Nanotechnology for Diabetes Management

Nanoscience & Nanotechnology Series

Editor-in-chief:
Nguyễn T. K. Thanh, *University College London, UK*

Series editors:
Gabriel Caruntu, *Central Michigan University, USA*
Shinya Maenosono, *Japan Advanced Institute of Science and Technology, Japan*
Neerish Revaprasadu, *University of Zululand, South Africa*

Titles in the series:
1: Nanotubes and Nanowires
2: Fullerenes: Principles and Applications
3: Nanocharacterisation
4: Atom Resolved Surface Reactions: Nanocatalysis
5: Biomimetic Nanoceramics in Clinical Use: From Materials to Applications
6: Nanofluidics: Nanoscience and Nanotechnology
7: Bionanodesign: Following Nature's Touch
8: Nano-Society: Pushing the Boundaries of Technology
9: Polymer-based Nanostructures: Medical Applications
10: Metallic and Molecular Interactions in Nanometer Layers, Pores and Particles: New Findings at the Yoctolitre Level
11: Nanocasting: A Versatile Strategy for Creating Nanostructured Porous Materials
12: Titanate and Titania Nanotubes: Synthesis, Properties and Applications
13: Raman Spectroscopy, Fullerenes and Nanotechnology
14: Nanotechnologies in Food
15: Unravelling Single Cell Genomics: Micro and Nanotools
16: Polymer Nanocomposites by Emulsion and Suspension
17: Phage Nanobiotechnology
18: Nanotubes and Nanowires, 2nd Edition
19: Nanostructured Catalysts: Transition Metal Oxides
20: Fullerenes: Principles and Applications, 2nd Edition
21: Biological Interactions with Surface Charge Biomaterials
22: Nanoporous Gold: From an Ancient Technology to a High-Tech Material
23: Nanoparticles in Anti-Microbial Materials: Use and Characterisation
24: Manipulation of Nanoscale Materials: An Introduction to Nanoarchitectonics
25: Towards Efficient Designing of Safe Nanomaterials: Innovative Merge of Computational Approaches and Experimental Techniques
26: Polymer–Graphene Nanocomposites
27: Carbon Nanotube-Polymer Composites
28: Nanoscience for the Conservation of Works of Art
29: Polymer Nanofibers: Building Blocks for Nanotechnology

30: Artificial Cilia
31: Nanodiamond
32: Nanofabrication and its Application in Renewable Energy
33: Semiconductor Quantum Dots: Organometallic and Inorganic Synthesis
34: Soft Nanoparticles for Biomedical Applications
35: Hierarchical Nanostructures for Energy Devices
36: Microfluidics for Medical Applications
37: Nanocharacterisation, 2nd Edition
38: Thermometry at the Nanoscale: Techniques and Selected Applications
39: Nanoceramics in Clinical Use: From Materials to Applications, 2nd Edition
40: Near-infrared Nanomaterials: Preparation, Bioimaging and Therapy Applications
41: Nanofluidics, 2nd Edition
42: Nanotechnologies in Food, 2nd Edition
43: ZnO Nanostructures: Fabrication and Applications
44: Diatom Nanotechnology: Progress and Emerging Applications
45: Nanostructured Materials for Type III Photovoltaics
46: Chemically Derived Graphene: Functionalization, Properties and Applications
47: Graphene-based Membranes for Mass Transport Applications
48: Carbon Nanostructures for Biomedical Applications
49: Surface Chemistry of Colloidal Nanocrystals
50: Reducing Agents in Colloidal Nanoparticle Synthesis
51: Carbon Nitride Nanostructures for Sustainable Energy Production and Environmental Remediation
52: Nanotubes and Nanowires, 3rd Edition
53: Bionanodesign: Old Forms for New Functions, 2nd Edition
54: Photothermal Nanomaterials
55: Concepts and Design of Materials Nanoarchitectonics
56: Anticorrosive Nanomaterials: Future Perspectives
57: Sustainable Nanotechnology
58: Anisotropic 2D Materials and Devices
59: Nanotechnology for Diabetes Management

How to obtain future titles on publication:
A standing order plan is available for this series. A standing order will bring delivery of each new volume immediately on publication.

For further information please contact:
Book Sales Department, Royal Society of Chemistry, Thomas Graham House, Science Park, Milton Road, Cambridge, CB4 0WF, UK
Telephone: +44 (0)1223 420066, Fax: +44 (0)1223 420247
Email: booksales@rsc.org
Visit our website at www.rsc.org/books

Nanotechnology for Diabetes Management

Edited by

Amar Abderrahmani
University of Lille, France
Email: amar.abderrahmani@univ-lille.fr

Sabine Szunerits
University of Lille, France
Email: sabine.szunerits@univ-lille.fr

Rabah Boukerroub
CNRS and University of Lille, France
Email: rabah.boukherroub@univ-lille.fr

and

Abdelfattah El Ouaamari
Robert Wood Johnson Medical School, USA
Email: ae353@rwjms.rutgers.edu

ROYAL SOCIETY
OF **CHEMISTRY**

Nanoscience & Nanotechnology Series No. 59

Print ISBN: 978-1-83916-470-5
PDF ISBN: 978-1-83916-549-8
EPUB ISBN: 978-1-83916-550-4
Print ISSN: 1757-7136
Electronic ISSN: 1757-7144

A catalogue record for this book is available from the British Library

The Royal Society of Chemistry is a charity, registered in England and Wales, Number 207890, and a company incorporated in England by Royal Charter (Registered No. RC000524), registered office: Burlington House, Piccadilly, London W1J 0BA, UK, Telephone: +44 (0) 20 7437 8656.

For further information see our web site at www.rsc.org

Printed in the United Kingdom by CPI Group (UK) Ltd, Croydon, CR0 4YY, UK

Preface

During the seminal lecture "*There is Plenty of Room at the Bottom*" in 1959, the 1965-nobelist physicist Prof. Richard Feynman claimed that the manipulation of atoms by science would be possible. Sixty years later, with the development of the science of objects at the nanometer scale, referred to as nanoscience, Feynman's vision becomes a reality and is radically changing the world. The number of daily-life products containing nanotechnologies is more and more impressively growing. These nanoproducts include not only cosmetics, furniture, coatings for cars and sport materials, but also therapeutics, anesthetics, imaging agents, vaccines and nutritional supplements. Nanotechnologies are now expected to offer great new perspectives as well as helping in combating some of the world's deadliest diseases which cause a huge socio-economic burden worldwide.

This book gathers together 11 outstanding chapters, presenting the evidence offered by nanotechnologies to operate a disruptive medicine for tackling the prevalence, disabilities and mortality caused by diabetes, a disease that affects men, women, children and adolescents. Diabetes is a complex disease, including type 2 (T2D) and type 1 diabetes (T1D), together representing approximately 95% of total diagnosed cases. The current antidiabetic medicines for the treatment of T2D and T1D, are mostly non-curative and, over the long term, they fail to slow and/or prevent the long-term lack of glycemic control and the progression of the disease toward life-threatening diseases. The chapters, including "*Optimizing the Current Type 2 Diabetes Antidiabetics with Nanotechnologies: Where do we Stand?*" (Chapter 3), "*Chemical Approaches for Beta-cell Biology*" (Chapter 1) and "*Amyloidosis Inhibition and Detection of Human Islet Amyloid Polypeptide with Nanomaterials*" (Chapter 4) show that nanotechnology can not only improve the efficacy and safety of current antidiabetic drugs, but also provide a

Nanoscience & Nanotechnology Series No. 59
Nanotechnology for Diabetes Management
Edited by Amar Abderrahmani, Sabine Szunerits, Rabah Boukerroub and Abdelfattah El Ouaamari

Published by the Royal Society of Chemistry, www.rsc.org

solution for engineering innovative curative antidiabetic drugs that will prevent the decline of pancreatic β-cells producing insulin, a key pathophysiological feature of T1D and T2D. Moreover, Chapter 5, entitled "*Phytochemicals in the Management of Diabetes*" reports a plethora of therapeutic candidates issued from plants that, if coupled to nanoparticles, might be a tremendous reservoir of future curative medicines. However, the search for innovative drugs requires high-throughput screening of hundreds of molecules, often achieved in non-human tissues. Chapter 7, entitled "*Microfluidic and Organ-on-a-chip-based Technologies for Diabetes Therapy and Research*" presents breakthroughs in 3D organ-on-chips (Ooc) microfluidic systems, made by nanomaterials, as fantastic support for human cell engineering by stem cells and for the culture of human tissues in an environmental medium that is very close to that of the human body. Ooc are anticipated to model human diabetes in a small dish and rapidly to ease the identification of the correct antidiabetic molecules for human therapy.

Nanotechnologies hold great promise in the transplantation of islets, the only curative treatment for T1D. This procedure is prescribed for patients with very unstable diabetes, severe hypoglycemia and without end-stage renal failure. Islet transplantation also enables the invasiveness and painful acts of glucose monitoring and insulin injections to be stopped. However, transplanted islets in the receipt can be rejected and/or progressively destroyed by immune cells attacks. In Chapter 8, "*3D Bioprinting of Islets*", the authors review the possibilities offered by nanoparticles for generating macrocapsules for the encapsulation and protection of islets against inflammation and other adverse factors ensuing during transplantation.

Complementary to islet transplantation, the artificial pancreas (AP) is expected to radically change diabetes management for a majority of patients with T1D. AP is formed by a continuous glucose monitor (CGM) coupled to a program connected to the insulin infusion pump. The CGM sticks to skin, and regulates blood glucose variations in an autonomous loop without user intervention. Chapter 10 "*Invasive and Implantable Glucose Sensors: Perspective for the Artificial Pancreas*" describes the Aps manufactured so far, and details the important challenges tackled by nanotechnologies for popularizing such a system. One of these is the development of more sensitive glucose sensors using a new generation of nanomaterial-based artificial enzymes enabling more rapid monitoring of glucose, and thereby achieving a perfect connection between glucose measurement and insulin pump functioning.

Other major achievements made by nanotechnologies are highlighted in Chapter 6 entitled "*Inhalation of Insulin for Diabetes Management*". This chapter deals with the release of a new generation of non-invasive, painless and easy mode of delivery of insulin, a hormone that is still mostly administrated via subcutaneous injection. In T1D, nanoparticle-based inhalation of insulin offers an unprecedented alternative for improving the comfort and compliance of patients, particularly those who have a needle phobia. The opportunity to adapt insulin therapy using inhaled insulin to a target patient is proof that nanotechnologies are contributing to make impactful advances into personalized diabetes medicine (PDM). Three chapters in this book illustrate this progress made by nanotechnologies in the set of PDM. In Chapter 9, entitled "*In vivo Imaging of the Pancreas and Gut Hormone Receptors*", the authors report the development of innovative imaging agents that will enable imaging of pancreatic islet β-cells from patients with T1D and T2D. The best radiolabeled imaging probes and the imaging techniques for detecting and quantifying β-cells are presented in this chapter. Nanotechnologies could be revolutionary in aiding the non-invasive delivery of these imaging agents as it is achieved for inhaled insulin. Quantification of β-cells, thanks to their imaging, is an invaluable opportunity for clinicians to prescribe and/or alternate the right therapeutics, to the right person, at the right time, while knowing the residual β-cells mass and function at the diagnosis time and during treatment. Chapter 11, entitled "*Artificial Intelligence for Diabetes-related Complications: The Eye as a Window to the Systemic Health*" underlines the importance of artificial intelligence (AI) and machine learning in PDM. Using the interpretation of retinal imaging data as a powerful predictor of cardiovascular disease risk, this chapter underlines the huge potential of AI and machine learning for incorporating and computing several biomarkers for giving the best diagnosis and prescription in clinical practice. To this end, the search for potent biomarkers is needed. Finally, Chapter 2 on "*Epigenomics of Type 2 Diabetes*", reports the progress made by epigenetics in identifying DNA biomarkers from blood cells, which could mirror the metabolic situation of several diseased organs and predict their improvement after therapy.

Nanotechnology for Diabetes Management is not only a book detailing the last progress made in nanotechnology for tackling diabetes and, thereby, stopping one of the fastest growing global health emergencies of the 21st century, it is also a seminal book that shows how much science can be disruptive, fruitful and groundbreaking when it is courageously and truly interdisciplinary.

Amar Abderrahmani
Sabine Szunerits
Abdelfattah El Ouaamari
Rabah Boukherroub

Contents

Nanoscience & Nanotechnology Series No. 59
Nanotechnology for Diabetes Management
Edited by Amar Abderrahmani, Sabine Szunerits, Rabah Boukerroub and Abdelfattah El Ouaamari

Published by the Royal Society of Chemistry, www.rsc.org

CHAPTER 1

Chemical Approaches for Beta-cell Biology

AMEDEO VETERE,[a] VISHAL S. PAREKH,[a] ASHLEY E. MODELL,[a,b,c] VERONIKA M. SHOBA,[a] AMIT CHOUDHARY[a,b,c] AND BRIDGET K. WAGNER*[a]

[a] Broad Institute, Chemical Biology & Therapeutics Science Program, 415 Main Street, Cambridge, MA 02142, USA; [b] Department of Medicine, Harvard Medical School, Boston, MA 02115, USA; [c] Divisions of Renal Medicine and Engineering, Brigham and Women's Hospital, Boston, MA 02115, USA
*Email: bwagner@broadinstitute.org

1.1 Introduction

Diabetes is a staggering epidemic that is constantly increasing. In its recent "Global Report on Diabetes," the World Health Organization (WHO) estimated that about 422 million adults were living with diabetes in 2016, compared to 108 million in 1980.[1,2] Despite this grim forecast, the pharmacologic toolkit available is somewhat limited. The stringent boundaries established by the FDA in terms of cardiovascular effects for any new antidiabetic drugs have made the development of novel chemical entities a "high-risk, low-reward" endeavor. Novel small-molecule approaches in islet biology, however, may prove fruitful in the future. In this chapter, we review the principles of chemical biology that enable small-molecule discovery, the current state of the art for therapeutic approaches, novel beta cell-focused approaches to use small molecules that improve phenotypes, and more recent efforts to deliver therapeutics selectively to the beta cell.

Nanoscience & Nanotechnology Series No. 59
Nanotechnology for Diabetes Management
Edited by Amar Abderrahmani, Sabine Szunerits, Rabah Boukerroub and Abdelfattah El Ouaamari

Published by the Royal Society of Chemistry, www.rsc.org

1.2 Principles of Chemical Biology

Chemistry and biology have long been considered independent disciplines, missing the important point that without chemistry we cannot have biological processes and *vice versa*. The emergence of the concept of "chemical biology" filled this gap. Chemical biology as a scientific discipline is a relatively new field; it only emerged about 20 years ago when scientists became interested in applying the principles of chemistry to study biological systems.[3] At its core, chemical biology is the process of discovering and using small molecules to modulate and understand how biological systems work. With this broad definition in mind, it is understandable to associate the process of small-molecule discovery with the concept of drug discovery. However, this is only part of the story. Small molecules as "drugs" have a very specific therapeutic scope and must meet stringent requirements in term of pharmacokinetics, safety, and bioavailability.[4] On the contrary, small molecules as "probes" can have more lenient pharmacokinetic requirements, but often even more stringent requirements for protein and cell selectivity. Chemical probes have the potential to provide important discovery tools to delve into biological phenomena.[5,6]

Chemical biology, as a discipline, and chemical probes, as its experimental tools, have provided an enormous boost to the expansion of the druggable genome. The completion of the Human Genome Project in 2003 coincided with the rise of chemical biology. The publication of a draft human genome sequence[7,8] provided the basis for understanding the roles of genes in normal physiology and in pathology. At the same time, chemical biology (or, as it is often referred to, chemical genetics) emerged as the way to modulate ideally each protein product emerging from all human genes.[9]

Chemical modulation of targets can be quite flexible, since it can provide narrow control of biological events on both temporal and concentration scales. At a temporal level, in the majority of cases, the effect of the chemical probe is rapid and reversible, as the effect of a chemical probe will last for the time it is present (covalent modulators are the exception). Further, the effects of small molecules are immediate, which helps determine the role of a cellular target. Data from knockout mice, on the other hand, can be confounded by developmental effects or compensation from a related isoform. At a concentration level, a key feature of chemical probes is their ability to show dose–response effects, where phenotypes are gradated by varying the concentration of the probe. Both of these properties are difficult to obtain by genetic manipulation. Temporal control of genetic manipulation would require conditional alleles, which are difficult to obtain and are prone to unwanted pleiotropic phenotypes.[10]

1.3 High-throughput Screening (HTS)

Genetics has been inspirational to the development of chemical biology as a modern discipline. From an experimental point of view, genetics studies are performed in a "forward genetics" or "reverse genetics" setting. In "forward genetics," random mutations are introduced in the experimental model and the subsequent phenotypes are observed. Subsequent sequencing analysis allows one to associate a mutation with a specific phenotype. In "reverse genetics," the mutations are introduced in a specific gene followed by analysis of the arising phenotype. Similarly, chemical genetics can be divided into "forward" and" reverse" chemical genetics. The procedural concept is similar. In "forward chemical genetics," the perturbation is obtained treating the experimental model with small molecules and the readout is the change of phenotype. With "reverse chemical genetics," a well-characterized and purified target protein is probed with small molecules followed by assessment of the specific activity of the target protein. A more general transposition of these concepts can be applied in chemical biology, and here we will discuss "target-based screening" and "phenotypic screening," for the concepts of "reverse chemical genetics" and "forward chemical genetics," respectively.

1.4 Target-based Screening

As the name suggests, in target-based screening a well-characterized protein target involved in a biological phenomenon or causative of a disease is used to interrogate chemical libraries to find high-affinity binders or modulators. Subsequently, active compounds are assessed in tissues and whole organisms. Target-based screening has been successfully used in both academic and industrial settings. The use of target-based screening has been particularly appealing to the drug discovery world since it seemed that targeting directly specific proteins would streamline the development of novel therapeutics. This concept stemmed from the observation that there were many examples of mutated proteins that were responsible for some cancers. Some notable examples of successful drugs that were discovered by target-based screening and that are in clinical use include inhibitors of Rho-dependent kinase (fasudil),[11] BCR-Abl (imatinib),[12] and EGFR (gefitinib).[13] Many more are approved or are at different stages of clinical trials[14] but the numbers are far from covering the entire known kinome. This is true not only for the kinome and but also for most of the druggable genome.

1.5 Phenotypic Screening

An alternative to target-based screening is phenotypic screening. In its general definition, phenotypic screening uses an agnostic approach by perturbing a phenotype with small molecules. Before the advancement of target-based drug discovery, new medicines were discovered using mainly phenotypic approaches. In an interesting analysis of drugs developed over the first decade of the 2000s, it emerged that 56% of the novel small-molecule approved drugs that constituted "new chemical entities" were discovered using phenotypic screening.[15] The past 20 years have seen an increase in this approach, with the discovery of first in class drugs for pathologies such as spinal muscular atrophy,[16] cystic fibrosis,[17] and hepatitis C.[18] Academic laboratories were also successful in identifying probes with novel mechanisms of action by using phenotypic screening as has been well summarized in a recent report.[19]

1.5.1 Hit Triage

Regardless of its final use (as a probe for basic research or as novel drug for clinical applications), a small molecule must satisfy basic criteria in terms of potency, solubility, toxicity, and specificity. Exhaustive descriptions of what constitutes a "good" probe have been recently reported.[5,20,21] Recently, the concept of "nuisance" has been added to the pool of factors to be considered in evaluating the quality of a probe.[22] Nuisance compounds are those that interfere with the assays used for the screening or, in more general terms, any compounds that show undesirable bioactivity. Nuisance compounds have always plagued phenotypic screening campaigns[23,24] but only recently has a systematic approach been pursued to categorize and characterize them.[22]

1.5.2 The Sources of Chemical Probes

1.5.2.1 Natural Products Libraries

Although the concept of chemical probes emerged as tools to realize the full potential of chemical biology, they were not strangers to either chemists or biologists. There are many examples of natural products whose discovery and use have enabled the understanding of important biological events. Some notable examples are colchicine, which helped study cytoskeleton organization,[25] FK506 and cyclosporine, which were instrumental to study immunosuppression,[26,27] rapamycin, which elucidated the role of mTOR,[28] and many others. Not surprisingly, the majority of the early studies focused on natural products. For centuries humankind has turned to natural compounds, mainly from plant extracts, as therapeutics. Far from being "old-fashioned," this trend continues and, not surprisingly, a great many FDA-approved drugs are natural products or their derivatives.[29] The search, purification, and eventually chemical synthesis of natural products, however, is an "art" that competes with the scope and need of chemical genetics, which aims to target every single human gene.[9,30–32]

1.5.2.2 Diversity-oriented Synthesis (DOS) Libraries

To increase the throughput of targeted genes or proteins it is necessary to access a wider and more diversified chemical space that can be achieved only by screening richer libraries of structurally diverse compounds. This goal has been approached by the introduction of diversity-oriented synthesis

(DOS) libraries of small molecules.[33–35] The idea behind DOS is to create, with few synthetic steps, rich libraries of small, conformationally diverse compounds. Part of what makes DOS compounds unique is the introduction of stereochemistry as a further level of diversity.[34,35] Since their inception, DOS libraries have been successful in identifying novel mechanism-of-action compounds and chemical probes. Some notable examples are the discovery of novel multistage antimalarial inhibitors,[36] a novel compound that protects cytokine-induced beta-cell apoptosis,[37] a highly specific inhibitor of the insulin-degrading enzyme,[38] and selective Cas9 inhibitors.[39]

1.5.2.3 Repurposing Collections

One of the paradigms of small-molecule discovery is to screen libraries of diverse compounds to identify those that show a high specificity toward the cellular target or phenotype. A more pragmatic approach has recently emerged with the introduction of the concept of "repurposing" screening. The idea behind repurposing screening is to exploit the relative promiscuity of well-annotated small molecules by using them to probe targets or phenotypes different from those for which they were originally developed. Repurposing screening represents an attractive and economical way to find novel therapeutic applications for compounds that had already gone through the painstaking process of different stages of preclinical development or even clinical trials. Although useful, large campaigns involving repurposing screening are still in their infancy. As recently reported,[40,41] the major hurdles to a widespread use of repurposing screening are (1) the lack of repositories of all available drugs, (2) low quality of drug annotations, and (3) the lack of clear readouts of drug activity. In the past few years there have been attempts to solve some of these problems. For example, today a few repurposing repositories are already available.[40–42]

1.5.2.4 DNA-encoded Libraries

A further improvement in the ability to identify novel compounds arose with the advent of DNA-encoded libraries (DEL).[43] The core concept behind DEL screening is to combine the diversity of organic chemistry with the uniqueness of DNA barcodes. The presence of unique DNA barcodes allows screening to be performed in a pooled fashion rather than in an arrayed manner. Pooled approaches have the advantage of allowing screening of millions, if not billions, of compounds in a single test tube.[44] The identification of hits is then achieved by evaluating the enrichment of unique DNA tags after PCR amplification and next-generation sequencing.[45] DEL screening represents a significant breakthrough in the field of chemical biology with the identification of novel protein ligands.[45] Despite its potential, DEL screening has some limitations, mainly the historical lack of structural diversity of chemical scaffolds and a chemical liability associated with the need for DNA-compatible chemistry.[46,47] In the past few years significant progress has been made to resolve both issues.[48,49]

1.5.2.5 *Molecular Glues*

A further breakthrough in the field of chemical biology arose with the realization of a phenomenon that was observed in the early 1990s, known as "chemical-induced proximity" or the "molecular glue" concept. As reported above, the discovery of cyclosporin A was instrumental to understanding immunosuppression, but it was years later that it was found that it acts in a unique mode by creating a protein–protein interaction, hence "molecular glue".[50] From there, chemical-induced proximity became the new frontier of chemical biology, and with all of its various applications, it is having profound effects on both biology and therapeutics.[51,52] The most recent example of chemical-induced proximity is the proteolysis targeting chimera (PROTAC) approach. PROTACs were first developed in the early 2000s.[53] In its original conception, it was based on the creation of a chimeric structure between a small molecule, specific for a target protein, and a peptide that had specificity toward E3 ligases. Thus, due to proximity, the lysines of the target protein are ubiquitinated by the E3 ligase complex and subsequently targeted to the proteasome to be degraded. Despite success in demonstrating proof-of-concept for PROTAC technology, peptide-based PROTACs had some disadvantages, such as high molecular weight, lability of peptide bonds, poor cell permeability, and low potency.[54] Over the last decade, we have witnessed a dramatic improvement in the PROTAC technology with the introduction of small molecule-based PROTACs.[55–57] The small molecule-based PROTAC technology relies on the construction of a heterobifunctional molecule from small-molecule ligands for the Von Hippel Lindau (VHL)[58,59] and cereblon (CRBN) E3 ligases[57] instead of peptide binders. We envision that the development of screening approaches and libraries for the discovery of such proximity inducers will tremendously benefit the field of drug discovery.

1.5.3 Target Identification and Mechanism-of-action Studies

The unbiased approach of phenotypic screening represents a powerful tool to identify compounds that affect a phenotype of interest, but a major challenge of this approach is the need to identify the target and mechanism of action for a new compound. Target identification is an important task needed to evaluate the specificity and selectivity of the probe and to, eventually, develop target-based screening strategies. Furthermore, target identification helps to better strategize medicinal chemistry efforts. Target identification is not an easy task but today there are at least three complementary approaches that are used for target identification: direct biochemical methods, genetic interaction methods, and computational inference methods.[60–64]

1.5.3.1 Label-based Target Identification

Direct biochemical methods rely on the affinity of the small molecule of interest for its protein target. This approach can be used to perform "label-based" or "label-free" target identification. Label-based approaches depend on a solid contribution from synthetic organic chemistry, since it typically requires the introduction of appropriate functional groups in the lead compound. Historically, this approach has involved immobilizing the small molecule of interest on a physical support that is used to "fish out" target proteins from a cellular extract,[65] whose identity can be detected by mass spectrometry.[61,66] This approach brings with it a risk of missing biologically significant but weak interactors, due to the stringent washes generally required by these protocols. To overcome these limitations, label-based approaches have shifted toward methods in which the small molecule of interest can establish covalent interactions with target protein(s) within a cellular context. This goal can be reached either by introducing electrophilic groups that can react with their nucleophilic counterparts in target proteins (activity-based probes), or by introduction of a photo-reactive group that can cross-link with the target protein upon UV-irradiation (photo-affinity probes).[67]

Activity-based probes possess common features, including (1) a reactive group for covalent binding, (2) a core structure that confers specificity, and (3) a reporter group (*i.e.*, biotin, or a generic fluorophore) for detection.[68–70] The reactive warhead can be designed to target key groups of active sites of a specific category of enzymes. The reporter group can be a constitutive part of the activity-based probe or can be added subsequently once the probe is covalently bound to its target.

To avoid high hit rates due to false positives, the majority of compounds used in high-throughput screening are based on chemistries that avoid covalent binding to target proteins. Nevertheless, covalent binding can become a useful instrument for target identification. One way to exploit this possibility is to introduce UV-reactive groups.[71,72] There are many factors that have to be considered before choosing the right photo-affinity probe,[72] but ultimately the nature of the core structure will dictate the choice based on the chemical manipulation it can withstand without losing activity and specificity. Label-based approaches have been used successfully for target identification deriving from unbiased phenotypic screening, as described recently.[73]

1.5.3.2 Label-free Target Identification

Label-based target identification approaches have undoubted advantages, but they typically require organic chemistry efforts that are not always available. To overcome this limitation, and to make target identification more accessible to biology-focused laboratories, many groups have embarked on efforts to develop label-free target identification methods. These approaches revolve around the ability of a small molecule, upon protein binding, to protect that protein from (1) protease digestion, (2) chemical denaturation, or (3) thermal denaturation; the respective methods are (1) drug affinity responsive target stability (DARTS), (2) stability of proteins from rates of oxidation (SPROX), and (3) cellular thermal shift assay (CETSA). Each of these methods has the option of using high-resolution mass spectrometry to determine and quantify the relative differences in proteins in compound-treated samples, with respect to control samples.

DARTS was developed based on the observation that when a small molecule interacts with its cognate protein target, the protein becomes less susceptible to the action of proteolytic enzymes.[74] Cell extracts are subjected to partial proteolysis after treatment with compound or DMSO. After SDS-PAGE, the bands differentially expressed in the treated sample *versus* the DMSO controls are excised and analyzed by mass spectrometry. SPROX uses a similar principle, but in this case is based on the susceptibility of exposed methionines on denatured proteins to oxidation by hydrogen peroxide.[75] Here, the assumption is that the presence of a small molecule will prevent denaturation and methionine oxidation. Cell extracts are treated with an increasing concentration of a denaturing agent, such as guanidinium chloride, and the exposed methionine residues are then oxidized with hydrogen peroxide. The samples are then digested, and peptides are analyzed by mass spectrometry. The readout is a shift of the so-called "transition midpoint," which is the concentration of denaturant required to reach 50% oxidized methionine and 50% native methionine.[75]

CETSA represents the latest approach introduced for target identification, with the first having been reported in 2013.[76] The principle behind CETSA is based on the well-appreciated understanding that the thermal stability of proteins can be modulated by a bound ligand. This observation forms the basis of the widely used biochemical thermal shift assay (TSA).[77] The major limitation of TSA is that it involves the use of purified protein. On the other hand, with CETSA, the concept of TSA was transposed to a cellular context. In its original iteration, CETSA was developed to study target engagement for known interactions in a cellular context.[76,78,79] Here, cells treated with the desired compound are subject to a brief thermal gradient. The cells are then lysed, and the soluble fraction is used to detect the desired target protein by western blotting. Plotting the band intensities as a function of temperature results in a melting profile, and the degree of shift in the melting temperature (T_m, corresponding to 50% denatured) in treated samples *versus* controls serves to evaluate the degree of target engagement. CETSA could not be applied to unbiased target identification until recently, when its principle was extended to proteome-wide applications with the introduction of thermal proteome profiling (TPP).[80] In a similar fashion as CETSA, cells treated with a specific compound are heated to increasing temperatures and then lysed, with the soluble fractions analyzed this time by mass spectrometry. The differential abundance of peptides in the treated *versus* the controls provides a melting profile for thousands of proteins and ultimately can help identify target protein(s). Since its first introduction, TPP has been used to identify novel targets for small molecules identified by phenotypic screening.[81–83]

1.5.3.3 Genetic Interaction Methods

An original goal set forth by leaders in chemical biology was to identify a chemical probe for each human protein.[9] Recent progress in functional genomics is bringing chemical biology closer to this goal. This is particularly true with the advent of CRISPR/Cas9 technology to manipulate the mammalian genome.[84–86] In brief, the methodology uses Cas9 as an effector protein that is "guided" to its target gene by the association of a guide RNA (sgRNA) complementary to the locus of interest. Cas9 introduces, then, a double-stranded DNA break, which triggers activation of the repair mechanisms of the cell. In the majority of cases, this results in frameshift mutations and subsequent loss of the corresponding protein.

Chemogenetic screens in the presence of genome-wide CRISPR can yield genes, whose corresponding protein knockout results in a loss of compound activity. With this approach, target cells are infected, in a pooled fashion, with lentivirus expressing Cas9 along with sgRNAs targeting each known gene. The infection is performed at a MOI that allows each cell to express only one sgRNA. In the setting of target identification, the pooled infection of CRISPR/Cas9/sgRNA is performed in the presence of a small molecule. This approach is particularly useful to identify targets of cytotoxic compounds, but the method is being expanded to other phenotypes as well. The outcome of the genetic perturbation may be the development of resistance to the compound (a suppressor interaction), or it may result in an increased susceptibility to the compound (an enhancer interaction). This approach has been used to identify novel targets of small molecules and directly correlate the gene to the effect of the small molecule.[64,87]

For cytotoxic small molecules, resistance screens also can be very valuable to identify targets. Here, the aim is to identify protective mutations in a specific gene that are induced after long-term treatment with a cytotoxic drug. In brief, cells are treated with a sublethal concentration of toxic compound until resistant clones emerge. After expansion, the transcriptomes of the resistant clones are then sequenced. The result of this approach is a series of clusters of genes of interest. CRISPR/Cas9 can then be used to deconvolute these clusters to pinpoint which mutation is essential for drug activity. This approach has been successfully used to identify novel target of small molecules.[87–89] As stated above, both chemogenetic and resistance screening approaches are particularly apt to identify the protein target of cytotoxic compounds. However, the phenotypic scale of application is rapidly expanding, and we expect that applications in the beta cell will arise in the near future.

1.5.3.4 Computational Inference Methods

Computational approaches are critical to a successful target identification campaign, regardless of the methods used. At the same time, computational approaches can serve as stand-alone methods for target identification. These approaches are based on the principle that structurally similar compounds, or phenotypically similar compounds, are likely to have similar mechanisms of action, and hence similar protein targets. In the past few years, many computational approaches have been successfully developed for target deconvolution.[90–92] Here we will discuss two methods, Cell Painting and the Connectivity Map.

Cell Painting is a target-naïve image-based profiling approach based on the analysis of up to 1500 morphological features (*i.e.*, size, shape, texture, intensity, *etc.*). These fingerprints are derived from six different fluorescent dyes specific for distinct cell organelles and compartments (nucleus, endoplasmic reticulum, mitochondria, cytoskeleton, Golgi apparatus), imaged with orthogonal fluorescent channels.[93] After treating with the compound of interest, cells are fixed, stained, and imaged. The feature pattern is compared with a reference database of previously acquired fingerprints from compounds with known mechanism of action, and pattern similarity allows inference of the targets for novel compounds.[94]

The Connectivity Map (CMap) takes a conceptual approach similar to Cell Painting. In this case, the readout is a gene-expression signature rather than a morphological fingerprint. CMap was first developed by using 164 distinct small molecules, selected among a broad range of activities (FDA-approved drugs, nondrug bioactive compounds). As a proof of concept, compounds sharing molecular targets were included to determine whether such compounds would share a molecular signature. Furthermore, the pool included compounds with the same clinical indication, to determine whether therapeutic class could determine connections.[95] These compounds were tested in MCF7 (breast cancer epithelial cell line), PC3 (prostate cancer epithelial cell line), and SKMEL5 (melanoma cell line) cell lines. More recently, transcriptional analysis was reduced to 1000 so-called "landmark" genes and termed an L1000 profile. It emerged that the L1000 profile was enough to cover 82% of the information of the entire transcriptome.[96] From a target deconvolution point of view, CMap is used by comparing the profiles of unknown compound with those derived from known perturbagens and has been successfully used to uncover novel mechanisms of action.[97]

Together, the judicious use of assay development, high-throughput screening, target identification, and mechanism-of-action studies provides a powerful blueprint for identifying new small molecules with beneficial activity in the islet or probes to further our understanding of disease. Further, these compounds can furnish the diabetes community with information about new cellular targets whose modulation may be of interest in islet biology. The application of these chemical biology approaches to islet biology and diabetes has become increasingly important to the advancement of the field.

1.6 The Chemical Biology of Diabetes

The development and use of chemical biology tools have deepened our knowledge of many diseases, including diabetes. Diabetes is one of the oldest reported human diseases, with the first reports emerging from Egypt around 1500 B.C.E.[98] Over time, there has been significant progress in understanding the pathology of diabetes, but perhaps not at the pace we would hope from a disease known for over 3000 years. Until the end of the 1980s, the basic idea about diabetes pathophysiology pivoted around the concept known as the triumvirate.[99] This concept stems from the basic knowledge of glucose homeostasis. Under normal conditions, after ingestion of a meal, the pancreas is primed to secrete insulin in response to hyperglycemia. The subsequent physiological hyperinsulinemia promotes uptake of glucose by the liver and muscle tissues and inhibits hepatic glucose synthesis. Therefore, the three pillars of glucose homeostasis are beta cell, liver, and muscle, hence the triumvirate model. An impairment at any of these levels can cause overt diabetes.[99] In real life, it was common to categorize diabetic patients based on these criteria. For example, individuals showing beta-cell dysfunction as causative of diabetes were typically the lean diabetic patients, while individuals whose diabetes started because of a reduced peripheral insulin sensitivity were typically the obese patients. Over the past 30 years, this vision of the pathogenesis of diabetes appeared reductive, and the triumvirate became the ominous octet.[100] Beta cells, liver, and muscle were joined by adipocytes, the gastrointestinal tract, alpha cells, the kidney, and the brain. Each of these can play an important role in the pathogenesis of diabetes.

Adipocytes are the major site of fatty-acid storage (as triglycerides) in the body. They are very efficient in their storage capacity in response to insulin, as long as the physiological conditions are normal. Obesity creates an imbalance where adipocytes enlarge and become insensitive to the antilipolytic effect of insulin. The excess of free fatty acids will, in turn, impair the ability of peripheral tissues to respond to insulin, and over time will promote beta-cell apoptosis.[101] Aside from its mechanical role, the gastrointestinal tract can be considered as an endocrine structure responsible for the synthesis and secretion of at least two key hormones, glucagon-like peptide-1 (GLP-1) and glucose-dependent insulinotropic peptide (GIP). GLP-1 is secreted by the L-cells of the distal small and large intestine, while GIP is secreted by the K-cells located in the early small intestine.[102] Both GLP-1 and GIP are at the forefront of insulin secretion, and collectively they are responsible for >70% of the insulin secreted after a meal. In overt diabetes, GLP-1 and GIP secretion are normal, but beta cells are insensitive to their action.[103] Like the gastrointestinal tract, the kidney plays a role in the regulation of metabolism. In physiological conditions, after an overnight fast, the kidney is responsible for ~20% of glucose production. This process is tightly negatively regulated by insulin itself. As in other peripheral tissues, the kidney becomes resistant to the effect of insulin in the case of diabetes. This loss of control increases the already present hyperglycemia, contributing to the severity of diabetes.[104]

The role of the brain in the pathogenesis of diabetes is not immediately evident, but not necessarily less important. Along with its direct effect on glucose uptake, insulin also has a potent effect on reducing food intake, at the level of hypothalamus. In the presence of obesity, which often accompanies diabetes, the hypothalamus becomes resistant to the anorectic effect of insulin, leading to an increase in food intake and contributing ultimately to the demise of beta cells.[105] Finally, alpha cells are rightly considered the counterpart of beta cells in the pancreatic islet. Alpha cells are responsible for the synthesis and secretion of glucagon, which is responsible for counteracting the effects of insulin and for inhibiting insulin secretion. In diabetic patients, glucagon remains elevated, even in presence of hyperglycemia.[106] Furthermore, elevated glucagon levels will prime the liver to start gluconeogenesis, contributing to hyperglycemia.

The pancreas is a mix of digestive enzyme-producing exocrine and hormone-producing islets. Both exocrine and endocrine compartments of the pancreas are essential for the homeostatic function. The major cells in the exocrine cells are acinar cells and ductal cells. Here, digestive enzymes produced in the acinar cells are drained *via* ducts into the duodenum. Pancreatic islets consist of four major endocrine cell types; beta cells, alpha cells, delta cells, and PP cells, which secrete insulin, glucagon, somatostatin, and pancreatic polypeptide hormone, respectively. Insulin and glucagon are major glucose-regulating hormones. Glucagon has a direct effect on the liver and stimulates gluconeogenesis under hypoglycemic conditions. The beta cells are the only cell types in the body to synthesize, process, store, and release insulin in response to glucose. Impaired or diminished insulin secretion leads to the development of impaired glucose tolerance and subsequently fasting hyperglycemia. Thus, insulin deficiency and fasting hyperglycemia are clinical features of diabetes and are related to low beta-cell mass, beta-cell dysfunction, and/or death. Thus, finding small molecules that impact on these phenotypes could have a major impact on diabetes understanding and treatment.

Diabetes is classified based on the clinical presentation, and there is considerable variation in disease progression due to its underlying etiology. The American Diabetes Association has classified diabetes into four categories: (1) type 1 diabetes (T1D), (2) type 2 diabetes (T2D), (3) gestational diabetes, and (4) specific types of diabetes due to other causes.[107] This last category includes neonatal diabetes, monogenic diabetes, disease relating to the exocrine pancreas (cystic fibrosis and pancreatitis), and chemical-induced diabetes, for example due to the use of glucocorticoids. T1D and T2D are the predominant diabetes types and are polygenic diseases; despite the baseline genetic risk, the numbers of individuals with T1D and T2D are rising due to changes in the environment and lifestyles.[108]

1.6.1 Type 1 Diabetes (T1D)

In type 1 diabetes, insulin deficiency is absolute, and it occurs as a result of autoimmune-mediated destruction of beta cells. Exogenous administration of insulin or insulin analogs is the only current therapy available for individuals with T1D. Several dogmas exist for the etiology and development of T1D.[109] For example, islet-intrinsic mechanisms[110] have been proposed as important to the development of T1D. Beta-cell stress and inflammatory cytokines trigger the generation of reactive oxygen species and nitrosative species, which lead to beta-cell dysfunction, attraction of immune cells, and beta-cell death. Islet-extrinsic mechanisms are associated with defective exocrine pancreatic function[111] and dysregulation in autoimmune factors.

The autoimmune basis of beta-cell death has gained much favor. As per this theory, naïve autoreactive T cells resident in the islet or islet lymph nodes are activated by beta cell-derived peptides. Autoreactive T cells transform B cells into plasma cells, which subsequently generate antibodies against one or more epitopes targeting beta cells. The predominant epitopes are insulin itself, glutamic acid decarboxylase (GAD65), islet antigen 2 (IA-2), and zinc transporter 8 (ZnT8, encoded by *SLC30A8*). Naïve autoreactive T cells also activate other T cells and B cells, which subsequently migrate to the islets – in a process known as insulitis – and release proinflammatory cytokines, which further increase beta-cell cytokine stress and subsequent killing by cytotoxic T cells.

Our early knowledge about the etiology of type 1 diabetes was enhanced by the use of the non-obese diabetic (NOD) mouse model, as well as immune profiling, and isolation of beta cells, islets, and pancreas from T1D individuals. It is interesting to note that recent studies indicate that beta-cell intrinsic mechanisms leading to overexpression of HLA1 and autoantigen presentation are likely involved in the recruitment and activation of CD4+ and CD8+ T cells.[110] While beta cell HLA1 overexpression is an important trigger,[112] a defining cause of T1D onset remains elusive.[113,114] Beta cells can express class II MHC molecules which are sufficient to interact with the circulating CD4+ T cells in islets at the onset of T1D.[115]

T1D has been defined as a polygenic disease, with underlying genetic risk factors (*e.g.*, *INS*), genetic linkages (to HLA1), and the presence of circulating autoantibodies (aAb) against beta-cell antigens.[116,117] However, several other hypotheses have been suggested to explain the increased incidence of T1D globally, including the decline of infectious diseases – the hygiene hypothesis[118] – and vitamin D deficiency.[119] Accumulating evidence indicates that loss of functionally mature beta cells precedes and may contribute to insulitis and the demise of beta cells themselves. For example, aberrant proinsulin processing has been observed in new-onset (<2 years) T1D,[120] and the presence of proinsulin-positive beta cells has been observed in long-standing T1D pancreata.[121] The acquisition of a senescence-associated secretory phenotype by beta cells[122] and activation of the HIF1a/PFKFB3 pathway[123] also argue for an early beta-cell dysfunction preceding the onset of T1D. Thus, protection of mature beta-cell function at the onset of, or during early stages of T1D, is an attractive therapeutic strategy for T1D. While previous studies have confirmed that a T1D-like cytokine storm can induce profound beta-cell stress, followed by autoimmune-mediated cell death, the auto-activation of oxidative stress and interferon signaling in beta cells *per se* is thought to provoke an antiviral response and trigger beta-cell death.[124]

Because T1D is considered an autoimmune-mediated irreversible loss of beta cells, efforts to treat T1D have been limited to modalities such as exogenous insulin, islet transplantation, or restoration of normal immune function. However, genome-wide association studies have shown that >50% of genes near loci associated with T1D are expressed in beta cells, indicating that beta cells can play an active role in the development of T1D.[125] Indeed, in prediabetic NOD mouse models, beta cells display activation of NF-κB along with morphologically stressed ER, and defective insulin secretion as revealed by elevation of the proinsulin:insulin ratio.[126] Targeting the ABL-Ire1a signaling pathway protected beta cells from ER stress, preserved beta-cell mass, and reversed autoimmune diabetes in NOD mice.[127] Additional studies in NOD mice showed that beta cell-specific deletion of Ire1a induced beta-cell dedifferentiation and reduced islet immune cell infiltration, due to diminished levels of MHC class I antigen and beta-cell autoantigens.[128] Consistent with this finding, recent studies observed that beta cells adopt a senescence-associated secretory phenotype in NOD mice, which, when eliminated, can reduce immune infiltration and protect the remaining healthy beta cells.[122] Interestingly, NOD mice can be protected from autoimmune diabetes by small-molecule inhibition of the JAK1/JAK2 pathway[129] or with the ER chaperone TUDCA.[130] Future work will be needed to perform high-throughput screening to discover small molecules that ensure the functional maturity and health of beta cells.

1.6.2 Type 2 Diabetes (T2D)

Like T1D, type 2 diabetes has a strong polygenic component that increases baseline risk. T2D is characterized by insulin resistance and impaired glucose-stimulated insulin secretion (GSIS), compromised pulsatility of insulin secretion, and amyloid deposits in the islet. Interestingly, many genetic loci linked with T2D are near known transcription factors or proteins involved in beta-cell development, maturation, or function. Despite this finding, the global trend in diabetes is increasing and strongly associated with obesity. In this case, beta cells compensate for increased peripheral insulin resistance by increasing insulin secretion, but prolonged and sustained insulin resistance increases the stress on beta cells, due to the increased need for insulin synthesis, trafficking, and secretion. As a result, beta-cell function deteriorates and fails to produce enough insulin to overcome insulin resistance, leading to the development of diabetes.[131]

Glucose and free fatty acids can directly induce insulin secretion. However, sustained and uncontrolled hyperglycemia, combined with elevated circulating levels of free saturated fatty acids (C16 or greater),[132] exhaust beta cells, resulting in impaired insulin secretion (beta-cell dysfunction) and beta-cell death. These effects of glucose and lipids in beta cells have been termed glucotoxicity,[133] lipotoxicity, or glucolipotoxicity. An alternative hypothesis has been posited to explain the growing number of diabetes cases in countries like India and China. Insulin is an important fetal growth factor, and the fetal insulin hypothesis suggests that impaired insulin secretion or increased insulin resistance results in low birth weights and contributes to impaired glucose tolerance, diabetes, and other cardiovascular complications.[134] Low birth size and subsequent obesity, which impacts on fatty acid and proinflammatory cytokines levels, can dramatically increase peripheral insulin resistance and the development of diabetes.[135] Together, there is good evidence that targeting the beta cell could be an important component to developing the future of diabetes therapeutics.

1.6.3 Modeling the Beta Cell for Small-molecule Discovery

1.6.3.1 Primary Human Islets

The availability of a good cellular model is critical for small-molecule discovery. Such models, however, have been difficult to obtain in the lab. Primary islets, isolated from the pancreata of cadaveric human donors, are the gold standard for *in vitro* beta-cell research, because any candidate hit from small-molecule screening may be more directly relevant to T1D or T2D. However, primary islets are fragile and difficult to culture for long periods of time, and in many parts of the world are not obtainable. Further, the quality of islet preparations has a high level of variation due to donor variability as well as potential differences in islet isolation procedures across clinical centers. Our group has developed an islet culture system suitable for high-throughput screening.[136,137] In our experience, using either dissociated human islet cells or small clusters has advanced our discovery activities.

1.6.3.2 Stem Cell-derived Beta Cells

Stem cell-derived islet-like aggregates are an attractive tool for screening a wide variety of compounds to study and/or improve beta-cell maturation, prevention of beta-cell degradation, and insulin secretion. In addition, pluripotent stem cell-derived beta cells offer expandability, allow genome editing, and enable transplantation in animal models of diabetes. Currently, several standard or reporter embryonic stem cell lines, with inherent genomic variability, are available in the islet research field.[138] On the contrary, induced pluripotent stem cells offer more flexibility in using patient-derived cells to derive beta cells that may more faithfully mimic the specific diabetes biology for drug discovery.[139]

1.6.3.3 Human Beta Cell Line

In stark contrast to the many immortalized cell lines available for cancer biology, it is only recently that an immortalized human beta-cell line (EndoC-βH1)[140] has become an alternative cell source for human beta-cell research. EndoC-βH1 expresses markers similar to native human beta cells, exhibits glucose-induced insulin secretion, and ameliorates chemical-induced diabetes *in vivo*. Human pancreatic buds were transduced with lentivirus encoding SV40 gene under the control of the insulin promoter and transplanted into SCID mice for fetal beta-cell maturation and subsequent generation of beta-cell teratomas. These cells were then transduced with lentivirus encoding human telomerase reverse transcriptase (hTERT) and re-transplanted into the SCID mouse for further expansion to generate *in vitro* cell lines. The insulin content and secretion were further improved in EndoC-βH2, a second-generation cell line where immortalizing transgenes were removed with a Cre expression system.[141] Electrophysiological processes, insulin exocytosis, and ultrastructural morphology of insulin granules of both EndoC-βH1 and EndoC-βH2 were similar to primary human beta cells.[142] EndoC-βH3, a third-generation cell line, enables excision of immortalizing transgenes using tamoxifen.[143] These human beta-cell lines offer expandability, which is necessary for screening, glucose-stimulated insulin secretion, and have a relatively similar genome to primary beta cells. The main limitation of these lines has been the challenge of culturing them in most labs; the cell lines grow quite slowly and require expensive specialty reagents. Overall, these cell lines are a very valuable and complementary resource for small-molecule and genetic discovery.

1.6.3.4 Insulinoma Cell Lines

Rodent insulinoma cell lines, such as INS-1E (selected through subcloning of original INS-1 cells),[144,145] INS-1 832/13 (derived by stable transfection with the human proinsulin gene),[146] MIN6,[147,148] and βTC6[149] have insulin secretion capacity similar to rodent islets, can respond to diabetogenic stress, and are cheaper alternatives for small-molecule screening. Careful consideration should be made, however, of the phenotype being measured. Rodent insulinoma lines are not good models for beta-cell proliferation, as there are vast differences between rodent and human beta cells in this process. Apoptosis and insulin secretion, on the other hand, are two phenotypes that are well-suited to the use of rodent cell lines. Indeed, we used the INS-1E cell line to generate a model for luminescence-based insulin secretion and high-throughput screening.[150] To achieve this, we inserted *Gaussia* luciferase into the C-peptide portion of proinsulin, allowing equimolar co-secretion of both insulin and luciferase, as we will discuss below. Together, the use of appropriate assay readouts (Table 1.1) on physiologically relevant cell models is an important component of chemical discovery in the beta cell.

1.7 Therapeutic Approaches in the Beta Cell

From this brief overview, it is clear that diabetes is a very complex disease with systemic complications. The landscape is further complicated by the fact that diabetes is a progressive disease that may require years to become overt. At the

Table 1.1 The assay models and readouts used for small-molecule discovery, including through high-throughput screening.

Model		Phenotype		Assay
T1D	One or more inflammatory cytokines (TNF-α, IFNα, IFNγ, IL-1β) or poly I:C	Regeneration	Beta cell mass/ replication	EdU, CldU, Ki67, PCNA, pH3
			Transdifferentiation (non-beta stem cells)/maturity	*PDX1, UCN3, FoxO1, MafA, NKX6.1*
		Protection		HLA1, HLA2, b2M, cleaved caspase-3, ATP, Annexin V, propidium iodide
		Health and function		ATP content, proinsulin: insulin ratio, GSIS
T2D	Glucotoxicity, lipotoxicity, glucolipotoxicity	Insulin secretion		Insulin ELISA, luciferase assay
		Beta-cell rest		Insulin ELISA, membrane potential
		Amyloid clearance		Thioflavin S
		Beta-cell protection		Cleaved caspase-3, ATP, Annexin V, propidium iodide
		Dedifferentiation/ maturity		*PDX1, UCN3, FoxO1, MafA, NKX6.1*

time of diagnosis (especially with T1D), the progression of the disease is already at an advanced state of beta-cell loss and dysfunction. Historically, the majority of therapeutic approaches have been aimed at lowering blood glucose by focusing on peripheral tissues (Figure 1.1). However, a few therapeutic strategies have focused on insulin secretion itself. An interesting summary of various chemical biology approaches applicable to the study of beta-cell function has been recently reported,[151] including a very thorough perspective of the chemical probes available to study islet function. Here, we will start with a description of current clinical strategies and follow that with a description of research approaches focused on the beta cell.

1.7.1 First-generation Therapeutics

1.7.1.1 Insulin Secretagogues

The pharmacological induction of insulin secretion has historically been obtained more with the use of a sledgehammer rather than a scalpel.

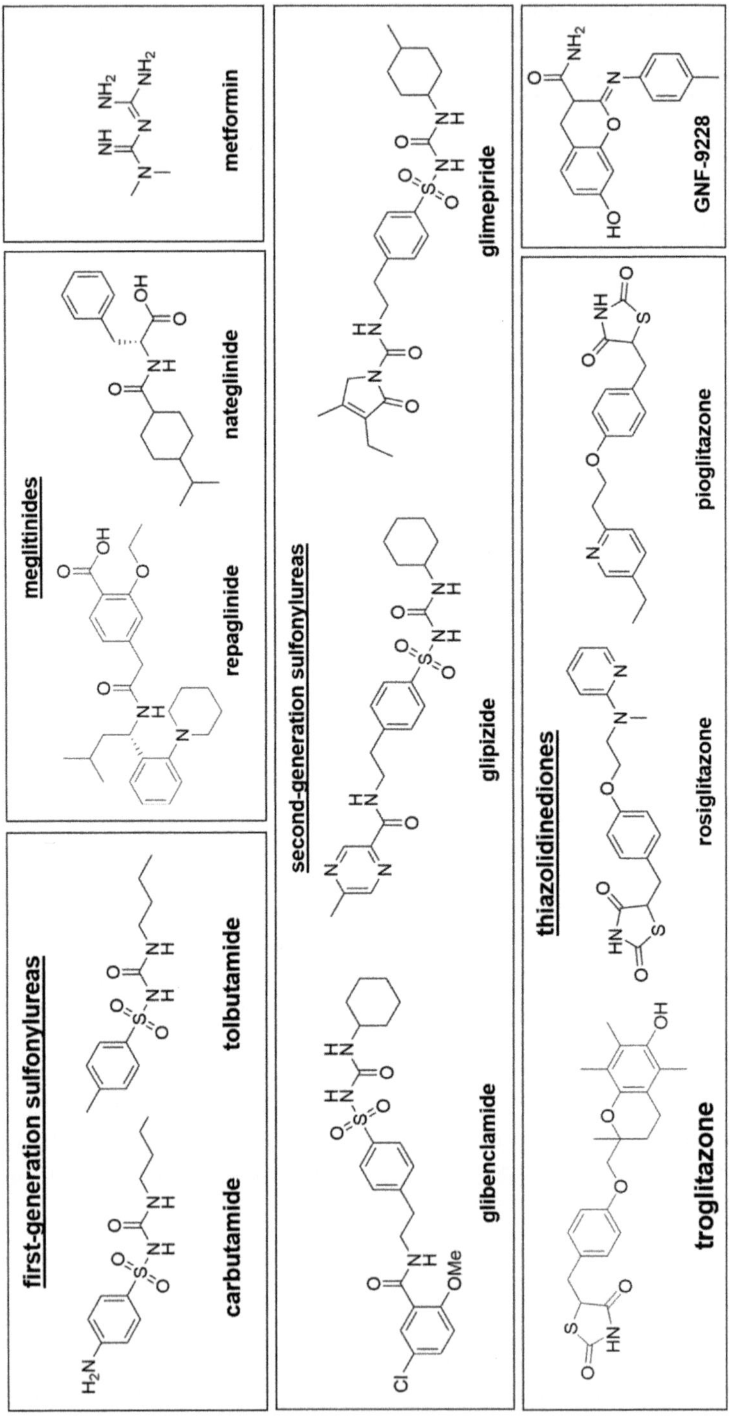

Figure 1.1 Chemical structures of current clinically used drugs for type 2 diabetes.

Therapeutic approaches promoting insulin secretion were (and still are today) based on sulfonylureas. In this process, insulin secretion from beta cells is induced by the increase in blood glucose. Oxidation of glucose drives an increase of ATP levels in the beta cell that in turn induces a closure of the ATP-sensitive potassium channel (K_{ATP}) and subsequent depolarization of cell membrane. This, in turn, induces calcium entry through L-type calcium channels and insulin secretion.[152] The sulfonylureas act by closing the K_{ATP} channel to induce insulin secretion.[153] The first generation of sulfonylureas is represented by tolbutamide and its analog carbutamide, introduced in the mid-1950s.[154] They are still available for prescription, but are no longer considered first-line therapeutics, due to side effects such as hypoglycemia, weight gain, hypersensitivity, and cardiovascular risks. Considering the high therapeutic value of sulfonylureas, much progress has been made to develop second-generation drugs that include, among others, glibenclamide, glipizide, and glimepiride. These second-generation sulfonylurea drugs were introduced around 1980. They are still widely prescribed although not as a first therapeutic approach. Second-generation sulfonylureas have increased potency but also suffer from side-effect profiles similar to the first-generation therapeutics, along with increased hepatotoxicity events.[155] Although sulfonylureas have been extensively used over many decades, their major liability is hypoglycemia. This effect is not surprising, since their mechanism of action is independent of both blood glucose levels and of glucose sensing by beta cells. The development of direct but glucose-dependent insulin secretagogues could have an important impact on treatment in the future, and as we will see below, the GLP-1-based incretin mimetics have an impact on this pathway.

Another class of insulin secretagogue is the meglitinides. Similar to the sulfonylureas, they act by blocking the K_{ATP} channel, but since they are short-acting, it is believed that they promote the first phase of insulin secretion. Physiologically, in response to glucose, beta cells secrete insulin in a biphasic mode. There is a first rapid secretion phase that occurs within minutes, and a second phase that occurs over a few hours.[156] Liver is more sensitive to the first phase of insulin secretion and in diabetes there is a marked loss of the first phase of insulin secretion.[157] Therefore, the idea behind the development of the meglitinides is that restoring the first phase of insulin secretion may create a therapeutic opportunity. In the clinic, the meglitinides appear to be efficacious both as monotherapy or in combination, and due to their fast-acting properties they are less prone to induce hypoglycemia.

1.7.1.2 Insulin Sensitizers

A second dimension of insulin-centric therapeutics is drugs that increase the insulin sensitivity of peripheral tissues. The archetype of this category is metformin. From a drug discovery point of view, metformin is quite unusual. Metformin is very poorly absorbed by the gut and therefore requires elevated doses to reach plasma efficacy (about 500–2500 mg per day), it is not metabolized, and it is excreted intact, with a half-life of about 5 hours.[158] Metformin was introduced as a medication in the United States in 1995. The anti-diabetic effects of metformin mainly work by increasing the sensitivity of tissues to insulin and by inhibiting hepatic gluconeogenesis.[159,160] Metformin is the most widely prescribed anti-diabetic drug, often as first-line therapy.[161] Metformin is well tolerated, can promote weight loss,[162] and does not show significant side effects.[158] Despite its intense use, the molecular mechanism of metformin remains elusive, although some reports indicate that metformin may indirectly promote activation of AMPK.[163]

In the same category of insulin sensitizers are the thiazolidinediones (troglitazone, rosiglitazone, and pioglitazone). Approved in the late 1990s, they work mainly by sensitizing adipose tissue to the effects of insulin. Thiazolidinediones activate PPARγ.[164] The PPAR nuclear receptor family is known to be involved in the regulation of fatty acid metabolism.[165] Not all thiazolidinediones were created equal, though. While troglitazone was efficacious as monotherapy or in combination with sulfonylureas or metformin, it showed clear signs of hepatoxicity[166] and was withdrawn from use in the United States in 2000. In terms of efficacy, both rosiglitazone and pioglitazone are comparable to troglitazone, and while less hepatotoxic, they showed a tendency to induce an increase in body weight.[167] Although quite efficacious and less hepatotoxic, the fate of rosiglitazone followed that of troglitazone. Rosiglitazone was withdrawn from the US market in 2007 after a meta-analysis suggested it may be associated with an increase in heart failure.[168]

1.7.2 Second-generation Therapeutics

This overview might appear to provide a gloomy vision of drug discovery for anti-diabetic drugs. Cardiovascular complications are strongly associated with diabetes, and they are probably the major cause of mortality. Having on the market drugs that improve glycemic control but that may cause cardiovascular complications became a significant concern to the FDA. Based on these concerns, the FDA issued guidelines in 2008 requiring that any novel anti-diabetic agent, regardless of the mechanism of action, should bring evidence showing that a new drug, if not actually beneficial for cardiovascular effects, should at least be neutral to cardiovascular risks (https://www.fda.gov/media/71297/download). These guidelines set the bar quite high, probably limiting the investment, in terms of funding and effort, that drug manufacturers were willing to dedicate for anti-diabetic research.

1.7.2.1 GLP-1 Receptor Agonists

Despite this outlook, a significant paradigm shift in the development of a novel anti-diabetic therapeutic arose from the discovery of the incretin hormones GIP and GLP-1.[169] Incretin hormones play an important role in promoting insulin secretion after ingestion of a meal.[102] GLP-1 stimulated particular interest due to its pleiotropic effects in the body. GLP-1 induces insulin secretion after oral ingestion of glucose and reduces glucagon secretion and glycogenolysis.[170] The GLP-1 receptor (GLP-1R) is a member of the class B of G-protein coupled receptors (GPCR). From a structural point of view, class B GPCRs are characterized by an N-terminal extracellular domain and a transmembrane domain that contain the GPCR motif (seven membrane-spanning α-helices). Signaling from class B GPCRs is mediated mainly by adenylate cyclase to increase the levels of intracellular cAMP.[171]

Class B GPCRs are famously resistant to drug discovery, which has hampered the identification of agonists; this has been true for GLP-1R as well. Efforts to identify small-molecule GLP-1R agonists in a drug-discovery setting were developed but have been slow to yield successes. There are numerous examples of small-molecule GLP-1R agonists reported in the literature,[172,173] but they have not yet reached clinical application. Nonetheless, intense efforts have been directed at developing biologic drugs based on the incretin biology. As a result, the landscape of GLP-1 analogs is quite impressive, with the introduction of first-in-class biologics such as exenatide and liraglutide. Both are GLP-1 analogs that have been engineered to have an extended half-life (2.5 hours, *vs.* 5 minutes for native GLP-1).[174] Aside from their ability to control glycemia, GLP-1 analogs are also showing some unexpected beneficial side effects, such as reduction of body weight and long-term cardiovascular benefits.

One of the basic concepts in the pathogenesis of diabetes is the failure of beta cells to accommodate increased insulin demand. The progressive decline in beta-cell function is accompanied by a loss of mass likely due to increased apoptosis. From a therapeutics point of view, it is emerging that inducing insulin secretion may be insufficient to prevent beta-cell exhaustion, and may even be harmful to the beta cell.[175] From these observations emerged the idea of "beta-cell rest." If driving insulin secretion over the long term leads to a deterioration in beta-cell function, then beta-cell rest would likely be beneficial. Therefore, any therapeutic agent that is able to promote both insulin secretion and beta-cell rest may have a beneficial effect in preventing beta-cell exhaustion. Currently, GLP-1R agonists seem to meet these criteria.

1.7.2.2 Sodium-glucose Co-transporter Inhibitors

Another novel class of anti-diabetic drugs with exceptional promise is the class of sodium-glucose co-transporter 2 (SGLT2) inhibitors. Under physiological conditions, excess plasma glucose is filtered through the renal glomerulus. From here, it is reabsorbed into the blood *via* the sodium-glucose co-transporter SGLT2.[176] The idea behind SGLT2 inhibitors is that blocking the re-absorption of glucose may induce glucosuria, an increase of excreted glucose in the urine. The story of the discovery of SGLT2 inhibitors is a "poster child" for chemical genetics, as defined in the introduction to this chapter. First, an "experiment of nature" showed that inhibiting SGLT2 induced glucosuria. It had been shown that autosomal recessive mutations in *SLC5A2* (the gene encoding SGLT2) are responsible for familial renal glucosuria. Even in its more extreme instances (complete absence of tubular glucose resorption), clinical manifestations are rare, and this condition is currently accepted to be relatively benign.[177] Second, phlorizin, a natural compound from the bark of pear, apple, and cherry trees and known to science since the late 1800s, was found to induce glucosuria and reduce plasma glucose. Mechanism-of-action studies showed that phlorizin acts as a competitive inhibitor of SGLT2.[178] As it stands, phlorizin lacks the properties of an ideal drug (*i.e.*, it has poor absorbance and low selectivity), but it provided an excellent starting point for a medicinal chemistry campaign. Over time, more successful derivatives of phlorizin were developed, such as canagliflozin, dapagliflozin, and empagliflozin, which reached the market around 2015.[178] Importantly, SGLT2 inhibitors have shown dramatic beneficial cardiovascular outcomes, including a reduced incidence of cardiovascular death and heart failure hospitalizations in people with and without diabetes, and those with and without prevalent heart failure.[179] These findings may lead to even wider clinical use of SGLT2 inhibitors and raise the bar even higher for the benefit that must be shown by novel therapeutic approaches.

1.7.3 New Therapeutic Approaches

The attention of the pharmaceutical industry has increasingly turned to the academic world to help de-risk novel potential targets for developing anti-diabetic agents. There are many examples of possible new avenues for the treatment of diabetes (Figure 1.2). Here, we will focus on beta cell-centered approaches, for which there are at least four major avenues being pursued: (1) glucose-dependent insulin secretion, (2) beta-cell proliferation, (3) beta-cell protection, and (4) beta-cell transdifferentiation.

1.7.3.1 Glucose-dependent Insulin Secretagogues

As we discussed above, GLP-1 analogs show glucose-dependent insulin secretion. In this competitive field, there are some novel opportunities that have been so far under-investigated, such as the quest for non-GLP-1R-agonist glucose-dependent insulin secretagogues. Surprisingly enough, there are limited data reported in the literature about screening campaigns to identify glucose-dependent insulin secretagogues.[180] In one instance, screening was performed in a low-throughput format, using insulin ELISA as readout and a collection of about 1200 small molecules from the Sigma library of pharmacologically active compounds. This effort identified several dopamine receptor antagonists that significantly enhanced glucose-dependent insulin secretion in rat islets.[180] Subsequent siRNA studies indicated that knockdown of *DRD2* (encoding dopamine receptor D2) had the same effect. However, these observations are in contrast with a body of literature showing that both pharmacological or genetic inactivation of *DRD2* impairs insulin secretion and causes glucose intolerance.[181,182]

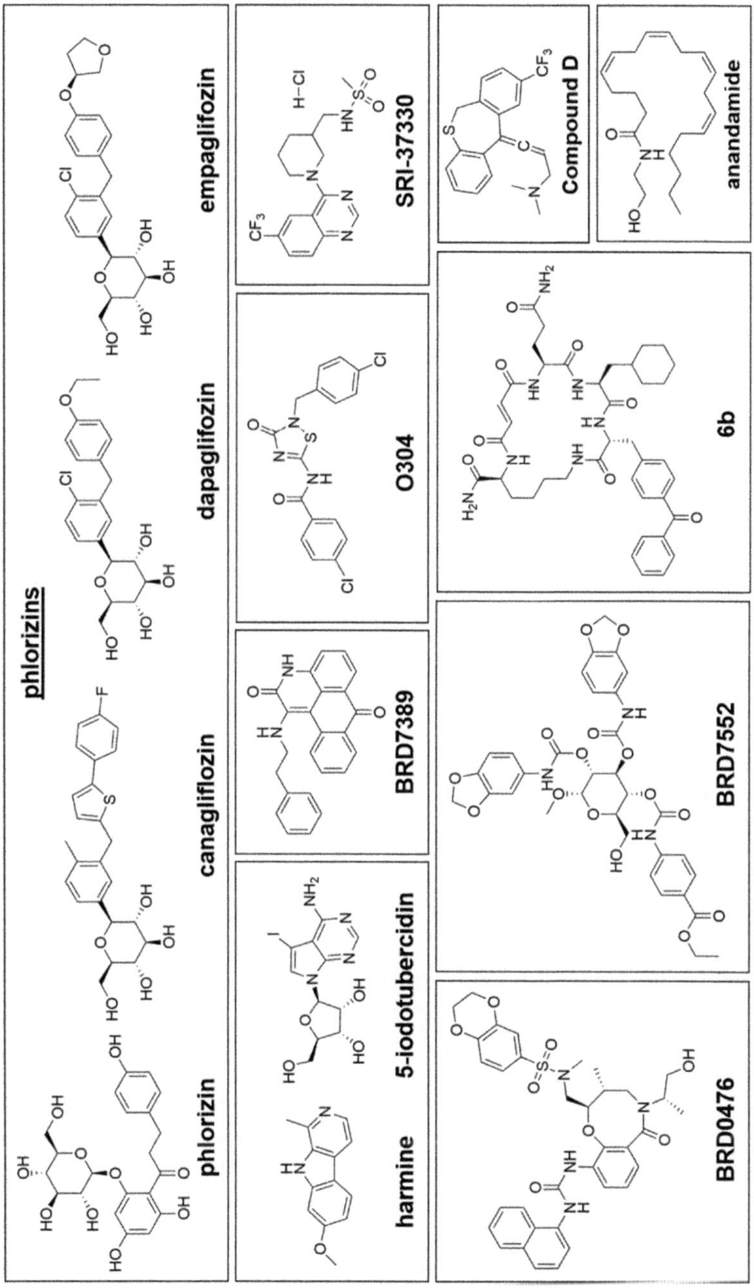

Figure 1.2 Chemical structures of small molecules emerging in new therapeutic approaches to type 1 and type 2 diabetes.

In recent years, our group has performed high-throughput phenotypic screening to identify glucose-dependent insulin secretagogues. First, as mentioned above, we developed a high-throughput luminescent reporter-based system.[150] This system is based on a construct in which the sequence of *Gaussia* luciferase is placed within the C-peptide portion of mouse proinsulin, a fragment that is normally cleaved within secretory vesicles by pH-sensitive prohormone convertase enzymes and co-secreted with mature insulin during exocytosis. With this construct, luciferase tracks to the vesicles but is kept in an inactive state until cleavage of the peptide occurs by PC2 in the acidified vesicle. The freed luciferase is then co-secreted with endogenous mature insulin exocytosis. Using a lentiviral vector, we created a stable integration of the construct within the INS-1E beta cell line. After stimulation with glucose or a secretagogue, luciferase activity is measured on a plate reader in the supernatant using the luciferase substrate coelenterazine. Using this system, we performed high-throughput screening of over 380 000 compounds from the NIH Molecular Libraries Probe Production Centers Network and identified a small molecule that induces insulin secretion in rat INS-1E cells and human islets in a glucose-dependent manner, with an EC50 of ∼1–2 μM. We also found that the compound strongly increases insulin secretion in islets from islet donors who had T2D, strengthening the notion that this compound could effectively modify human disease. These results indicate that novel glucose-dependent insulin secretagogues can be identified by high-throughput screening, and future work to determine the mechanism of action may have an impact on small-molecule discovery in this area.

1.7.3.2 Beta-cell Proliferation

Loss of beta-cell number and function is a key landmark of the pathology of diabetes.[183] Therefore, ways to expand pancreatic beta-cell mass and function could be an effective therapeutic approach. This goal is based on the premise that type 1 diabetes patients with long-standing disease exhibit residual beta-cell mass.[184–186] Small molecule-induced beta-cell proliferation has long been sought after, and there are some interesting examples in the literature.[187–189] Among the most notable examples are the discovery of dual-specificity tyrosine-regulated kinase1a (DYRK1A) inhibitors harmine and 5-iodotubercidine (5-IT) as potent inducers of human beta-cell proliferation,[190–192] including a series of aminopyrazines that were developed to increase potency in promoting human beta-cell proliferation.[192] Interestingly, this class of compounds showed inhibitory effects not only toward DYRK1A but also to GSK3β. Although GSK3β is a well-known target of beta-cell (but also general cell) proliferation,[193,194] it emerged that the simultaneous inhibition of DYRK1A and GSK3β proved more effective in inducing beta-cell proliferation than inhibition of either alone.[192] Additional studies on DYRK1A have shown that simultaneous inhibition of DYRK1A and DYRK1B is necessary for activity;[195] as it turns out, most current chemical matter inhibits both kinases. These results indicate that polypharmacology may be a superior avenue to induce beta-cell proliferation.

In line with these observations, a recent report showed that the simultaneous pharmacological inhibition of DYRK1A and TGFβ[196] and DYRK1A and GLP-1R agonists[197] induces synergistic beta-cell proliferation in human cells, as inhibition of the latter also promotes beta-cell proliferation.[198] Although important as a proof of concept, targeting DYRK1A may have some disadvantages from a pharmacological point of view. First, DYRK1A has a broad tissue expression; therefore, any compound targeting it may induce undesirable proliferation events in peripheral tissues. Second, just about all DYRK1A inhibitors so far described are limited by a lack of specificity. In fact, both 5-IT and harmine target not only DYRK1A but also a subset of other kinases, comprising DYRK1B, DYRK2, DYRK3, DYRK4, CLK1, CLK2, CLK3, and CLK4.[190] Inhibiting these kinases may have as-yet unknown effects in humans.

Overall, though, these discoveries are paving the way to identify compounds with novel mechanisms of action with high selectivity for beta cells. For example, recently, GNF-9228 was discovered to promote DYRK1A-independent human beta-cell proliferation.[199] Along this line is the discovery that heterozygous inactivation of the Na^{+}/Ca^{2+} exchanger (isoform 1; NCX1) increases beta-cell proliferation in mice.[200] Based on these observations, recently the same group reported that chemical inhibition of the same target using phenoxybenzamyl-derivatives small molecules promoted beta-cell proliferation in mice.[201] The use of a more stringent approach resulted in the identification of glucocorticoids as inducers of beta-cell proliferation in rats and humans.[202] In this particular instance, beta cells from adult rats and humans were FACS purified and treated with a collection of 1280 bioactive compounds (LOPAC) for 6 days. At least four compounds were able to induce beta-cell proliferation, and these were glucocorticoid receptor ligands. Among those hits, hydrocortisone and methylprednisolone almost doubled the number of beta cells in 2 weeks. Glucocorticoids, however, induce insulin resistance on their own,[203] but in this case, the authors aimed to expand beta cells *ex vivo* for transplantation, so this systemic effect may not be as much of an issue.

1.7.3.3 Beta-cell Protection

Another approach to deal with the loss of beta-cell mass during diabetes is represented by protection from cell death. This kind of approach must be tailored based on the type of cellular insult imposed on the beta cell. While in T1D, the immune system recognizes the beta cell as foreign, resulting in destruction of the beta-cell population, in T2D, the insulin resistance that precedes overt disease will drive beta cells to exhaustion and ultimately cell death. In the case of immune attack, a model frequently used involves treatment with inflammatory cytokines such as interleukin-1β (IL-1β), interferon-γ (IFN-γ), and tumor necrosis factor-α (TNF-α) produced by macrophages and

T cells.[204,205] IL-1β and TNF-α induce NFκB expression, with downstream gene expression regulated by nitric oxide (NO) signaling, which increases endoplasmic reticulum stress-response pathways and decreases beta cell-specific function.[204,206] NFκB activation, along with IFN-γ-induced STAT1 signaling,[207,208] work together to abolish insulin secretion and to induce beta-cell apoptosis.[209] Efforts to inhibit individual cytokines with protein-based receptor antagonists had previously progressed to clinical trials, but did not lead to approved therapies.[210] For these reasons, we used Cas9 to engineer IL-10 to be secreted, along with insulin, beta cells.[211] On the other hand, small-molecule suppressors of cytokine-induced beta-cell apoptosis could lead to entirely new avenues for therapeutic intervention.

Studies have described small molecules, such as extracts from *Artemisia capillaris*[212] and St. John's wort,[213] that have protective effects against cytokines. Many of these natural products were discovered because of their general antioxidant or anti-inflammatory properties, but some of them have shown inhibitory activity toward JAK-STAT[214] or NFκB signaling.[215] Aside from these serendipitous approaches, more systematic ones were described. For example, we reported the identification of BRD0476 from high-throughput screening efforts to discover suppressors of beta-cell apoptosis.[216] BRD0476 inhibits IFN-γ-induced JAK2 and signal transducer and activator of transcription 1 (STAT1) signaling to promote beta-cell survival. Interestingly, BRD0476 does not have kinase inhibitory activity but we found that it works through an unusual mechanism of action by targeting the deubiquitinase ubiquitin-specific peptidase 9X (USP9X). In particular, we found that a competition between phosphorylation and ubiquitination on JAK2 explains the ability of BRD0476 to protect beta cells from death.[37] More recently, *TYK2* (tyrosine kinase 2) has been proposed as a genetic association with T1D,[217] and preclinical evaluation of TYK2 inhibitors for treatment of T1D is underway.[218] In this case, TYK2 activation occurs through IFN-α signaling in the beta cell, and these compounds show great promise.

The fate of the beta cell in overt T2D is no less dramatic. Glucotoxicity, lipotoxicity, and glucolipotoxicity are considered the main driving factors of beta-cell death in T2D. The underlying concept is that once the primary pathogenesis of diabetes is established, hyperglycemia and hyperlipidemia ensue, and thereafter exert additional damaging or toxic effects on the beta cells. In terms of small-molecule discovery, glucolipotoxicity-induced beta-cell death has been relatively neglected. At least two reports of high-throughput screening campaigns have been described.[219,220] The first report focused more on the protection from lipotoxicity. The rat insulinoma cell line INS-1 was exposed to palmitate to induce cell death in the presence of compounds. The collection comprised about 2000 compounds encompassing kinase inhibitors and other bioactive compounds. Hit prioritization and *in silico* optimization ultimately identified MAP4K4 inhibitors and the endocannabinoid anandamide as effective compounds at inhibiting palmitate-induced apoptosis in INS-1 cells as well as in rat and human islets.[219]

The second report focused on a more comprehensive approach aimed at protecting beta cells from glucolipotoxicity. Again, the INS-1E cell line was used as a model, and the compound collection included ~300 000 compounds. The readout, Caspase-Glo, was tailored to identify compounds that specifically prevent apoptosis rather than generic cell death. Further hit prioritizations identified six scaffolds that specifically protected beta cells from glucolipotoxicity-induced apoptosis.[220] Following systematic medicinal chemistry efforts, Compound D was obtained as the most potent and selective. Furthermore, mechanism-of-action studies indicated that Compound D works through a reduction of calcium influx, although definitive target proteins were not identified.

In terms of more general protection from beta-cell death, the thioredoxin-interacting protein (TXNIP) has emerged as a promising therapeutic target. TXNIP was found to be the top glucose-induced gene in a human pancreatic islet gene-expression microarray.[221] Its increased expression is associated with diabetes and beta-cell apoptosis, while its absence in beta cells is accompanied by a significant decrease in apoptosis and with an overall protective effect from diabetes. Interrogation of a 300 000-compound library identified SRI-37330 as a non-toxic small molecule, which effectively rescued mice from obesity-induced diabetes.[222] Moreover, SRI-37330 inhibited glucagon secretion, reduced hepatic glucose production, and reversed hepatic steatosis. The continued growth of this area will ensure that the validity of this approach will become more thoroughly vetted.

1.7.3.4 Beta-cell Transdifferentiation

The concept behind the idea of transdifferentiation is the plasticity of cell state. Advances in developing iPS cells have shown that cellular reprogramming no longer belongs to the realm of speculation. Furthermore, in the context of islet biology, many examples of spontaneous reprogramming have been described.[223] These findings support the idea that small-molecule methods may be used to reprogram cells to a beta-like state.

An intuitive starting point for beta-cell transdifferentiation is the alpha cell, since both share a common developmental lineage.[224] There are different examples of successful screening campaigns to identify small molecules promoting alpha- to beta-cell transdifferentiation. One of the first was reported in 2010 when BRD7389 was identified from a phenotypic screen of about 30 000 compounds.[225] From a molecular point of view, BRD7389 was identified as having a strong inhibitory effect on the RSK kinase family. Subsequently, another report showed that chronic treatment with GABA could induce alpha- to beta- transdifferentiation *in vivo*.[226] Similar results were obtained by another group, who also identified artemisinins as potentiators of GABA signaling by stabilizing gephyrin.[227] Following these reports, an effort rebutting these findings was reported, when lineage-tracing experiments failed to recapitulate the results, putting into question whether these agents may represent novel diabetes therapies.[228]

There is a long controversy around the idea that new beta cells may derive not only from replication but also from progenitors or stem cells.[229] In recent years, the scientific community seemed to reach the conclusion that these events are not mutually exclusive. From a therapeutic point of view, this opens more opportunities and chemical biology has taken advantage. A report described high-throughput screening of ~60 000 compounds. The human ductal cell line PANC-1 was used as a model and the readout was expression of *PDX1*, a master regulator of both beta-cell development and insulin expression. This effort identified BRD7552 as an inducer of PDX1 mRNA and protein expression in primary ductal cells. Furthermore, prolonged compound treatment induced both insulin mRNA and protein expression.[230]

1.7.4 Emerging Therapeutic Approaches

We have attempted to provide an overview of the most systematic unbiased approaches underway in four major phenotypic areas of intervention: glucose-dependent insulin secretion, beta-cell proliferation, beta-cell protection, and transdifferentiation. The complexity of beta-cell biology, though, means many more avenues may be open for small-molecule discovery as well. Here, we will provide some of the more recent and more relevant examples of possible therapeutic avenues that stem from basic science.

As described above, one of the most successful anti-diabetic drugs still prescribed is metformin. Although the precise mechanism of action of metformin remains mysterious, a body of evidence seems to indicate that, ultimately, metformin may work by activating AMPK, a key enzyme controlling glucose uptake and metabolism by peripheral tissues.[231] These premises were used to identify the compound O304 as a pan-AMPK activator. O304 is a small heterocyclic compound developed by Betagenon AB through rationale design for AMPK activators from an early hit molecule identified in a cellular screen. O304 showed beneficial effects in increasing glucose uptake, reducing beta-cell stress, and promoting beta-cell rest.[232] Moreover, O304 showed good performance in reducing fasting plasma glucose in patients with T2D and, interestingly, O304 improved peripheral microvascular complications.

Among the major players regarding anti-diabetic therapy, a central role is played by insulin itself. Since its discovery in 1922, insulin remains a life-saving drug for many. Therefore, it is no surprise that many approaches have aimed at improving the performance of endogenous insulin. One of the most notable examples is the targeting of insulin degrading enzyme (IDE), a ubiquitous zinc-metalloprotease responsible for the degradation of insulin, glucagon, and amyloid beta-protein.[233] Again, the proof of concept that IDE may be an interesting target for anti-diabetic drugs arose from the observation that a non-protein inhibitor of IDE extracted from liver was able to increase the effect of insulin *in vivo*.[234] Despite these observations, the discovery of small-molecule inhibitors of IDE proved quite challenging.

In the first report from 2014, inhibitors were obtained by screening a DNA-templated library of 13 824 synthetic macrocycles.[235] From this screen, six candidate IDE-binding molecules were identified, and among those, the compound 6b had the highest inhibitory effect on IDE. *In vivo* studies showed 6b was able to improve glucose tolerance upon oral glucose administration. 6b, however, also prevented glucagon degradation, which is an undesirable effect for an effective anti-diabetic drug. In a subsequent iteration, the same group described novel small-molecule IDE inhibitors identified in a screen of ~17 000 compounds.[38] The top compound there showed greater efficacy when insulin was the substrate, as opposed to glucagon.

When administered exogenously, as often happens with diabetic patients, insulin is very effective, but it lacks the physiological control of being secreted only in the presence of elevated glucose. Although insulin analogs have improved glycemic control, glycemic excursions still cause major health problems and complications. Many attempts have been made over the years to develop so-called "smart" insulins, also known as glucose-responsive insulins (GRIs). The concept of GRIs is not new and was introduced for the first time in 1979.[236] So far, excellent GRIs can be obtained by embedding insulin in polymeric matrices containing glucose-responsive elements (*i.e.*, lectin), glucose oxidase, or phenylboronic acids (PBAs). PBAs are small molecules with high affinity for diols such as sugars. Since glucose is the most abundant sugar in the blood, PBAs are particularly apt to develop GRIs. The most recent progress in the development of GRI has been well described.[237]

1.8 Selectively Targeting the Beta Cell

1.8.1 Need for Beta-cell Targeting Moieties

As we have discussed, an attractive way to combat the reduced beta-cell mass in T1D or T2D is to develop small molecules that induce beta-cell regeneration. There is an abundance of chemical matter targeting DYRK1A, which is a kinase that, when inhibited, increases beta-cell proliferation. Efforts in the past several years of beta-cell biology have led to the development of leucettine, harmine, GNF4877, 5-iodotubercidin (5-IT), CC-401, and others.[190,191,238–245] Most of these compounds induce only 1–3% of beta cells to proliferate. More recently, DYRK1A inhibitors were combined with TGFβ inhibitors to have a synergistic mechanism of action to proliferate beta cells, resulting in a 5–8% increase.[196] However, DYRK1A and TGFβ receptors are ubiquitous and not specific. Additionally, DYRK1A inhibitors were used in conjunction with GLP-1R agonists and shown to increase beta-cell proliferation to 5–6%, but again, DYRK1A inhibitors are not specific to beta cells. It has been predicted that 10% of proliferating beta cells may be enough to maintain glucose homeostasis.[197] While these combinations increase the efficacy of beta-cell proliferation, they do not address the problem of off-targets and specificity.

Thus, current beta-cell mitogens have therapeutic limitations because they have a low therapeutic window. Not only do they have low efficiencies in beta cells, but they are also not specific and proliferate other cells as well. Furthermore, the pancreas is composed of a heterogeneous mixture of cells (*e.g.*, alpha, beta, delta). One way to increase the therapeutic window of these compounds is by increasing the localization of the cargo to beta cells, or by creating pro-drugs for the specific release of the cargo in beta cells. In this chapter, we define "cargo" as a small molecule-based entity for delivery. Here, we will review compounds that take advantage of the unique features of beta cells. We focus on examples that (1) conjugate cargo to beta-cell surface markers, and (2) create pro-drugs that take advantage of the high levels of zinc in beta cells. Finally, we conclude with workflows to find new beta-cell markers to further increase specificity.

1.8.2 Beta-cell Targeting

There are several relatively specific cell-surface markers that could be used for beta-cell targeting (Figure 1.3). Among them, GLP-1R, VMAT2, dopamine receptor D2, and SUR1 are the most well-studied. One approach for cargo delivery stems from the concept of antibody–drug conjugates: tethering cargo to antibodies that are highly specific to cancer markers on a desired cell type is able to localize the cargo to the cell. Over the past few decades, several groups have developed radiolabeled or fluorescent ligands to help identify and sort beta cells. The logical next step is to conjugate these probes to cargo to specifically target them to beta cells.

A number of probes targeting cell surface markers exist (Table 1.2), including exendin-4, which targets GLP-1R; dihydrotetrabenazine (DTBZ), which targets VMAT2; (+)-4-propyl-9-hydroxynaphthoxazine (PHNO) and fallypride, which

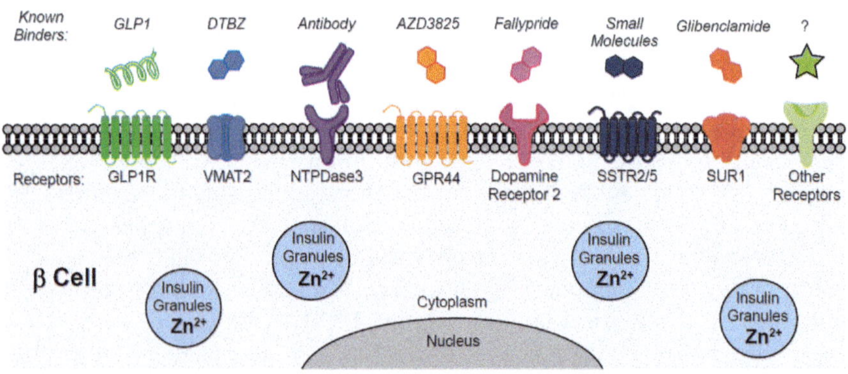

Figure 1.3 Cell schematic of the beta cell, with promising surface markers indicated. The top row lists known binders, and whether they are peptides (green helix), small molecules (double hexagons), or proteins (purple antibody). The bottom row lists the names of the candidate surface receptors.

Table 1.2 Beta-cell markers that can be leveraged for cell-specific therapeutic delivery. Note that not all markers are surface markers (*e.g.*, zinc).

Marker	Binder	Specificity and possible off-targets	Applications
GLP1 receptor	Peptide GLP-1(7–36)-NH_2	Relatively specific, but can be found in other tissues	Co-delivery of glucagon, amylin, cholecytokinin, gastrin, and nuclear hormones in the form of hybrid peptides, delivery of antisense oligonucleotides, imaging of transplanted islets in humans
VMAT2	DTBZ	Expressed in beta cells and neuroendocrine cells. Not found in rat pancreas	Imaging of endogenous pancreatic beta-cell mass in human, used for delivery of proliferating reagent to mice pancreas with uncleavable linker
Dopamine receptor 2	Fallypride and (+)-4-propyl-9-hydroxynaphthoxazine (PHNO)	Expressed in brain, endocrine tissue, pancreas	[^{18}F]-Fallypride used in imaging of transplanted islets in mice, [^{11}C]-PHNO showed promise in differentiation between T1DM and health subjects
SUR1	Glibenclamide	Pancreatic beta cells and in neurons, astrocytes, oligodendrocytes, endothelial cells	Glibenclamide-based dendrimers were used to label human islets *in vitro*
GPR44	AZD3825	Identified as a beta cell-specific biomarker through a proteomic screen, but its ligand accumulates in small intestines and spleen	Used to stain human pancreas, and *in vivo* in pigs and non-human primates
NTPDase 3	Antibody	Somewhat specific to beta cells, found on delta cells, but not on alpha cells	NTPDase3 antibody enabled sorting of beta cells and *in vivo* imaging
Somatostatin receptor SSTR2/5	Beta-naphthylalanine, several other binders reported	SSTR2 in both alpha and beta cells, SSTR5 expressed in β-cells only	Higher uptake in pancreas demonstrated in mice
LAT	5-Hydroxy-tryptophan, DOPA	Localized to islets, absent in exocrine pancreas. Found in brain, liver, kidney, *etc.*	Estimation of beta-cell mass in humans
GLUT2	Fluorodeoxyglucose	Expressed in liver, endocrine pancreas, and kidneys	Imaging in clinical islet transplantation
VDCC	Manganese-based probes	Ubiquitously expressed in various *cell* types throughout the body	Used to distinguish between non-diabetic and type 2 diabetic patients by MRI
Zinc	Zinc chelators	Functional beta cells have over a million-fold higher zinc ion (Zn^{2+}) concentration than other cell types and plasma	Imaging and sorting of beta cells, more recently, *in vitro* delivery of mitogens

target D2; and glibenclamide, which targets the K_{ATP} channel. Radiolabeled probes based on their specific ligands were used for PET imaging of the pancreas in humans. Other markers, such as GRP44, somatostatin receptor SSTR2/5, and NTPDase3, are less studied, but still hold promise. GPR44 was identified as a beta cell-specific marker through a proteomic screen,[246] and its ligand AZD3825 demonstrated ten times higher endocrine-to-exocrine binding ratio than for DTBZ. In contrast to SSTR2, which can be found on the surface of both alpha and beta cells, SSTR5 was shown to be almost exclusively expressed in beta cells.[247] Furthermore, an increased pancreatic uptake of its radiolabeled ligand beta-naphthylalanine was demonstrated in mice.[248] Besides beta-naphthylalanine, several small-molecule binders of SSTR5 were discovered and can be leveraged for the development of beta cell-specific probes.[249–251] Recently, NTPDase3 was identified as a highly specific marker of beta cells.[252] Immunohistochemistry on pancreatic tissue sections from human donors showed high expression of NTPDase3 on the surface of beta cells, some expression on delta cells, but none on the surface of alpha cells. Using NTPDase3 antibody, isolation and *in vivo* imaging of beta cells were performed. In addition to all of these mentioned candidates, voltage-gated calcium channel (VDCC), large neutral amino acid transporter (LAT), and glucose transporter 2 (GLUT2) were reported to be used to image beta cells. For example, the correlation between C-peptide production and ^{11}C-5-hydroxy-tryptophan uptake points to the possibility of LAT exploration in beta cell-targeted delivery.[253]

1.8.2.1 Leveraging GLP-1R for Beta-cell Targeting

As we have discussed, the glucagon-like peptide-1 receptor (GLP-1R) is a GPCR that is highly expressed on beta cells, although there is also some distribution on delta cells, found in the islet, and other cell types.[254] When GLP-1 binds its receptor, it stimulates a cascade that ultimately triggers insulin release. Thus, several peptide-based receptor agonists have been developed to trigger insulin release, but there are practical side-effects to using these in the clinic. More recently, non-peptidic, orally available GLP1-R agonists have been developed, and while clinical development is underway, they can serve now as a potential probe to which to tether cargo.[255] Several groups have already conjugated GLP-1 peptides to cargo and demonstrated their specific delivery to beta cells. These have included ^{125}I-radiolabeled GLP-1 analogs to visualize insulinomas *in vivo*.[256] Following this work, an ^{11}In-labeled exendin analog was developed and used for visualization of beta-cell mass in rats, mice, and recently humans by single photon emission computed tomography.[257–259] Other efforts developed a near-infrared probe based on extendin-4 and VivoTag-680 fluorochrome. Three minutes after intravenous administration of this probe to mice, pancreatic islets were detectable by intravital microscopy and the signal peaked after 20 minutes, with 6 : 1 target-to-background selectivity.[260] Soon after this discovery, the group improved selectivity of imaging up to 64 : 1 using VivoTag-750.[261] Besides applications in beta-cell imaging, GLP-1R was also used for the delivery of therapeutic agents. For example, estrogen conjugated to GLP-1 was used as a targeted therapy to reverse metabolic syndrome in rodents.[262] Further, anti-sense oligonucleotides conjugated to a GLP-1 peptide showed increased uptake in beta cells.[263,264] As an alternative approach, GLP-1 and glucagon co-agonists have been developed, as well as GLP-1 and GIP co-agonists. These examples illustrate how GLP-1R-targeting moieties could be used to deliver active cargo (drugs) to beta cells.

1.8.2.2 Leveraging VMAT2 for Beta-cell Targeting

The first attempt to perform beta cell-specific delivery of mitogens covalently linked the beta-cell proliferating agent aminopyrazine (AP) to the VMAT2 ligand DTBZ.[265] The AP–DTBZ conjugate retained its VMAT2 binding affinity but demonstrated a four-fold reduction in activity to induce proliferation. *In vivo* tissue distribution in Balb/C mice (by LC-MS) demonstrated a promising enrichment of the conjugate in the pancreas, with a tissue-to-plasma ratio of fifteen. However, a significant amount of AP–DTBZ was also found in the liver and kidneys, with tissue-to-plasma ratios of 2.1 and 4.9, respectively. The authors speculated that such low specificity can be attributed to the unsuccessful choice of animal model, since VMAT2 expression in rodent beta cells has been reported to be much lower than in other species. After this publication, there were no follow-up attempts to improve the efficiency of the system or test it in a more suitable model – pig or primate.

1.8.2.3 Leveraging Zinc for Beta-cell Targeting

In addition to cell-surface markers discussed above, beta cells have an incredibly high level of Zn^{2+} ions in their insulin vesicles, with concentrations reaching up to 30 mM in the "tightly bound" form and 100 μM in the "free or loosely bound" form.[266] By contrast, most other cell types and plasma have no more than 1 nM of Zn^{2+}. Thus, beta cells have over a million-fold higher free Zn^{2+} concentration than other cell types and plasma, allowing Zn^{2+} indicator dyes to be employed in beta-cell sorting and islet transplantation imaging.[267,268] Recently, zinc-based targeting of the beta cell was further extended to specific delivery of proliferation-inducing agents. For example, covalent attachment of the dipicolylamine zinc chelator to the beta-cell mitogen GNF4877 resulted in enhanced accumulation of the hybrid molecule in the beta cell.[269] Treatment of human islets with chelator-modified GNF4877 led to only 1.3-fold selectivity in the induction of proliferation of beta cells *versus* non-beta cells. One of the possible explanations for low specificity is a conjugation-related reduction in the potency of the mitogen. A similar drop in potency (four-fold) was observed when dihydrotetrabenazine (DTBZ) was covalently conjugated to the beta-cell proliferating agent aminopyrazine in the VMAT2-based targeting strategy discussed above.

The specificity of beta-cell targeting can be further improved with the use of a pro-drug approach, in which an administered compound is converted to an active drug only at the site of action. In this approach, an active molecule can be masked by conjugation with a binder of a beta-cell marker (*e.g.*, Zn^{2+} or VMAT2), which leads to not only the enrichment of drug in beta cells but also unmasking of the active compound following reaction-based activation steps. Zn^{2+} is a powerful Lewis acid, allowing it to be used for the development of highly sensitive and selective reaction-based probes to image Zn^{2+} in tissues.[246,250] In these probes, coordination of the Zn^{2+} ion to its chelator is followed by ligand-directed Zn^{2+}-catalyzed ester/carbamate hydrolysis and release of fluorescent compound. In our proof-of-concept studies, molecules able to release ZP1 and BODIPY selectively in beta cells were designed and applied for beta-cell sorting and islet imaging.[270,271] Furthermore, second-generation conjugates of zinc chelator with proliferating reagents (GNF4877 and harmine) were developed.[271,272] In contrast to the first-generation molecules developed, new hybrid compounds are based on a pro-drug strategy, thus allowing beta cell-specific delivery of unmodified proliferation-inducing agents with preserved potency. There is still much room left for improvement of beta cell-targeting pro-drugs based on the Lewis acidity of zinc, and new assays that could allow rapid screening of known zinc chelators, cargos (mitogens, immunosuppressants, *etc.*), and the nature of chemical bonds between them, need to be developed.

1.8.3 Next Steps Toward Selective Targeting

In addition to the markers mentioned above, there are several underexplored proteins which could be leveraged for the development of beta cell-specific therapies. For example, TMEM27 and its protease BACE2 are enriched in beta cells, and radio- and fluorophore-labeled TMEM27 antibodies were used for beta-cell imaging.[273,274] However, the development of specific small-molecule binders for TMEM27 will be required to increase the therapeutic potential of this target. TAK-875, a binder of another promising target, GRP40, was recently used for the design of novel probes for beta-cell imaging.[275] In the case of zinc-based probes and pro-drugs, specificity can be further improved with the use of novel zinc binders, such as low-affinity hydroxyphenylbenzoxazole or benzimidazole-based ligands.[276] However, despite the incredible potential of these candidate markers, specificity still might be a limiting factor on the way to therapeutic application. Thus far, we have been unable to single out even one marker that does not have off-target expression in pancreas or other tissues. Nevertheless, this limitation can be addressed *via* the careful design of agents with several layers of specificity. For example, one group synthesized a triagonist:GLP-1/GIP/glucagon,[277] and we predict that combinations like these will continue to emerge to compensate for the lack of beta cell-specific therapeutic agents. Furthermore, any of the chemical biology approaches described throughout this chapter (*e.g.*, DELs, PROTACs, CRISPR screening, photo-cross-linking, and target identification) can be used to develop probes or drugs for the understanding and treatment of diabetes.

Abbreviations

CETSA	cellular thermal shift assay
CMap	Connectivity Map
DARTS	drug affinity responsive target stability
DEL	DNA-encoded library
DOS	diversity-oriented synthesis
GIP	glucose-dependent insulinotropic peptide
GLP-1	glucagon-like peptide-1
GPCR	G-protein coupled receptor
GRI	glucose-responsive insulin
GSIS	glucose-stimulated insulin secretion
HTS	high-throughput screening
NOD	non-obese diabetic mouse model
PROTAC	proteolysis targeting chimera
SGLT2	sodium-glucose co-transporter 2
SPROX	stability of proteins from rates of oxidation
T1D	type 1 diabetes
T2D	type 2 diabetes
TPP	thermal proteome profiling
TSA	thermal shift assay
WHO	World Health Organization

References

1. W. H. Herman and P. Zimmet, *Diabetes Care*, 2012, **35**, 943–944.
2. G. Roglic, *Global Report on Diabetes*, World Health Organization, 2016.
3. P. Workman and I. Collins, *Chem. Biol.*, 2010, **17**, 561–577.
4. R. M. Rydzewski, *Real World Drug Discovery: A Chemist's Guide to Biotech and Pharmaceutical Research*, Elsevier, 2008.
5. B. K. Wagner, *Introduction to Chemical Probes*, Royal Societry of Chemistry, 2021.
6. C. J. Gerry and S. L. Schreiber, *Nat. Rev. Drug Discovery*, 2018, **17**, 333–352.
7. E. S. Lander, L. M. Linton, B. Birren, C. Nusbaum, M. C. Zody, J. Baldwin, K. Devon, K. Dewar, M. Doyle, W. FitzHugh, R. Funke, D. Gage, K. Harris, A. Heaford, J. Howland, L. Kann, J. Lehoczky, R. LeVine, P. McEwan, K. McKernan, J. Meldrim, J. P. Mesirov, C. Miranda, W. Morris, J. Naylor, C. Raymond, M. Rosetti, R. Santos, A. Sheridan, C. Sougnez, Y. Stange-Thomann, N. Stojanovic, A. Subramanian, D. Wyman, J. Rogers, J. Sulston, R. Ainscough, S. Beck, D. Bentley, J. Burton, C. Clee, N. Carter, A. Coulson, R. Deadman, P. Deloukas, A. Dunham, I. Dunham, R. Durbin, L. French, D. Grafham, S. Gregory, T. Hubbard, S. Humphray, A. Hunt, M. Jones, C. Lloyd, A. McMurray, L. Matthews, S. Mercer, S. Milne,

J. C. Mullikin, A. Mungall, R. Plumb, M. Ross, R. Shownkeen, S. Sims, R. H. Waterston, R. K. Wilson, L. W. Hillier, J. D. McPherson, M. A. Marra, E. R. Mardis, L. A. Fulton, A. T. Chinwalla, K. H. Pepin, W. R. Gish, S. L. Chissoe, M. C. Wendl, K. D. Delehaunty, T. L. Miner, A. Delehaunty, J. B. Kramer, L. L. Cook, R. S. Fulton, D. L. Johnson, P. J. Minx, S. W. Clifton, T. Hawkins, E. Branscomb, P. Predki, P. Richardson, S. Wenning, T. Slezak, N. Doggett, J. F. Cheng, A. Olsen, S. Lucas, C. Elkin, E. Uberbacher, M. Frazier, R. A. Gibbs, D. M. Muzny, S. E. Scherer, J. B. Bouck, E. J. Sodergren, K. C. Worley, C. M. Rives, J. H. Gorrell, M. L. Metzker, S. L. Naylor, R. S. Kucherlapati, D. L. Nelson, G. M. Weinstock, Y. Sakaki, A. Fujiyama, M. Hattori, T. Yada, A. Toyoda, T. Itoh, C. Kawagoe, H. Watanabe, Y. Totoki, T. Taylor, J. Weissenbach, R. Heilig, W. Saurin, F. Artiguenave, P. Brottier, T. Bruls, E. Pelletier, C. Robert, P. Wincker, D. R. Smith, L. Doucette-Stamm, M. Rubenfield, K. Weinstock, H. M. Lee, J. Dubois, A. Rosenthal, M. Platzer, G. Nyakatura, S. Taudien, A. Rump, H. Yang, J. Yu, J. Wang, G. Huang, J. Gu, L. Hood, L. Rowen, A. Madan, S. Qin, R. W. Davis, N. A. Federspiel, A. P. Abola, M. J. Proctor, R. M. Myers, J. Schmutz, M. Dickson, J. Grimwood, D. R. Cox, M. V. Olson, R. Kaul, C. Raymond, N. Shimizu, K. Kawasaki, S. Minoshima, G. A. Evans, M. Athanasiou, R. Schultz, B. A. Roe, F. Chen, H. Pan, J. Ramser, H. Lehrach, R. Reinhardt, W. R. McCombie, M. de la Bastide, N. Dedhia, H. Blocker, K. Hornischer, G. Nordsiek, R. Agarwala, L. Aravind, J. A. Bailey, A. Bateman, S. Batzoglou, E. Birney, P. Bork, D. G. Brown, C. B. Burge, L. Cerutti, H. C. Chen, D. Church, M. Clamp, R. R. Copley, T. Doerks, S. R. Eddy, E. E. Eichler, T. S. Furey, J. Galagan, J. G. Gilbert, C. Harmon, Y. Hayashizaki, D. Haussler, H. Hermjakob, K. Hokamp, W. Jang, L. S. Johnson, T. A. Jones, S. Kasif, A. Kaspryzk, S. Kennedy, W. J. Kent, P. Kitts, E. V. Koonin, I. Korf, D. Kulp, D. Lancet, T. M. Lowe, A. McLysaght, T. Mikkelsen, J. V. Moran, N. Mulder, V. J. Pollara, C. P. Ponting, G. Schuler, J. Schultz, G. Slater, A. F. Smit, E. Stupka, J. Szustakowki, D. Thierry-Mieg, J. Thierry-Mieg, L. Wagner, J. Wallis, R. Wheeler, A. Williams, Y. I. Wolf, K. H. Wolfe, S. P. Yang, R. F. Yeh, F. Collins, M. S. Guyer, J. Peterson, A. Felsenfeld, K. A. Wetterstrand, A. Patrinos, M. J. Morgan, P. de Jong, J. J. Catanese, K. Osoegawa, H. Shizuya, S. Choi, Y. J. Chen, J. Szustakowki and International Human Genome Sequencing Consortium, *Nature*, 2001, **409**, 860–921.

8. J. C. Venter, M. D. Adams, E. W. Myers, P. W. Li, R. J. Mural, G. G. Sutton, H. O. Smith, M. Yandell, C. A. Evans, R. A. Holt, J. D. Gocayne, P. Amanatides, R. M. Ballew, D. H. Huson, J. R. Wortman, Q. Zhang, C. D. Kodira, X. H. Zheng, L. Chen, M. Skupski, G. Subramanian, P. D. Thomas, J. Zhang, G. L. Gabor Miklos, C. Nelson, S. Broder, A. G. Clark, J. Nadeau, V. A. McKusick, N. Zinder, A. J. Levine, R. J. Roberts, M. Simon, C. Slayman, M. Hunkapiller, R. Bolanos, A. Delcher, I. Dew, D. Fasulo, M. Flanigan, L. Florea, A. Halpern, S. Hannenhalli, S. Kravitz, S. Levy, C. Mobarry, K. Reinert, K. Remington, J. Abu-Threideh, E. Beasley,

K. Biddick, V. Bonazzi, R. Brandon, M. Cargill, I. Chandramouliswaran, R. Charlab, K. Chaturvedi, Z. Deng, V. Di Francesco, P. Dunn, K. Eilbeck, C. Evangelista, A. E. Gabrielian, W. Gan, W. Ge, F. Gong, Z. Gu, P. Guan, T. J. Heiman, M. E. Higgins, R. R. Ji, Z. Ke, K. A. Ketchum, Z. Lai, Y. Lei, Z. Li, J. Li, Y. Liang, X. Lin, F. Lu, G. V. Merkulov, N. Milshina, H. M. Moore, A. K. Naik, V. A. Narayan, B. Neelam, D. Nusskern, D. B. Rusch, S. Salzberg, W. Shao, B. Shue, J. Sun, Z. Wang, A. Wang, X. Wang, J. Wang, M. Wei, R. Wides, C. Xiao, C. Yan, A. Yao, J. Ye, M. Zhan, W. Zhang, H. Zhang, Q. Zhao, L. Zheng, F. Zhong, W. Zhong, S. Zhu, S. Zhao, D. Gilbert, S. Baumhueter, G. Spier, C. Carter, A. Cravchik, T. Woodage, F. Ali, H. An, A. Awe, D. Baldwin, H. Baden, M. Barnstead, I. Barrow, K. Beeson, D. Busam, A. Carver, A. Center, M. L. Cheng, L. Curry, S. Danaher, L. Davenport, R. Desilets, S. Dietz, K. Dodson, L. Doup, S. Ferriera, N. Garg, A. Gluecksmann, B. Hart, J. Haynes, C. Haynes, C. Heiner, S. Hladun, D. Hostin, J. Houck, T. Howland, C. Ibegwam, J. Johnson, F. Kalush, L. Kline, S. Koduru, A. Love, F. Mann, D. May, S. McCawley, T. McIntosh, I. McMullen, M. Moy, L. Moy, B. Murphy, K. Nelson, C. Pfannkoch, E. Pratts, V. Puri, H. Qureshi, M. Reardon, R. Rodriguez, Y. H. Rogers, D. Romblad, B. Ruhfel, R. Scott, C. Sitter, M. Smallwood, E. Stewart, R. Strong, E. Suh, R. Thomas, N. N. Tint, S. Tse, C. Vech, G. Wang, J. Wetter, S. Williams, M. Williams, S. Windsor, E. Winn-Deen, K. Wolfe, J. Zaveri, K. Zaveri, J. F. Abril, R. Guigo, M. J. Campbell, K. V. Sjolander, B. Karlak, A. Kejariwal, H. Mi, B. Lazareva, T. Hatton, A. Narechania, K. Diemer, A. Muruganujan, N. Guo, S. Sato, V. Bafna, S. Istrail, R. Lippert, R. Schwartz, B. Walenz, S. Yooseph, D. Allen, A. Basu, J. Baxendale, L. Blick, M. Caminha, J. Carnes-Stine, P. Caulk, Y. H. Chiang, M. Coyne, C. Dahlke, A. Mays, M. Dombroski, M. Donnelly, D. Ely, S. Esparham, C. Fosler, H. Gire, S. Glanowski, K. Glasser, A. Glodek, M. Gorokhov, K. Graham, B. Gropman, M. Harris, J. Heil, S. Henderson, J. Hoover, D. Jennings, C. Jordan, J. Jordan, J. Kasha, L. Kagan, C. Kraft, A. Levitsky, M. Lewis, X. Liu, J. Lopez, D. Ma, W. Majoros, J. McDaniel, S. Murphy, M. Newman, T. Nguyen, N. Nguyen, M. Nodell, S. Pan, J. Peck, M. Peterson, W. Rowe, R. Sanders, J. Scott, M. Simpson, T. Smith, A. Sprague, T. Stockwell, R. Turner, E. Venter, M. Wang, M. Wen, D. Wu, M. Wu, A. Xia, A. Zandieh and X. Zhu, *Science*, 2001, **291**, 1304–1351.

9. S. L. Schreiber, *Bioorg. Med. Chem.*, 1998, **6**, 1127–1152.
10. Z. A. Knight and K. M. Shokat, *Cell*, 2007, **128**, 425–430.
11. H. Nagumo, Y. Sasaki, Y. Ono, H. Okamoto, M. Seto and Y. Takuwa, *Am. J. Physiol.: Cell Physiol.*, 2000, **278**, C57–65.
12. B. J. Druker and N. B. Lydon, *J. Clin. Invest.*, 2000, **105**, 3–7.
13. M. Ranson, W. Mansoor and G. Jayson, *Expert Rev. Anticancer Ther.*, 2002, **2**, 161–168.
14. K. S. Bhullar, N. O. Lagaron, E. M. McGowan, I. Parmar, A. Jha, B. P. Hubbard and H. P. V. Rupasinghe, *Mol. Cancer*, 2018, **17**, 48.
15. D. C. Swinney, *Clin. Pharmacol. Ther.*, 2013, **93**, 299–301.

16. S. Campagne, S. Boigner, S. Rudisser, A. Moursy, L. Gillioz, A. Knorlein, J. Hall, H. Ratni, A. Clery and F. H. Allain, *Nat. Chem. Biol.*, 2019, **15**, 1191–1198.
17. M. P. Boyle and K. De Boeck, *Lancet Respir. Med.*, 2013, **1**, 158–163.
18. J. A. Lemm, D. O'Boyle, 2nd, M. Liu, P. T. Nower, R. Colonno, M. S. Deshpande, L. B. Snyder, S. W. Martin, D. R. St Laurent, M. H. Serrano-Wu, J. L. Romine, N. A. Meanwell and M. Gao, *J. Virol.*, 2010, **84**, 482–491.
19. B. K. Wagner, *Phenotypic Screening: Methods and Protocols*, Springer, 2018.
20. F. Vincent, P. M. Loria, A. D. Weston, C. M. Steppan, R. Doyonnas, Y. M. Wang, K. L. Rockwell and M. C. Peakman, *Cell Chem. Biol.*, 2020, **27**, 1332–1346.
21. C. H. Arrowsmith, J. E. Audia, C. Austin, J. Baell, J. Bennett, J. Blagg, C. Bountra, P. E. Brennan, P. J. Brown, M. E. Bunnage, C. Buser-Doepner, R. M. Campbell, A. J. Carter, P. Cohen, R. A. Copeland, B. Cravatt, J. L. Dahlin, D. Dhanak, A. M. Edwards, M. Frederiksen, S. V. Frye, N. Gray, C. E. Grimshaw, D. Hepworth, T. Howe, K. V. Huber, J. Jin, S. Knapp, J. D. Kotz, R. G. Kruger, D. Lowe, M. M. Mader, B. Marsden, A. Mueller-Fahrnow, S. Muller, R. C. O'Hagan, J. P. Overington, D. R. Owen, S. H. Rosenberg, B. Roth, R. Ross, M. Schapira, S. L. Schreiber, B. Shoichet, M. Sundstrom, G. Superti-Furga, J. Taunton, L. Toledo-Sherman, C. Walpole, M. A. Walters, T. M. Willson, P. Workman, R. N. Young and W. J. Zuercher, *Nat. Chem. Biol.*, 2015, **11**, 536–541.
22. J. L. Dahlin, D. S. Auld, I. Rothenaigner, S. Haney, J. Z. Sexton, J. W. M. Nissink, J. Walsh, J. A. Lee, J. M. Strelow, F. S. Willard, L. Ferrins, J. B. Baell, M. A. Walters, B. K. Hua, K. Hadian and B. K. Wagner, *Cell Chem. Biol.*, 2021, **28**, 356–370.
23. J. L. Dahlin, K. M. Nelson, J. M. Strasser, D. Barsyte-Lovejoy, M. M. Szewczyk, S. Organ, M. Cuellar, G. Singh, J. H. Shrimp, N. Nguyen, J. L. Meier, C. H. Arrowsmith, P. J. Brown, J. B. Baell and M. A. Walters, *Nat. Commun.*, 2017, **8**, 1527.
24. J. L. Dahlin, J. W. Nissink, J. M. Strasser, S. Francis, L. Higgins, H. Zhou, Z. Zhang and M. A. Walters, *J. Med. Chem.*, 2015, **58**, 2091–2113.
25. B. Dasgeb, D. Kornreich, K. McGuinn, L. Okon, I. Brownell and D. L. Sackett, *Br. J. Dermatol.*, 2018, **178**, 350–356.
26. J. F. Borel, C. Feurer, H. U. Gubler and H. Stahelin, *Agents Actions*, 1976, **6**, 468–475.
27. T. Goto, T. Kino, H. Hatanaka, M. Nishiyama, M. Okuhara, M. Kohsaka, H. Aoki and H. Imanaka, *Transplant. Proc.*, 1987, **19**, 4–8.
28. E. J. Brown, M. W. Albers, T. B. Shin, K. Ichikawa, C. T. Keith, W. S. Lane and S. L. Schreiber, *Nature*, 1994, **369**, 756–758.
29. D. J. Newman and G. M. Cragg, *J. Nat. Prod.*, 2020, **83**, 770–803.
30. C. J. O'Connor, L. Laraia and D. R. Spring, *Chem. Soc. Rev.*, 2011, **40**, 4332–4345.

31. D. R. Spring, *Chem. Soc. Rev.*, 2005, **34**, 472–482.
32. A. J. Carter, O. Kraemer, M. Zwick, A. Mueller-Fahrnow, C. H. Arrowsmith and A. M. Edwards, *Drug Discovery Today*, 2019, **24**, 2111–2115.
33. M. D. Burke and S. L. Schreiber, *Angew. Chem., Int. Ed.*, 2004, **43**, 46–58.
34. S. L. Schreiber, *Science*, 2000, **287**, 1964–1969.
35. C. J. Gerry and S. L. Schreiber, *Curr. Opin. Chem. Biol.*, 2020, **56**, 1–9.
36. N. Kato, E. Comer, T. Sakata-Kato, A. Sharma, M. Sharma, M. Maetani, J. Bastien, N. M. Brancucci, J. A. Bittker, V. Corey, D. Clarke, E. R. Derbyshire, G. L. Dornan, S. Duffy, S. Eckley, M. A. Itoe, K. M. Koolen, T. A. Lewis, P. S. Lui, A. K. Lukens, E. Lund, S. March, E. Meibalan, B. C. Meier, J. A. McPhail, B. Mitasev, E. L. Moss, M. Sayes, Y. Van Gessel, M. J. Wawer, T. Yoshinaga, A. M. Zeeman, V. M. Avery, S. N. Bhatia, J. E. Burke, F. Catteruccia, J. C. Clardy, P. A. Clemons, K. J. Dechering, J. R. Duvall, M. A. Foley, F. Gusovsky, C. H. Kocken, M. Marti, M. L. Morningstar, B. Munoz, D. E. Neafsey, A. Sharma, E. A. Winzeler, D. F. Wirth, C. A. Scherer and S. L. Schreiber, *Nature*, 2016, **538**, 344–349.
37. D. H. Chou, A. Vetere, A. Choudhary, S. S. Scully, M. Schenone, A. Tang, R. Gomez, S. M. Burns, M. Lundh, T. Vital, E. Comer, P. W. Faloon, V. Dancik, C. Ciarlo, J. Paulk, M. Dai, C. Reddy, H. Sun, M. Young, N. Donato, J. Jaffe, P. A. Clemons, M. Palmer, S. A. Carr, S. L. Schreiber and B. K. Wagner, *J. Am. Chem. Soc.*, 2015, **137**, 7929–7934.
38. J. P. Maianti, G. A. Tan, A. Vetere, A. J. Welsh, B. K. Wagner, M. A. Seeliger and D. R. Liu, *Nat. Chem. Biol.*, 2019, **15**, 565–574.
39. B. Maji, S. A. Gangopadhyay, M. Lee, M. Shi, P. Wu, R. Heler, B. Mok, D. Lim, S. U. Siriwardena, B. Paul, V. Dancik, A. Vetere, M. F. Mesleh, L. A. Marraffini, D. R. Liu, P. A. Clemons, B. K. Wagner and A. Choudhary, *Cell*, 2019, **177**, 1067–1079 e1019.
40. S. M. Corsello, J. A. Bittker, Z. Liu, J. Gould, P. McCarren, J. E. Hirschman, S. E. Johnston, A. Vrcic, B. Wong, M. Khan, J. Asiedu, R. Narayan, C. C. Mader, A. Subramanian and T. R. Golub, *Nat. Med.*, 2017, **23**, 405–408.
41. J. Janes, M. E. Young, E. Chen, N. H. Rogers, S. Burgstaller-Muehlbacher, L. D. Hughes, M. S. Love, M. V. Hull, K. L. Kuhen, A. K. Woods, S. B. Joseph, H. M. Petrassi, C. W. McNamara, M. S. Tremblay, A. I. Su, P. G. Schultz and A. K. Chatterjee, *Proc. Natl. Acad. Sci. U. S. A.*, 2018, **115**, 10750–10755.
42. Y. Cha, T. Erez, I. J. Reynolds, D. Kumar, J. Ross, G. Koytiger, R. Kusko, B. Zeskind, S. Risso, E. Kagan, S. Papapetropoulos, I. Grossman and D. Laifenfeld, *Br. J. Pharmacol.*, 2018, **175**, 168–180.
43. S. Brenner and R. A. Lerner, *Proc. Natl. Acad. Sci. U. S. A.*, 1992, **89**, 5381–5383.
44. R. A. Goodnow, Jr., C. E. Dumelin and A. D. Keefe, *Nat. Rev. Drug Discovery*, 2017, **16**, 131–147.
45. D. Neri and R. A. Lerner, *Annu. Rev. Biochem.*, 2018, **87**, 479–502.
46. P. Dickson and T. Kodadek, *Org. Biomol. Chem.*, 2019, **17**, 4676–4688.
47. R. M. Franzini and C. Randolph, *J. Med. Chem.*, 2016, **59**, 6629–6644.

48. C. J. Gerry, M. J. Wawer, P. A. Clemons and S. L. Schreiber, *J. Am. Chem. Soc.*, 2019, **141**, 10225–10235.
49. C. J. Gerry, Z. Yang, M. Stasi and S. L. Schreiber, *Org. Lett.*, 2019, **21**, 1325–1330.
50. S. L. Schreiber and G. R. Crabtree, *Immunol. Today*, 1992, **13**, 136–142.
51. S. L. Schreiber, *Cell*, 2021, **184**, 3–9.
52. B. Z. Stanton, E. J. Chory and G. R. Crabtree, *Science*, 2018, **359**, eaao5902.
53. K. M. Sakamoto, K. B. Kim, A. Kumagai, F. Mercurio, C. M. Crews and R. J. Deshaies, *Proc. Natl. Acad. Sci. U. S. A.*, 2001, **98**, 8554–8559.
54. A. C. Lai and C. M. Crews, *Nat. Rev. Drug Discovery*, 2017, **16**, 101–114.
55. D. L. Buckley, J. L. Gustafson, I. Van Molle, A. G. Roth, H. S. Tae, P. C. Gareiss, W. L. Jorgensen, A. Ciulli and C. M. Crews, *Angew. Chem., Int. Ed.*, 2012, **51**, 11463–11467.
56. D. L. Buckley, I. Van Molle, P. C. Gareiss, H. S. Tae, J. Michel, D. J. Noblin, W. L. Jorgensen, A. Ciulli and C. M. Crews, *J. Am. Chem. Soc.*, 2012, **134**, 4465–4468.
57. G. E. Winter, D. L. Buckley, J. Paulk, J. M. Roberts, A. Souza, S. Dhe-Paganon and J. E. Bradner, *Science*, 2015, **348**, 1376–1381.
58. D. L. Buckley, K. Raina, N. Darricarrere, J. Hines, J. L. Gustafson, I. E. Smith, A. H. Miah, J. D. Harling and C. M. Crews, *ACS Chem. Biol.*, 2015, **10**, 1831–1837.
59. D. P. Bondeson, A. Mares, I. E. Smith, E. Ko, S. Campos, A. H. Miah, K. E. Mulholland, N. Routly, D. L. Buckley, J. L. Gustafson, N. Zinn, P. Grandi, S. Shimamura, G. Bergamini, M. Faelth-Savitski, M. Bantscheff, C. Cox, D. A. Gordon, R. R. Willard, J. J. Flanagan, L. N. Casillas, B. J. Votta, W. den Besten, K. Famm, L. Kruidenier, P. S. Carter, J. D. Harling, I. Churcher and C. M. Crews, *Nat. Chem. Biol.*, 2015, **11**, 611–617.
60. S. Ziegler, V. Pries, C. Hedberg and H. Waldmann, *Angew. Chem., Int. Ed.*, 2013, **52**, 2744–2792.
61. M. Schenone, V. Dancik, B. K. Wagner and P. A. Clemons, *Nat. Chem. Biol.*, 2013, **9**, 232–240.
62. K. Kubota, M. Funabashi and Y. Ogura, *Biochim. Biophys. Acta, Proteins Proteomics*, 2019, **1867**, 22–27.
63. H. Park, J. Ha and S. B. Park, *Curr. Opin. Chem. Biol.*, 2019, **50**, 66–72.
64. Q. T. L. Pasquer, I. A. Tsakoumagkos and S. Hoogendoorn, *Molecules*, 2020, **25**, 5702.
65. B. J. Leslie and P. J. Hergenrother, *Chem. Soc. Rev.*, 2008, **37**, 1347–1360.
66. U. Rix and G. Superti-Furga, *Nat. Chem. Biol.*, 2009, **5**, 616–624.
67. X. Chen and Y. W. Wu, *Org. Biomol. Chem.*, 2016, **14**, 5417–5439.
68. L. I. Willems, W. A. van der Linden, N. Li, K. Y. Li, N. Liu, S. Hoogendoorn, G. A. van der Marel, B. I. Florea and H. S. Overkleeft, *Acc. Chem. Res.*, 2011, **44**, 718–729.
69. B. F. Cravatt, A. T. Wright and J. W. Kozarich, *Annu. Rev. Biochem.*, 2008, **77**, 383–414.
70. D. S. Hewings, J. A. Flygare, M. Bogyo and I. E. Wertz, *FEBS J.*, 2017, **284**, 1555–1576.

71. H. A. Flaxman, D. K. Miyamoto and C. M. Woo, *Curr. Protoc. Chem. Biol.*, 2019, **11**, e75.
72. E. Smith and I. Collins, *Future Med. Chem.*, 2015, **7**, 159–183.
73. L. P. Conway, W. Li and C. G. Parker, *Cell Chem. Biol.*, 2021, **28**, 371–393.
74. B. Lomenick, R. Hao, N. Jonai, R. M. Chin, M. Aghajan, S. Warburton, J. Wang, R. P. Wu, F. Gomez, J. A. Loo, J. A. Wohlschlegel, T. M. Vondriska, J. Pelletier, H. R. Herschman, J. Clardy, C. F. Clarke and J. Huang, *Proc. Natl. Acad. Sci. U. S. A.*, 2009, **106**, 21984–21989.
75. E. C. Strickland, M. A. Geer, D. T. Tran, J. Adhikari, G. M. West, P. D. DeArmond, Y. Xu and M. C. Fitzgerald, *Nat. Protoc.*, 2013, **8**, 148–161.
76. D. Martinez Molina, R. Jafari, M. Ignatushchenko, T. Seki, E. A. Larsson, C. Dan, L. Sreekumar, Y. Cao and P. Nordlund, *Science*, 2013, **341**, 84–87.
77. F. H. Niesen, H. Berglund and M. Vedadi, *Nat. Protoc.*, 2007, **2**, 2212–2221.
78. R. Jafari, H. Almqvist, H. Axelsson, M. Ignatushchenko, T. Lundback, P. Nordlund and D. Martinez Molina, *Nat. Protoc.*, 2014, **9**, 2100–2122.
79. D. Martinez Molina and P. Nordlund, *Annu. Rev. Pharmacol. Toxicol.*, 2016, **56**, 141–161.
80. M. M. Savitski, F. B. Reinhard, H. Franken, T. Werner, M. F. Savitski, D. Eberhard, D. Martinez Molina, R. Jafari, R. B. Dovega, S. Klaeger, B. Kuster, P. Nordlund, M. Bantscheff and G. Drewes, *Science*, 2014, **346**, 1255784.
81. J. M. Dziekan, H. Yu, D. Chen, L. Dai, G. Wirjanata, A. Larsson, N. Prabhu, R. M. Sobota, Z. Bozdech and P. Nordlund, *Sci. Transl. Med.*, 2019, **11**, eaau3174.
82. M. Kitagawa, P. J. Liao, K. H. Lee, J. Wong, S. C. Shang, N. Minami, O. Sampetrean, H. Saya, D. Lingyun, N. Prabhu, G. K. Diam, R. Sobota, A. Larsson, P. Nordlund, F. McCormick, S. Ghosh, D. M. Epstein, B. W. Dymock and S. H. Lee, *Nat. Commun.*, 2017, **8**, 2200.
83. S. Vartanian, T. P. Ma, J. Lee, P. M. Haverty, D. S. Kirkpatrick, K. Yu and D. Stokoe, *Mol. Cell. Proteomics*, 2016, **15**, 1220–1231.
84. P. D. Hsu, E. S. Lander and F. Zhang, *Cell*, 2014, **157**, 1262–1278.
85. H. Wang, M. La Russa and L. S. Qi, *Annu. Rev. Biochem.*, 2016, **85**, 227–264.
86. A. V. Wright, J. K. Nunez and J. A. Doudna, *Cell*, 2016, **164**, 29–44.
87. M. Jost and J. S. Weissman, *ACS Chem. Biol.*, 2018, **13**, 366–375.
88. C. Kasap, O. Elemento and T. M. Kapoor, *Nat. Chem. Biol.*, 2014, **10**, 626–628.
89. J. E. Neggers, B. Kwanten, T. Dierckx, H. Noguchi, A. Voet, L. Bral, K. Minner, B. Massant, N. Kint, M. Delforge, T. Vercruysse, E. Baloglu, W. Senapedis, M. Jacquemyn and D. Daelemans, *Nat. Commun.*, 2018, **9**, 502.
90. T. Katsila, G. A. Spyroulias, G. P. Patrinos and M. T. Matsoukas, *Comput. Struct. Biotechnol. J.*, 2016, **14**, 177–184.
91. G. B. Li, Z. J. Yu, S. Liu, L. Y. Huang, L. L. Yang, C. T. Lohans and S. Y. Yang, *J. Chem. Inf. Model.*, 2017, **57**, 1640–1651.

92. N. S. Madhukar, P. K. Khade, L. Huang, K. Gayvert, G. Galletti, M. Stogniew, J. E. Allen, P. Giannakakou and O. Elemento, *Nat. Commun.*, 2019, **10**, 5221.
93. S. M. Gustafsdottir, V. Ljosa, K. L. Sokolnicki, J. Anthony Wilson, D. Walpita, M. M. Kemp, K. Petri Seiler, H. A. Carrel, T. R. Golub, S. L. Schreiber, P. A. Clemons, A. E. Carpenter and A. F. Shamji, *PLoS One*, 2013, **8**, e80999.
94. T. Schneidewind, A. Brause, A. Pahl, A. Burhop, T. Mejuch, S. Sievers, H. Waldmann and S. Ziegler, *ChemBioChem*, 2020, **21**, 3197–3207.
95. J. Lamb, E. D. Crawford, D. Peck, J. W. Modell, I. C. Blat, M. J. Wrobel, J. Lerner, J. P. Brunet, A. Subramanian, K. N. Ross, M. Reich, H. Hieronymus, G. Wei, S. A. Armstrong, S. J. Haggarty, P. A. Clemons, R. Wei, S. A. Carr, E. S. Lander and T. R. Golub, *Science*, 2006, **313**, 1929–1935.
96. A. Subramanian, R. Narayan, S. M. Corsello, D. D. Peck, T. E. Natoli, X. Lu, J. Gould, J. F. Davis, A. A. Tubelli, J. K. Asiedu, D. L. Lahr, J. E. Hirschman, Z. Liu, M. Donahue, B. Julian, M. Khan, D. Wadden, I. C. Smith, D. Lam, A. Liberzon, C. Toder, M. Bagul, M. Orzechowski, O. M. Enache, F. Piccioni, S. A. Johnson, N. J. Lyons, A. H. Berger, A. F. Shamji, A. N. Brooks, A. Vrcic, C. Flynn, J. Rosains, D. Y. Takeda, R. Hu, D. Davison, J. Lamb, K. Ardlie, L. Hogstrom, P. Greenside, N. S. Gray, P. A. Clemons, S. Silver, X. Wu, W. N. Zhao, W. Read-Button, X. Wu, S. J. Haggarty, L. V. Ronco, J. S. Boehm, S. L. Schreiber, J. G. Doench, J. A. Bittker, D. E. Root, B. Wong and T. R. Golub, *Cell*, 2017, **171**, 1437–1452 e1417.
97. A. C. Ravindranath, N. Perualila-Tan, A. Kasim, G. Drakakis, S. Liggi, S. C. Brewerton, D. Mason, M. J. Bodkin, D. A. Evans, A. Bhagwat, W. Talloen, H. W. Gohlmann, Z. Shkedy, A. Bender and Q. Consortium, *Mol. BioSyst.*, 2015, **11**, 86–96.
98. M. Karamanou, A. Protogerou, G. Tsoucalas, G. Androutsos and E. Poulakou-Rebelakou, *World J. Diabetes*, 2016, **7**, 1–7.
99. R. A. DeFronzo, *Diabetes*, 1988, **37**, 667–687.
100. R. A. Defronzo, *Diabetes*, 2009, **58**, 773–795.
101. R. Lupi, F. Dotta, L. Marselli, S. Del Guerra, M. Masini, C. Santangelo, G. Patane, U. Boggi, S. Piro, M. Anello, E. Bergamini, F. Mosca, U. Di Mario, S. Del Prato and P. Marchetti, *Diabetes*, 2002, **51**, 1437–1442.
102. D. J. Drucker, *Cell Metab.*, 2006, **3**, 153–165.
103. M. A. Nauck, I. Vardarli, C. F. Deacon, J. J. Holst and J. J. Meier, *Diabetologia*, 2011, **54**, 10–18.
104. J. E. Gerich, *Diabetes Med.*, 2010, **27**, 136–142.
105. M. W. Schwartz, S. C. Woods, D. Porte, Jr., R. J. Seeley and D. G. Baskin, *Nature*, 2000, **404**, 661–671.
106. A. D. Baron, L. Schaeffer, P. Shragg and O. G. Kolterman, *Diabetes*, 1987, **36**, 274–283.
107. A. American Diabetes, *Diabetes Care*, 2020, **43**, S14–S31.

108. J. S. Skyler, G. L. Bakris, E. Bonifacio, T. Darsow, R. H. Eckel, L. Groop, P. H. Groop, Y. Handelsman, R. A. Insel, C. Mathieu, A. T. McElvaine, J. P. Palmer, A. Pugliese, D. A. Schatz, J. M. Sosenko, J. P. Wilding and R. E. Ratner, *Diabetes*, 2017, **66**, 241–255.
109. A. Petrelli, M. A. Atkinson, M. Pietropaolo and N. Giannoukakis, *Diabetes*, 2021, **70**, 313–322.
110. C. Bender, S. Rajendran and M. G. von Herrath, *Front. Endocrinol.*, 2020, **11**, 606434.
111. M. R. Rickels, A. W. Norris and R. L. Hull, *Diabetologia*, 2020, **63**, 2030–2039.
112. S. J. Richardson, T. Rodriguez-Calvo, I. C. Gerling, C. E. Mathews, J. S. Kaddis, M. A. Russell, M. Zeissler, P. Leete, L. Krogvold, K. Dahl-Jorgensen, M. von Herrath, A. Pugliese, M. A. Atkinson and N. G. Morgan, *Diabetologia*, 2016, **59**, 2448–2458.
113. M. Lundberg, L. Krogvold, E. Kuric, K. Dahl-Jorgensen and O. Skog, *Diabetes*, 2016, **65**, 3104–3110.
114. L. Marroqui, R. S. Dos Santos, A. Op de Beeck, A. Coomans de Brachene, L. Marselli, P. Marchetti and D. L. Eizirik, *Diabetologia*, 2017, **60**, 656–667.
115. M. A. Russell, S. D. Redick, D. M. Blodgett, S. J. Richardson, P. Leete, L. Krogvold, K. Dahl-Jorgensen, R. Bottino, M. Brissova, J. M. Spaeth, J. A. B. Babon, R. Haliyur, A. C. Powers, C. Yang, S. C. Kent, A. G. Derr, A. Kucukural, M. G. Garber, N. G. Morgan and D. M. Harlan, *Diabetes*, 2019, **68**, 988–1001.
116. F. Pociot, B. Akolkar, P. Concannon, H. A. Erlich, C. Julier, G. Morahan, C. R. Nierras, J. A. Todd, S. S. Rich and J. Nerup, *Diabetes*, 2010, **59**, 1561–1571.
117. M. J. Redondo, A. K. Steck and A. Pugliese, *Pediatr. Diabetes*, 2018, **19**, 346–353.
118. J. F. Bach and L. Chatenoud, *Cold Spring Harb. Perspect. Med.*, 2012, **2**, a007799.
119. L. Adorini, G. Penna, N. Giarratana and M. Uskokovic, *J. Cell. Biochem.*, 2003, **88**, 227–233.
120. P. Leete, R. A. Oram, T. J. McDonald, B. M. Shields, C. Ziller, T. s. team, A. T. Hattersley, S. J. Richardson and N. G. Morgan, *Diabetologia*, 2020, **63**, 1258–1267.
121. E. K. Sims, F. Syed, J. Nyalwidhe, H. T. Bahnson, L. Haataja, C. Speake, M. A. Morris, A. N. Balamurugan, R. G. Mirmira, J. Nadler, T. L. Mastracci, P. Arvan, C. J. Greenbaum and C. Evans-Molina, *Transl. Res.*, 2019, **213**, 90–99.
122. P. J. Thompson, A. Shah, V. Ntranos, F. Van Gool, M. Atkinson and A. Bhushan, *Cell Metab.*, 2019, **29**, 1045–1060 e1010.
123. H. Nomoto, L. Pei, C. Montemurro, M. Rosenberger, A. Furterer, G. Coppola, B. Nadel, M. Pellegrini, T. Gurlo, P. C. Butler and S. Tudzarova, *Diabetologia*, 2020, **63**, 149–161.
124. G. F. Bottazzo, *Diabetes Med.*, 1986, **3**, 119–130.

125. S. A. Soleimanpour and D. A. Stoffers, *Trends Endocrinol. Metab.*, 2013, **24**, 324–331.
126. S. A. Tersey, Y. Nishiki, A. T. Templin, S. M. Cabrera, N. D. Stull, S. C. Colvin, C. Evans-Molina, J. L. Rickus, B. Maier and R. G. Mirmira, *Diabetes*, 2012, **61**, 818–827.
127. S. Morita, S. A. Villalta, H. C. Feldman, A. C. Register, W. Rosenthal, I. T. Hoffmann-Petersen, M. Mehdizadeh, R. Ghosh, L. Wang, K. Colon-Negron, R. Meza-Acevedo, B. J. Backes, D. J. Maly, J. A. Bluestone and F. R. Papa, *Cell Metab.*, 2017, **25**, 1207.
128. H. Lee, Y. S. Lee, Q. Harenda, S. Pietrzak, H. Z. Oktay, S. Schreiber, Y. Liao, S. Sonthalia, A. E. Ciecko, Y. G. Chen, S. Keles, R. Sridharan and F. Engin, *Cell Metab.*, 2020, **31**, 822–836 e825.
129. P. M. Trivedi, K. L. Graham, N. A. Scott, M. R. Jenkins, S. Majaw, R. M. Sutherland, S. Fynch, A. M. Lew, C. J. Burns, B. Krishnamurthy, T. C. Brodnicki, S. I. Mannering, T. W. Kay and H. E. Thomas, *Diabetes*, 2017, **66**, 1650–1660.
130. F. Engin, A. Yermalovich, T. Nguyen, S. Hummasti, W. Fu, D. L. Eizirik, D. Mathis and G. S. Hotamisligil, *Sci. Transl. Med.*, 2013, **5**, 211ra156.
131. F. M. Ashcroft and P. Rorsman, *Cell*, 2012, **148**, 1160–1171.
132. Y. S. Oh, G. D. Bae, D. J. Baek, E. Y. Park and H. S. Jun, *Front. Endocrinol.*, 2018, **9**, 384.
133. N. Kaiser, G. Leibowitz and R. Nesher, *J. Pediatr. Endocrinol. Metab.*, 2003, **16**, 5–22.
134. A. T. Hattersley and J. E. Tooke, *Lancet*, 1999, **353**, 1789–1792.
135. C. S. Yajnik, *Nutr. Rev.*, 2001, **59**, 1–9.
136. D. Walpita, T. Hasaka, J. Spoonamore, A. Vetere, K. K. Takane, D. Fomina-Yadlin, N. Fiaschi-Taesch, A. Shamji, P. A. Clemons, A. F. Stewart, S. L. Schreiber and B. K. Wagner, *J. Biomol. Screen*, 2012, **17**, 509–518.
137. D. Walpita and B. K. Wagner, *Curr. Protoc. Chem. Biol.*, 2014, **6**, 157–168.
138. D. Balboa, R. B. Prasad, L. Groop and T. Otonkoski, *Diabetologia*, 2019, **62**, 1329–1336.
139. K. G. Maxwell and J. R. Millman, *Cell Rep. Med.*, 2021, **2**, 100238.
140. P. Ravassard, Y. Hazhouz, S. Pechberty, E. Bricout-Neveu, M. Armanet, P. Czernichow and R. Scharfmann, *J. Clin. Invest.*, 2011, **121**, 3589–3597.
141. R. Scharfmann, S. Pechberty, Y. Hazhouz, M. von Bulow, E. Bricout-Neveu, M. Grenier-Godard, F. Guez, L. Rachdi, M. Lohmann, P. Czernichow and P. Ravassard, *J. Clin. Invest.*, 2014, **124**, 2087–2098.
142. B. Hastoy, M. Godazgar, A. Clark, V. Nylander, I. Spiliotis, M. van de Bunt, M. V. Chibalina, A. Barrett, C. Burrows, A. I. Tarasov, R. Scharfmann, A. L. Gloyn and P. Rorsman, *Sci. Rep.*, 2018, **8**, 16994.
143. M. Benazra, M. J. Lecomte, C. Colace, A. Muller, C. Machado, S. Pechberty, E. Bricout-Neveu, M. Grenier-Godard, M. Solimena, R. Scharfmann, P. Czernichow and P. Ravassard, *Mol. Metab.*, 2015, **4**, 916–925.
144. M. Asfari, D. Janjic, P. Meda, G. Li, P. A. Halban and C. B. Wollheim, *Endocrinology*, 1992, **130**, 167–178.

145. A. Merglen, S. Theander, B. Rubi, G. Chaffard, C. B. Wollheim and P. Maechler, *Endocrinology*, 2004, **145**, 667–678.
146. H. E. Hohmeier, H. Mulder, G. Chen, R. Henkel-Rieger, M. Prentki and C. B. Newgard, *Diabetes*, 2000, **49**, 424–430.
147. J. Miyazaki, K. Araki, E. Yamato, H. Ikegami, T. Asano, Y. Shibasaki, Y. Oka and K. Yamamura, *Endocrinology*, 1990, **127**, 126–132.
148. H. Ishihara, T. Asano, K. Tsukuda, H. Katagiri, K. Inukai, M. Anai, M. Kikuchi, Y. Yazaki, J. I. Miyazaki and Y. Oka, *Diabetologia*, 1993, **36**, 1139–1145.
149. V. Poitout, L. E. Stout, M. B. Armstrong, T. F. Walseth, R. L. Sorenson and R. P. Robertson, *Diabetes*, 1995, **44**, 306–313.
150. S. M. Burns, A. Vetere, D. Walpita, V. Dancik, C. Khodier, J. Perez, P. A. Clemons, B. K. Wagner and D. Altshuler, *Cell Metab.*, 2015, **21**, 126–137.
151. J. Huey, K. Keutler and C. Schultz, *Cell Chem. Biol.*, 2020, **27**, 1015–1031.
152. J. E. Campbell and C. B. Newgard, *Nat. Rev. Mol. Cell Biol.*, 2021, **22**, 142–158.
153. P. Proks, F. Reimann, N. Green, F. Gribble and F. Ashcroft, *Diabetes*, 2002, **51**(Suppl 3), S368–376.
154. F. R. Brown, Jr, R. W. Friskey, L. Grindle, L. W. Kinsell and S. Splitter, *Calif. Med.*, 1956, **85**, 285–288.
155. J. H. Hoofnagle, *LiverTox: Clinical and Research Information on Drug-Induced Liver Injury*, Bethesda, MD, 2012.
156. P. Rorsman, L. Eliasson, E. Renstrom, J. Gromada, S. Barg and S. Gopel, *News Physiol. Sci.*, 2000, **15**, 72–77.
157. J. E. Gerich, *Diabetes*, 2002, **51**(Suppl 1), S117–121.
158. G. G. Graham, J. Punt, M. Arora, R. O. Day, M. P. Doogue, J. K. Duong, T. J. Furlong, J. R. Greenfield, L. C. Greenup, C. M. Kirkpatrick, J. E. Ray, P. Timmins and K. M. Williams, *Clin. Pharmacokinet.*, 2011, **50**, 81–98.
159. R. S. Hundal, M. Krssak, S. Dufour, D. Laurent, V. Lebon, V. Chandramouli, S. E. Inzucchi, W. C. Schumann, K. F. Petersen, B. R. Landau and G. I. Shulman, *Diabetes*, 2000, **49**, 2063–2069.
160. A. Klip and L. A. Leiter, *Diabetes Care*, 1990, **13**, 696–704.
161. J. Flory and K. Lipska, *JAMA*, 2019, **321**, 1926–1927.
162. A. Golay, *Int. J. Obes.*, 2008, **32**, 61–72.
163. L. G. Fryer, A. Parbu-Patel and D. Carling, *J. Biol. Chem.*, 2002, **277**, 25226–25232.
164. M. J. Nanjan, M. Mohammed, B. R. Prashantha Kumar and M. J. N. Chandrasekar, *Bioorg. Chem.*, 2018, **77**, 548–567.
165. V. Dubois, J. Eeckhoute, P. Lefebvre and B. Staels, *J. Clin. Invest.*, 2017, **127**, 1202–1214.
166. J. H. Hoofnagle, *LiverTox: Clinical and Research Information on Drug-Induced Liver Injury*, Bethesda, MD, 2012.
167. A. H. Barnett, *Vasc. Health Risk Manag.*, 2009, **5**, 141–151.
168. S. E. Nissen and K. Wolski, *N. Engl. J. Med.*, 2007, **356**, 2457–2471.

169. J. J. Holst, *Front. Endocrinol.*, 2019, **10**, 260.
170. J. J. Holst, *Physiol. Rev.*, 2007, **87**, 1409–1439.
171. S. R. Hoare, *Drug Discovery Today*, 2005, **10**, 417–427.
172. X. Deng, M. S. Tavallaie, R. Sun, J. Wang, Q. Cai, J. Shen, S. Lei, L. Fu and F. Jiang, *Bioorg. Chem.*, 2020, **99**, 103810.
173. L. L. Baggio and D. J. Drucker, *Mol. Metab.*, 2021, **46**, 101090.
174. M. A. Nauck and J. J. Meier, *Eur. J. Endocrinol.*, 2019, **181**, R211–R234.
175. K. Aston-Mourney, J. Proietto, G. Morahan and S. Andrikopoulos, *Diabetologia*, 2008, **51**, 540–545.
176. E. M. Wright, D. D. Loo and B. A. Hirayama, *Physiol. Rev.*, 2011, **91**, 733–794.
177. R. Santer and J. Calado, *Clin. J. Am. Soc. Nephrol.*, 2010, **5**, 133–141.
178. R. F. Rosenwasser, S. Sultan, D. Sutton, R. Choksi and B. J. Epstein, *Diabetes Metab. Syndr. Obes.*, 2013, **6**, 453–467.
179. D. K. McGuire, W. J. Shih, F. Cosentino, B. Charbonnel, D. Z. I. Cherney, S. Dagogo-Jack, R. Pratley, M. Greenberg, S. Wang, S. Huyck, I. Gantz, S. G. Terra, U. Masiukiewicz and C. P. Cannon, *JAMA Cardiol.*, 2021, **6**, 148–158.
180. W. Wu, J. Shang, Y. Feng, C. M. Thompson, S. Horwitz, J. R. Thompson, E. D. MacIntyre, N. A. Thornberry, K. Chapman, Y. P. Zhou, A. D. Howard and J. Li, *J. Biomol. Screen.*, 2008, **13**, 128–134.
181. J. E. de Leeuw van Weenen, E. T. Parlevliet, P. Maechler, L. M. Havekes, J. A. Romijn, D. M. Ouwens, H. Pijl and B. Guigas, *Biochem. Pharmacol.*, 2010, **79**, 1827–1836.
182. I. Garcia-Tornadu, A. M. Ornstein, A. Chamson-Reig, M. B. Wheeler, D. J. Hill, E. Arany, M. Rubinstein and D. Becu-Villalobos, *Endocrinology*, 2010, **151**, 1441–1450.
183. E. Ferrannini, *Cell Metab.*, 2010, **11**, 349–352.
184. H. A. Keenan, J. K. Sun, J. Levine, A. Doria, L. P. Aiello, G. Eisenbarth, S. Bonner-Weir and G. L. King, *Diabetes*, 2010, **59**, 2846–2853.
185. R. A. Oram, T. J. McDonald, B. M. Shields, M. M. Hudson, M. H. Shepherd, S. Hammersley, E. R. Pearson, A. T. Hattersley and U. Team, *Diabetes Care*, 2015, **38**, 323–328.
186. J. L. Sherr, T. Ghazi, A. Wurtz, L. Rink and K. C. Herold, *Diabetes Metab. Res. Rev.*, 2014, **30**, 154–162.
187. X. Ma and S. Zhu, *Acta Biochim. Biophys. Sin*, 2017, **49**, 289–301.
188. S. P. Singh and N. Ninov, *Int. J. Dev. Biol.*, 2018, **62**, 453–464.
189. A. Vetere, A. Choudhary, S. M. Burns and B. K. Wagner, *Nat. Rev. Drug Discovery*, 2014, **13**, 278–289.
190. E. Dirice, D. Walpita, A. Vetere, B. C. Meier, S. Kahraman, J. Hu, V. Dancik, S. M. Burns, T. J. Gilbert, D. E. Olson, P. A. Clemons, R. N. Kulkarni and B. K. Wagner, *Diabetes*, 2016, **65**, 1660–1671.
191. P. Wang, J. C. Alvarez-Perez, D. P. Felsenfeld, H. Liu, S. Sivendran, A. Bender, A. Kumar, R. Sanchez, D. K. Scott, A. Garcia-Ocana and A. F. Stewart, *Nat. Med.*, 2015, **21**, 383–388.
192. W. Shen, B. Taylor, Q. Jin, V. Nguyen-Tran, S. Meeusen, Y. Q. Zhang, A. Kamireddy, A. Swafford, A. F. Powers, J. Walker, J. Lamb,

B. Bursalaya, M. DiDonato, G. Harb, M. Qiu, C. M. Filippi, L. Deaton, C. N. Turk, W. L. Suarez-Pinzon, Y. Liu, X. Hao, T. Mo, S. Yan, J. Li, A. E. Herman, B. J. Hering, T. Wu, H. Martin Seidel, P. McNamara, R. Glynne and B. Laffitte, *Nat. Commun.*, 2015, **6**, 8372.

193. H. Liu, M. S. Remedi, K. L. Pappan, G. Kwon, N. Rohatgi, C. A. Marshall and M. L. McDaniel, *Diabetes*, 2009, **58**, 663–672.
194. R. Mussmann, M. Geese, F. Harder, S. Kegel, U. Andag, A. Lomow, U. Burk, D. Onichtchouk, C. Dohrmann and M. Austen, *J. Biol. Chem.*, 2007, **282**, 12030–12037.
195. C. Ackeifi, E. Swartz, K. Kumar, H. Liu, S. Chalada, E. Karakose, D. K. Scott, A. Garcia-Ocana, R. Sanchez, R. J. DeVita, A. F. Stewart and P. Wang, *JCI Insight*, 2020, **5**, e132594.
196. P. Wang, E. Karakose, H. Liu, E. Swartz, C. Ackeifi, V. Zlatanic, J. Wilson, B. J. Gonzalez, A. Bender, K. K. Takane, L. Ye, G. Harb, F. Pagliuca, D. Homann, D. Egli, C. Argmann, D. K. Scott, A. Garcia-Ocana and A. F. Stewart, *Cell Metab.*, 2019, **29**, 638–652 e635.
197. C. Ackeifi, P. Wang, E. Karakose, J. E. Manning Fox, B. J. Gonzalez, H. Liu, J. Wilson, E. Swartz, C. Berrouet, Y. Li, K. Kumar, P. E. MacDonald, R. Sanchez, B. Thorens, R. DeVita, D. Homann, D. Egli, D. K. Scott, A. Garcia-Ocana and A. F. Stewart, *Sci. Transl. Med.*, 2020, **12**, eaaw9996.
198. S. Dhawan, E. Dirice, R. N. Kulkarni and A. Bhushan, *Diabetes*, 2016, **65**, 1208–1218.
199. H. E. Hohmeier, L. Zhang, B. Taylor, S. Stephens, D. Lu, P. McNamara, B. Laffitte and C. B. Newgard, *PLoS One*, 2020, **15**, e0224344.
200. E. Nguidjoe, S. Sokolow, S. Bigabwa, N. Pachera, E. D'Amico, F. Allagnat, J. M. Vanderwinden, A. Sener, M. Manto, M. Depreter, J. Mast, G. Joanny, E. Montanya, J. Rahier, A. K. Cardozo, D. L. Eizirik, S. Schurmans and A. Herchuelz, *Diabetes*, 2011, **60**, 2076–2085.
201. J. Papin, F. P. Zummo, N. Pachera, C. Guay, R. Regazzi, A. K. Cardozo and A. Herchuelz, *J. Endocr. Soc.*, 2018, **2**, 631–645.
202. Z. Assefa, S. Akbib, A. Lavens, G. Stange, Z. Ling, K. H. Hellemans and D. Pipeleers, *Am. J. Physiol.: Endocrinol. Metab.*, 2016, **311**, E698–E705.
203. C. Beaupere, A. Liboz, B. Feve, B. Blondeau and G. Guillemain, *Int. J. Mol. Sci.*, 2021, **22**, 623.
204. M. Cnop, N. Welsh, J. C. Jonas, A. Jorns, S. Lenzen and D. L. Eizirik, *Diabetes*, 2005, **54**(Suppl 2), S97–107.
205. D. L. Eizirik and T. Mandrup-Poulsen, *Diabetologia*, 2001, **44**, 2115–2133.
206. M. I. Darville and D. L. Eizirik, *Diabetologia*, 1998, **41**, 1101–1108.
207. H. I. Callewaert, C. A. Gysemans, L. Ladriere, W. D'Hertog, J. Hagenbrock, L. Overbergh, D. L. Eizirik and C. Mathieu, *Diabetes*, 2007, **56**, 2169–2173.
208. C. A. Gysemans, L. Ladriere, H. Callewaert, J. Rasschaert, D. Flamez, D. E. Levy, P. Matthys, D. L. Eizirik and C. Mathieu, *Diabetes*, 2005, **54**, 2396–2403.

209. L. G. Grunnet, R. Aikin, M. F. Tonnesen, S. Paraskevas, L. Blaabjerg, J. Storling, L. Rosenberg, N. Billestrup, D. Maysinger and T. Mandrup-Poulsen, *Diabetes*, 2009, **58**, 1807–1815.
210. C. M. Larsen, M. Faulenbach, A. Vaag, A. Volund, J. A. Ehses, B. Seifert, T. Mandrup-Poulsen and M. Y. Donath, *N. Engl. J. Med.*, 2007, **356**, 1517–1526.
211. D. Lim, V. Sreekanth, K. J. Cox, B. K. Law, B. K. Wagner, J. M. Karp and A. Choudhary, *Nat. Commun.*, 2020, **11**, 4043.
212. E. K. Kim, K. B. Kwon, M. J. Han, M. Y. Song, J. H. Lee, N. Lv, K. B. Choi, D. G. Ryu, K. S. Kim, J. W. Park and B. H. Park, *Int. J. Mol. Med.*, 2007, **19**, 535–540.
213. M. Menegazzi, M. Novelli, P. Beffy, V. D'Aleo, E. Tedeschi, R. Lupi, E. Zoratti, P. Marchetti, H. Suzuki and P. Masiello, *Int. J. Biochem. Cell Biol.*, 2008, **40**, 1509–1521.
214. T. Matsuda, K. Ferreri, I. Todorov, Y. Kuroda, C. V. Smith, F. Kandeel and Y. Mullen, *Endocrinology*, 2005, **146**, 175–185.
215. M. Y. Song, E. K. Kim, W. S. Moon, J. W. Park, H. J. Kim, H. S. So, R. Park, K. B. Kwon and B. H. Park, *Toxicol. Appl. Pharmacol.*, 2009, **235**, 57–67.
216. D. H. Chou, J. R. Duvall, B. Gerard, H. Liu, B. A. Pandya, B. C. Suh, E. M. Forbeck, P. Faloon, B. K. Wagner and L. A. Marcaurelle, *ACS Med. Chem. Lett.*, 2011, **2**, 698–702.
217. S. Nagafuchi, Y. Kamada-Hibio, K. Hirakawa, N. Tsutsu, M. Minami, A. Okada, K. Kai, M. Teshima, A. Moroishi, Y. Murakami, Y. Umeno, Y. Yokogawa, K. Kogawa, K. Izumi, K. Anzai, R. Iwakiri, K. Hamaguchi, N. Sasaki, S. Nohara, E. Yoshida, M. Harada, K. Akashi, T. Yanase, J. Ono, T. Okeda, R. Fujimoto, K. Ihara, T. Hara, Y. Kikuchi, M. Iwase, T. Kitazono, F. Kojima, S. Kono, H. Kurisaki, S. Kondo and H. Katsuta, *EBioMedicine*, 2015, **2**, 744–749.
218. A. Coomans de Brachene, A. Castela, A. Op de Beeck, R. G. Mirmira, L. Marselli, P. Marchetti, C. Masse, W. Miao, S. Leit, C. Evans-Molina and D. L. Eizirik, *Diabetes, Obes. Metab.*, 2020, **22**, 1827–1836.
219. S. H. Lee, D. Cunha, C. Piermarocchi, G. Paternostro, A. Pinkerton, L. Ladriere, P. Marchetti, D. L. Eizirik, M. Cnop and F. Levine, *Biochem. Pharmacol.*, 2017, **138**, 140–149.
220. J. Vogel, J. Yin, L. Su, S. X. Wang, R. Zessis, S. Fowler, C. H. Chiu, A. C. Wilson, A. Chen, F. Zecri, G. Turner, T. M. Smith, B. DeChristopher, H. Xing, D. M. Rothman, X. Cai and A. Berdichevsky, *Diabetes*, 2020, **69**, 1032–1041.
221. A. Shalev, C. A. Pise-Masison, M. Radonovich, S. C. Hoffmann, B. Hirshberg, J. N. Brady and D. M. Harlan, *Endocrinology*, 2002, **143**, 3695–3698.
222. L. A. Thielen, J. Chen, G. Jing, O. Moukha-Chafiq, G. Xu, S. Jo, T. B. Grayson, B. Lu, P. Li, C. E. Augelli-Szafran, M. J. Suto, M. Kanke, P. Sethupathy, J. K. Kim and A. Shalev, *Cell Metab.*, 2020, **32**, 353–365 e358.

223. H. S. Spijker, R. B. Ravelli, A. M. Mommaas-Kienhuis, A. A. van Apeldoorn, M. A. Engelse, A. Zaldumbide, S. Bonner-Weir, T. J. Rabelink, R. C. Hoeben, H. Clevers, C. L. Mummery, F. Carlotti and E. J. de Koning, *Diabetes*, 2013, **62**, 2471–2480.
224. P. Collombat, A. Mansouri, J. Hecksher-Sorensen, P. Serup, J. Krull, G. Gradwohl and P. Gruss, *Genes Dev.*, 2003, **17**, 2591–2603.
225. D. Fomina-Yadlin, S. Kubicek, D. Walpita, V. Dancik, J. Hecksher-Sorensen, J. A. Bittker, T. Sharifnia, A. Shamji, P. A. Clemons, B. K. Wagner and S. L. Schreiber, *Proc. Natl. Acad. Sci. U. S. A.*, 2010, **107**, 15099–15104.
226. N. Ben-Othman, A. Vieira, M. Courtney, F. Record, E. Gjernes, F. Avolio, B. Hadzic, N. Druelle, T. Napolitano, S. Navarro-Sanz, S. Silvano, K. Al-Hasani, A. Pfeifer, S. Lacas-Gervais, G. Leuckx, L. Marroqui, J. Thevenet, O. D. Madsen, D. L. Eizirik, H. Heimberg, J. Kerr-Conte, F. Pattou, A. Mansouri and P. Collombat, *Cell*, 2017, **168**, 73–85 e11.
227. J. Li, T. Casteels, T. Frogne, C. Ingvorsen, C. Honore, M. Courtney, K. V. M. Huber, N. Schmitner, R. A. Kimmel, R. A. Romanov, C. Sturtzel, C. H. Lardeau, J. Klughammer, M. Farlik, S. Sdelci, A. Vieira, F. Avolio, F. Briand, I. Baburin, P. Majek, F. M. Pauler, T. Penz, A. Stukalov, M. Gridling, K. Parapatics, C. Barbieux, E. Berishvili, A. Spittler, J. Colinge, K. L. Bennett, S. Hering, T. Sulpice, C. Bock, M. Distel, T. Harkany, D. Meyer, G. Superti-Furga, P. Collombat, J. Hecksher-Sorensen and S. Kubicek, *Cell*, 2017, **168**, 86–100 e115.
228. A. M. Ackermann, N. G. Moss and K. H. Kaestner, *Cell Metab.*, 2018, **28**, 787–792 e783.
229. A. Granger and J. A. Kushner, *J. Intern. Med.*, 2009, **266**, 325–338.
230. Y. Yuan, K. Hartland, Z. Boskovic, Y. Wang, D. Walpita, P. A. Lysy, C. Zhong, D. W. Young, Y. K. Kim, N. J. Tolliday, E. M. Sokal, S. L. Schreiber and B. K. Wagner, *Chem. Biol.*, 2013, **20**, 1513–1522.
231. N. Musi and L. J. Goodyear, *Acta Physiol. Scand.*, 2003, **178**, 337–345.
232. P. Steneberg, E. Lindahl, U. Dahl, E. Lidh, J. Straseviciene, F. Backlund, E. Kjellkvist, E. Berggren, I. Lundberg, I. Bergqvist, M. Ericsson, B. Eriksson, K. Linde, J. Westman, T. Edlund and H. Edlund, *JCI Insight*, 2018, **3**, e99114.
233. W. Farris, S. Mansourian, Y. Chang, L. Lindsley, E. A. Eckman, M. P. Frosch, C. B. Eckman, R. E. Tanzi, D. J. Selkoe and S. Guenette, *Proc. Natl. Acad. Sci. U. S. A.*, 2003, **100**, 4162–4167.
234. I. A. Mirsky and G. Perisutti, *Science*, 1955, **122**, 559–560.
235. J. P. Maianti, A. McFedries, Z. H. Foda, R. E. Kleiner, X. Q. Du, M. A. Leissring, W. J. Tang, M. J. Charron, M. A. Seeliger, A. Saghatelian and D. R. Liu, *Nature*, 2014, **511**, 94–98.
236. M. Brownlee and A. Cerami, *Science*, 1979, **206**, 1190–1191.
237. N. K. Rege, N. F. B. Phillips and M. A. Weiss, *Curr. Opin. Endocrinol., Diabetes Obes.*, 2017, **24**, 267–278.
238. Y. Abdolazimi, Z. Zhao, S. Lee, H. Xu, P. Allegretti, T. M. Horton, B. Yeh, H. P. Moeller, R. J. Nichols, D. McCutcheon, A. Shalizi, M. Smith, N. A. Armstrong and J. P. Annes, *Endocrinology*, 2018, **159**, 3143–3157.

239. M. Debdab, F. Carreaux, S. Renault, M. Soundararajan, O. Fedorov, P. Filippakopoulos, O. Lozach, L. Babault, T. Tahtouh, B. Baratte, Y. Ogawa, M. Hagiwara, A. Eisenreich, U. Rauch, S. Knapp, L. Meijer and J. P. Bazureau, *J. Med. Chem.*, 2011, **54**, 4172–4186.
240. H. Falke, A. Chaikuad, A. Becker, N. Loaëc, O. Lozach, S. Abu Jhaisha, W. Becker, P. G. Jones, L. Preu, K. Baumann, S. Knapp, L. Meijer and C. Kunick, *J. Med. Chem.*, 2015, **58**, 3131–3143.
241. Y. A. Liu, Q. Jin, Y. Zou, Q. Ding, S. Yan, Z. Wang, X. Hao, B. Nguyen, X. Zhang, J. Pan, T. Mo, K. Jacobsen, T. Lam, T. Y. Wu, H. M. Petrassi, B. Bursulaya, M. DiDonato, W. P. Gordon, B. Liu, J. Baaten, R. Hill, V. Nguyen-Tran, M. Qiu, Y. Q. Zhang, A. Kamireddy, S. Espinola, L. Deaton, S. Ha, G. Harb, Y. Jia, J. Li, W. Shen, A. M. Schumacher, K. Colman, R. Glynne, S. Pan, P. McNamara, B. Laffitte, S. Meeusen, V. Molteni and J. Loren, *J. Med. Chem.*, 2020, **63**, 2958–2973.
242. K. Kumar, C. Suebsuwong, P. Wang, A. Garcia-Ocana, A. F. Stewart and R. J. DeVita, *J. Med. Chem.*, 2021, **64**, 2901–2922.
243. K. Kumar, P. Wang, A. S. E. S. Khamrui, C. Secor, B. L. M. R. F. S. A. Sanchez and R. J. DeVita, *Molecules*, 2020, **25**, 1983.
244. K. Kumar, P. Wang, R. Sanchez, E. A. Swartz, A. F. Stewart and R. J. DeVita, *J. Med. Chem.*, 2018, **61**, 7687–7699.
245. T. Tahtouh, J. M. Elkins, P. Filippakopoulos, M. Soundararajan, G. Burgy, E. Durieu, C. Cochet, R. S. Schmid, D. C. Lo, F. Delhommel, A. E. Oberholzer, L. H. Pearl, F. Carreaux, J. P. Bazureau, S. Knapp and L. Meijer, *J. Med. Chem.*, 2012, **55**, 9312–9330.
246. E. Hellström-Lindahl, A. Danielsson, F. Ponten, P. Czernichow, O. Korsgren, L. Johansson and O. Eriksson, *Acta Diabetol.*, 2016, **53**, 413–421.
247. M. Braun, *Vitam. Horm.*, 2014, **95**, 165–193.
248. J. K. Amartey, C. Esguerra, I. Al-Jammaz, R. S. Parhar and B. Al-Otaibi, *Appl. Radiat. Isot.*, 2006, **64**, 769–777.
249. R. E. Martin, L. G. Green, W. Guba, N. Kratochwil and A. Christ, *J. Med. Chem.*, 2007, **50**, 6291–6294.
250. H. Hirose, T. Yamasaki, M. Ogino, R. Mizojiri, Y. Tamura-Okano, H. Yashiro, Y. Muraki, Y. Nakano, J. Sugama, A. Hata, S. Iwasaki, M. Watanabe, T. Maekawa and S. Kasai, *Bioorg. Med. Chem.*, 2017, **25**, 4175–4193.
251. A. Negi, J. Zhou, S. Sweeney and P. V. Murphy, *Eur. J. Med. Chem.*, 2019, **163**, 148–159.
252. D. C. Saunders, M. Brissova, N. Phillips, S. Shrestha, J. T. Walker, R. Aramandla, G. Poffenberger, D. K. Flaherty, K. P. Weller, J. Pelletier, T. Cooper, M. T. Goff, J. Virostko, A. Shostak, E. D. Dean, D. L. Greiner, L. D. Shultz, N. Prasad, S. E. Levy, R. H. Carnahan, C. Dai, J. Sevigny and A. C. Powers, *Cell Metab.*, 2019, **29**, 745–754 e744.
253. O. Eriksson, D. Espes, R. K. Selvaraju, E. Jansson, G. Antoni, J. Sörensen, M. Lubberink, A. R. Biglarnia, J. W. Eriksson, A. Sundin, H. Ahlström, B. Eriksson, L. Johansson, P. O. Carlsson and O. Korsgren, *Diabetes*, 2014, **63**, 3428–3437.

254. A. Segerstolpe, A. Palasantza, P. Eliasson, E. M. Andersson, A. C. Andreasson, X. Sun, S. Picelli, A. Sabirsh, M. Clausen, M. K. Bjursell, D. M. Smith, M. Kasper, C. Ammala and R. Sandberg, *Cell Metab.*, 2016, **24**, 593–607.
255. H. J. Choe and Y. M. Cho, *Endocrinol. Metab.*, 2021, **36**, 22–29.
256. M. Gotthardt, M. Fischer, I. Naeher, J. B. Holz, H. Jungclas, H. W. Fritsch, M. Béhé, B. Göke, K. Joseph and T. M. Behr, *Eur. J. Nucl. Med. Mol. Imaging*, 2002, **29**, 597–606.
257. M. Brom, W. Woliner-van der Weg, L. Joosten, C. Frielink, T. Bouckenooghe, P. Rijken, K. Andralojc, B. J. Göke, M. de Jong, D. L. Eizirik, M. Béhé, T. Lahoutte, W. J. Oyen, C. J. Tack, M. Janssen, O. C. Boerman and M. Gotthardt, *Diabetologia*, 2014, **57**, 950–959.
258. L. Joosten, M. Brom, H. Peeters, D. Bos, E. Himpe, L. Bouwens, O. Boerman and M. Gotthardt, *Mol. Pharm.*, 2019, **16**, 4024–4030.
259. G. M. Thurber, *J. Nucl. Med.*, 2021, **62**, 745–747.
260. T. Reiner, R. H. Kohler, C. W. Liew, J. A. Hill, J. Gaglia, R. N. Kulkarni and R. Weissleder, *Bioconjug. Chem.*, 2010, **21**, 1362–1368.
261. T. Reiner, G. Thurber, J. Gaglia, C. Vinegoni, C. W. Liew, R. Upadhyay, R. H. Kohler, L. Li, R. N. Kulkarni, C. Benoist, D. Mathis and R. Weissleder, *Proc. Natl. Acad. Sci. U. S. A.*, 2011, **108**, 12815–12820.
262. B. Finan, B. Yang, N. Ottaway, K. Stemmer, T. D. Muller, C. X. Yi, K. Habegger, S. C. Schriever, C. Garcia-Caceres, D. G. Kabra, J. Hembree, J. Holland, C. Raver, R. J. Seeley, W. Hans, M. Irmler, J. Beckers, M. H. de Angelis, J. P. Tiano, F. Mauvais-Jarvis, D. Perez-Tilve, P. Pfluger, L. Zhang, V. Gelfanov, R. D. DiMarchi and M. H. Tschop, *Nat. Med.*, 2012, **18**, 1847–1856.
263. C. Ämmälä, W. J. Drury, 3rd, L. Knerr, I. Ahlstedt, P. Stillemark-Billton, C. Wennberg-Huldt, E. M. Andersson, E. Valeur, R. Jansson-Löfmark, D. Janzén, L. Sundström, J. Meuller, J. Claesson, P. Andersson, C. Johansson, R. G. Lee, T. P. Prakash, P. P. Seth, B. P. Monia and S. Andersson, *Sci. Adv.*, 2018, **4**, eaat3386.
264. L. Knerr, T. P. Prakash, R. Lee, W. J. Drury Iii, M. Nikan, W. Fu, E. Pirie, L. Maria, E. Valeur, A. Hayen, M. Olwegard-Halvarsson, J. Broddefalk, C. Ammala, M. E. Ostergaard, J. Meuller, L. Sundstrom, P. Andersson, D. Janzen, R. Jansson-Lofmark, P. P. Seth and S. Andersson, *J. Am. Chem. Soc.*, 2021, **143**, 3416–3429.
265. M. K. Schäfer, N. R. Hartwig, N. Kalmbach, M. Klietz, M. Anlauf, L. E. Eiden and E. Weihe, *Diabetologia*, 2013, **56**, 1047–1056.
266. Y. V. Li, *Endocrine*, 2014, **45**, 178–189.
267. H. Ichii, L. Inverardi, A. Pileggi, R. D. Molano, O. Cabrera, A. Caicedo, S. Messinger, Y. Kuroda, P. O. Berggren and C. Ricordi, *Am. J. Transplant.*, 2005, **5**, 1635–1645.
268. B. Lukowiak, B. Vandewalle, R. Riachy, J. Kerr-Conte, V. Gmyr, S. Belaich, J. Lefebvre and F. Pattou, *J. Histochem. Cytochem.*, 2001, **49**, 519–528.
269. T. M. Horton, P. A. Allegretti, S. Lee, H. P. Moeller, M. Smith and J. P. Annes, *Cell Chem. Biol.*, 2019, **26**, 213–222 e216.

270. S. Kahraman, D. Manna, E. Dirice, B. Maji, J. Small, B. K. Wagner, A. Choudhary and R. N. Kulkarni, *Life Sci. Alliance*, 2021, **4**, e202000840.
271. M. Lee, B. Maji, D. Manna, S. Kahraman, R. M. Elgamal, J. Small, P. Kokkonda, A. Vetere, J. M. Goldberg, S. J. Lippard, R. N. Kulkarni, B. K. Wagner and A. Choudhary, *J. Am. Chem. Soc.*, 2020, **142**, 6477–6482.
272. K. Yang, M. Lee, P. A. Jones, S. S. Liu, A. Zhou, J. Xu, V. Sreekanth, J. L. Y. Wu, L. Vo, E. A. Lee, R. Pop, Y. Lee, B. K. Wagner, D. A. Melton, A. Choudhary and J. M. Karp, *Sci. Adv.*, 2020, **6**, 33208361.
273. D. Vats, H. Wang, D. Esterhazy, K. Dikaiou, C. Danzer, M. Honer, F. Stuker, H. Matile, C. Migliorini, E. Fischer, J. Ripoll, R. Keist, W. Krek, R. Schibli, M. Stoffel and M. Rudin, *Diabetologia*, 2012, **55**, 2407–2416.
274. D. Esterházy, I. Stützer, H. Wang, M. P. Rechsteiner, J. Beauchamp, H. Döbeli, H. Hilpert, H. Matile, M. Prummer, A. Schmidt, N. Lieske, B. Boehm, L. Marselli, D. Bosco, J. Kerr-Conte, R. Aebersold, G. A. Spinas, H. Moch, C. Migliorini and M. Stoffel, *Cell Metab.*, 2011, **14**, 365–377.
275. R. Bertrand, A. Wolf, Y. Ivashchenko, M. Löhn, M. Schäfer, M. Brönstrup, M. Gotthardt, V. Derdau and O. Plettenburg, *ACS Chem. Biol.*, 2016, **11**, 1745–1754.
276. X. Yan, J. J. Kim, H. S. Jeong, Y. K. Moon, Y. K. Cho, S. Ahn, S. B. Jun, H. Kim and Y. You, *Inorg. Chem.*, 2017, **56**, 4332–4346.
277. T. D. Muller, C. Clemmensen, B. Finan, R. D. DiMarchi and M. H. Tschop, *Pharmacol. Rev.*, 2018, **70**, 712–746.

CHAPTER 2

Epigenomics of Type 2 Diabetes

A. NAZIA PARVEEN AND B. SANGEETA DHAWAN*

Department of Translational Research and Cellular Therapeutics, Arthur Riggs Diabetes and Metabolism Research Institute, Beckman Research Institute of City of Hope, Duarte, CA 91010, USA
*Email: sdhawan@coh.org

2.1 Introduction

Type 2 diabetes (T2D) is a disease that impairs the body's ability to process blood glucose due to an insufficiency of insulin, a hormone produced by pancreatic beta-cells, resulting in chronic hyperglycemia. Reduced insulin availability, and impaired insulin action due to insulin resistance in the peripheral tissues are two key physiological aspects of T2D. Insulin resistance can result from obesity, metabolic syndrome, and chronic over-nutrition, disrupting glucose uptake and storage by peripheral tissues. Failure of the insulin-producing beta-cells to compensate for insulin resistance is central to the pathogenesis of T2D, and involves the progressive impairment of beta-cell function, followed by beta-cell death. Collectively, this results in chronic hyperglycemia, which can ultimately lead to multiple, and often life-threatening health complications such as nephropathy, retinopathy, neuropathy, and cardiovascular disease.

T2D is fast emerging as a global pandemic, with a drastic increase in disease incidence in the past few decades. Of particular concern is the recent prevalence of early-onset T2D, and the number of children and young adults receiving a T2D diagnosis is rapidly rising.[1] A family history of T2D is a major, well-recognized risk factor for disease development, with ~25% of affected individuals having at least one diagnosed family member.[2–5] This has spurred

Nanoscience & Nanotechnology Series No. 59
Nanotechnology for Diabetes Management
Edited by Amar Abderrahmani, Sabine Szunerits, Rabah Boukerroub and Abdelfattah El Ouaamari

Published by the Royal Society of Chemistry, www.rsc.org

an intensive search for genetic factors that contribute to T2D risk. Genome-wide association studies (GWAS) have led to the identification of several genetic variants associated with diabetes, and provided valuable insights into the pathogenesis of such a complex disease.[6,7] However, genetic factors contribute to only a small percentage of diabetes associated with family history. Moreover, genetic variants alone cannot account for the sharp increase in T2D incidence in recent times, as the human genome is unlikely to have undergone any large-scale changes in such a short span of time. Collectively, this suggests an influence of environmental factors on the genome.[8] The contribution of non-genetic risk factors for T2D, such as sedentary lifestyle, over-nutrition, obesity, and aging, is now well recognized, and further underscores the importance of the gene–environment interaction. The effect of environmental variations on gene expression is mediated by epigenetic mechanisms, which enable a cell to respond and adapt to its environment.[9,10] An epigenetic change can be defined as a temporary or permanent, heritable change in chromatin structure and function, that leads to a phenotypic change, without any change in the underlying nucleotide sequence.[11] There are multiple layers of epigenetic regulation such as DNA methylation, covalent modifications of histone proteins, chromatin remodeling, and non-coding RNAs that facilitate the modulation of chromatin accessibility and gene expression.[12] The growth and homeostasis of metabolic organs are regulated by epigenetic mechanisms in response to environmental cues such as morphogens, nutrients, metabolism, and cellular stress.[13] Nutritional homeostasis and metabolism have a direct link with epigenetic regulation, given that many of the epigenetic modifiers require metabolites as cofactors. Accordingly, changes in cellular metabolism and nutrient quality and accessibility can change the activity of epigenetic regulatory enzymes, thus impacting chromatin accessibility, gene expression, and cellular phenotype.[14–16] Thus, one can imagine that scenarios such as over- or under-nutrition and hyperglycemia can disrupt the epigenetic program and function of metabolic organs, and contribute to T2D risk, and disease development, and complications. Similarly, adverse environmental conditions during the intrauterine growth phase or early postnatal life can impact the differentiation, growth, and function of metabolic organs, and increase the risk of T2D in adulthood.[17–19] Therefore, environmental factors such as nutrition, obesity, and the complex interplay between genes and the environment account for the diabetes progression[9,10] (Figure 2.1).

A large number of recent epigenome profiling studies in different metabolic organs have provided important insights into the pathophysiology of T2D, and have not only revealed the epigenetic underpinnings of beta-cell failure and peripheral insulin resistance, but also helped explain vulnerability to the numerous vascular complications.[20] These epigenome-wide association studies (EWAS) have explained certain disease traits that cannot be described by GWAS.[20,21] This chapter will focus on the epigenome-wide association studies in different metabolic organs such as islets, liver, adipose tissue, *etc.* in T2D, and delve into the contribution of environmental and epigenetic factors in T2D risk and disease pathogenesis. It will also

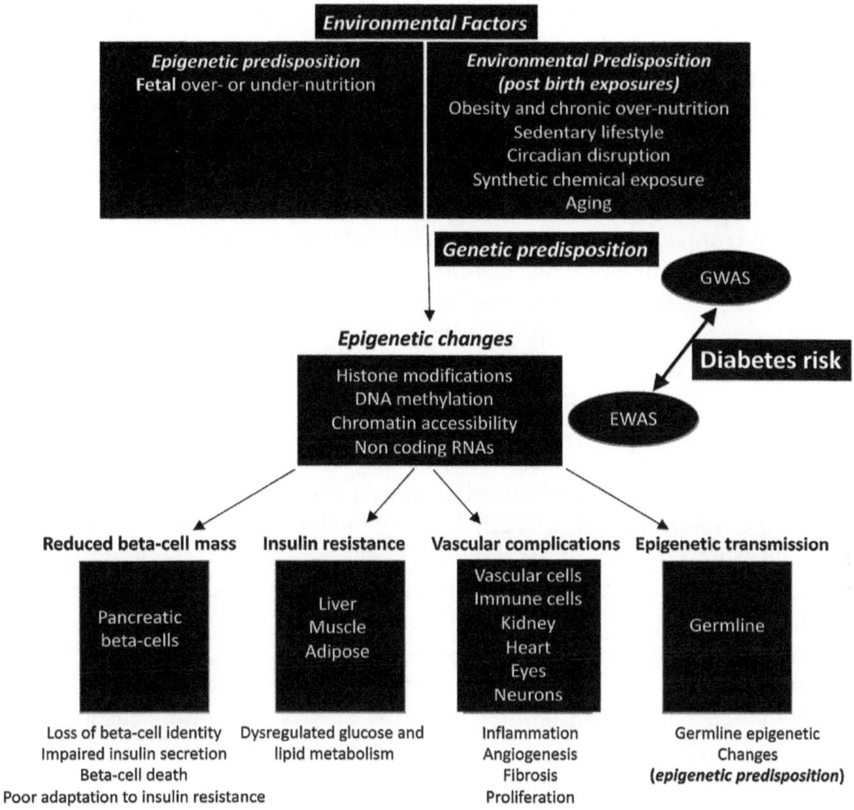

Figure 2.1 Environmental factors such as conditions during fetal growth, or adverse changes in the environment during adult life, can create a state of systemic epigenetic vulnerability. Either alone, or in combination with genetic factors, such epigenetic dysregulation can impact the function of metabolic tissues and lead to vascular damage, and contribute to increased diabetes risk. Environmentally induced epigenetic changes can also occur in the germline and have the potential of transmitting disease risk to subsequent generations. Understanding epigenome-wide changes in diabetes, alone or in relation to genome-wide association studies of diabetes risk, can provide mechanistic insights into disease pathogenesis, as well as generate novel avenues for therapeutic intervention.

discuss the nanobiology of cell–cell communication and its impact on the cellular epigenome in the context of diabetes pathogenesis, and highlight recent breakthroughs in epigenetic engineering that are likely to open new avenues for the development of therapeutic approaches for T2D.

2.2 Epigenetics, Gene–Environment Interactions, and Diabetes Risk

Environmental factors such as changes in dietary quality and quantity have long been known to impact metabolic health and glucose homeostasis.

Obesity, recognized as the biggest modifiable risk factor for T2D, is associated with poor dietary habits and lack of physical activity.[22] In addition to a high dietary carbohydrate and fat content increasing the risk of diabetes, high total protein intake has also been associated with reduced insulin sensitivity and increased incidence of T2D.[23,24] Emerging evidence suggests that dietary routine can affect the epigenetic pathways underlying metabolic regulation. For instance, dietary fat content has been shown to alter the adipose and muscle epigenome, especially at loci known to be involved in the regulation of metabolism in humans.[25,26] Similarly, a high-fat (HF) diet in mice leads to aberrant DNA methylation and expression of genes relevant to obesity and diabetes in the liver, coinciding with altered glucose homeostasis.[27] Dietary composition has a direct effect on the epigenome, given that many metabolites serve as key cofactors for various enzymatic reactions involved in epigenetic modifications.[14–16] For instance, changes in glucose or fat metabolism can change the levels of metabolites such as acetyl-coA and α-ketoglutarate (α-KG), which are essential for histone acetylation and histone/DNA demethylation reactions, respectively. Similarly, the quality and quantity of dietary protein can alter the levels of amino-acid methionine and dictate the levels of methyl-donor S-adenosylmethionine (SAM), a substrate for DNA methylation. This intricate link between dietary content, metabolism, and the epigenome serves as a way for different metabolic organs to adapt their function to the changing milieu.

Besides diet, other environmental factors such as exercise, seasonal variations, and sleep can also impact the epigenome and metabolic homeostasis. Exercise, both acute and chronic, results in the reprogramming of DNA methylation patterns in the subcutaneous white adipose tissue (sWAT), and skeletal muscle in sedentary humans, and impacts the expression of genes involved in adipogenesis, mitochondrial function, muscle contraction, and inflammation.[28–30] Exercise has been shown to improve metabolic health in lean as well as obese individuals. This suggests that the beneficial effects may occur *via* mechanisms other than simply reducing obesity,[22] likely *via* epigenetic changes in pathways involved in adipogenesis and improved muscle insulin sensitivity. Seasonal variations such as temperature changes can have a profound effect on adipose tissue function. Heat or cold exposure have been shown to alter the epigenome, gene-expression patterns, and phenotype of beige adipocytes, allowing metabolic adaptation to temperature changes.[31] Cold exposure has also been shown to induce epigenetic re-programming in sperm, facilitating improved adaptation to over-nutrition and hypothermia in the offspring.[32] Sleep disruption and circadian misalignment are now recognized as major risk factors for T2D.[33] Circadian rhythms directly impact the cellular epigenome, as illustrated by the histone acetyltransferase CLOCK, which serves as a core component of the circadian clock.[34] The circadian feeding–fasting behavior establishes another key link between the circadian clock, metabolism, and the epigenome, supporting systemic metabolic adaptation to diurnal changes in nutrient availability, by regulating circadian gene-expression patterns (reviewed in ref. 34). For instance, feeding–fasting behavior has been shown to epigenetically modulate the rhythmicity of expression of several glucose and lipid metabolism genes in the liver, *via* accumulation of an RNA-binding protein NONO triggered by feeding.[35] This suggests that the timing of nutrient availability is critical for epigenetic regulation of metabolic homeostasis. In agreement with this hypothesis, time-restricted feeding has been shown to prevent metabolic syndrome in mice harboring disruptions in the clock machinery.[36] Thus, circadian disruption can alter the cellular metabolic and epigenetic landscape, and consequently impair adaptation to nutrient availability, predisposing to an increased risk of T2D. Together, these studies show that adverse environmental and lifestyle changes can contribute to T2D pathogenesis.[37]

Type 2 diabetes is a multifactorial disease, involving a combination of genetic and environmental factors that contribute to diabetes risk and disease development. In animal models of diabetes, genetic background has been shown to influence the outcome of diabetogenic stimuli, such as a high-fat diet.[38,39] In humans, diabetes risk and associated contributing factors such as insulin sensitivity, obesity, and beta-cell mass, *etc.* vary between different ethnic and geographical backgrounds.[40,41] However, while there are rare cases of diabetes due to single gene mutations (*e.g.* maturity-onset diabetes of the young), the majority of T2D-associated genetic variants identified to date contribute only to a small percentage of total T2D cases, and mostly increase disease risk rather than being singularly causal. Collectively, these observations point to the significance of gene–environment interactions. Indeed, environmental factors can modify the disease risk posed by genetic variations. Emerging data suggest that the risk posed by genetic variants associated with T2D can be mitigated by increased physical activity and a healthy diet, while a sedentary lifestyle and Western diet can increase the risk. For instance, a low glycemic index diet has been shown to reduce the T2D risk due to variants of the *TCF7L2* locus.[22] Similarly, higher body mass index (BMI, a measure of obesity/adiposity) and Western diet have been shown to increase the risk of T2D conferred by associated genetic variants.[22] A clearer understanding of such gene–environment interactions is likely to advance the development of personalized patient care approaches using lifestyle interventions such as optimized dietary and physical activity regimens.

2.3 Epigenetics and the Developmental Basis of Diabetes Risk

The strongest and most convincing evidence for the role of environmental factors in determining diabetes risk comes from studies focusing on early developmental exposures. A large body of work spanning many vertebrate species has identified poor maternal nutrition and intrauterine growth retardation as major risk factors for diabetes development in offspring, firmly pointing to the involvement of epigenetic factors.[42] Epidemiological studies on the Dutch Hunger Winter, a major famine in 1944 in The Netherlands, have demonstrated that adverse nutritional circumstances during fetal development lead to increased risk for developing diabetes in subsequent generations.[43] Data from such studies have led to the "thrifty phenotype hypothesis," which proposes that poor nutrition during early life leads to lasting changes in glucose homeostasis.[17] These studies also revealed critical developmental time windows that are highly vulnerable to nutritional restriction in the context of diabetes risk, such that those individuals exposed during early fetal life had higher rates of obesity and heart disease. On the other hand, while exposure during late gestation led to low birth weight, these individuals continued to maintain lower weight throughout life compared to those born before and after the famine.[44] These studies also showed that nutritional deprivation in early development is particularly maladaptive in the context of exposure to plentiful nutrition in later life. The developing embryo undergoes adaptations termed "predictive adaptive responses (PARs)" that ensure its immediate survival in response to adverse intrauterine exposures, and future success upon similar exposures in later life.[45] However, a postnatal environment that contrasts the *in utero* circumstances, can lead to metabolic maladaptation and predispose to higher risk of metabolic syndrome. Collectively, this large body of work has formed the basis of "Developmental Origins of Health and Disease," or DoHaD, a concept that is now well-recognized, and underscores the epigenetic basis of metabolic disease.

As the above discussion suggests, maternal malnutrition is a major risk factor for poor metabolic health in the offspring. Dietary protein restriction is often used to model for maternal malnutrition, and has been shown to cause global DNA hypermethylation and locus-specific DNA methylation changes in livers of the offspring in rodents (reviewed in ref. 46). Similarly, intrauterine growth restriction (IUGR) in rodent models leads to impaired pancreatic development, reduced functional beta-cell mass, and higher risk for diabetes,[47,48] associated with disruption of DNA methylation signatures essential for beta-cell function and identity.[42] On the other hand, maternal over-nutrition and gestational diabetes can also have an adverse impact on the metabolic fitness of the offspring.[42] Epidemiological data show that exposure to maternal obesity or diabetes during fetal life predisposes the offspring to insulin resistance, impaired beta-cell function, and high risk of diabetes.[49,50] Maternal high-fat diet and obesity in mice lead to aberrant DNA methylation of genes involved in glucose homeostasis in the liver,[51] as well as opioid and dopamine signaling in the hypothalamus,[52] resulting in altered hepatic glucose homeostasis, insulin resistance, and altered feeding behavior. Therefore, over- and under-nutrition during embryonic development and postnatal growth can compromise the development and function of metabolic organs, predisposing the offspring to diabetes.[10,42] Impaired metabolic homeostasis in either parent can alter the metabolic program in the offspring concurrent with specific epigenetic changes, and points to the involvement of epigenetics in the transmission of diabetes risk.[43,53,54] Environmental exposure can also cause aberrant epigenetic changes in the germline, which may be transmitted to subsequent generations and potentially contribute to the inheritance of T2D risk.[53,55–58] Altogether, these studies show that parental metabolic health and lifestyle choices can induce transgenerational changes in the epigenome and metabolic fitness.

The rising popularity of dietary supplements in recent years is another emergent challenge with potential epigenetic consequences relevant to metabolic health. Optimal availability of micronutrients is critical for programming and maintaining the epigenetic landscape. For example, the DNA methylation reaction requires *S*-adenosyl methionine (SAM) as a co-factor, the levels of which are determined by methyl-donors such as folate and vitamin B12, the amino acid methionine, as well as flux through the TCA cycle.[59] Aberrant levels of these micro- and macro-nutrients during embryonic development can therefore impact the epigenetic pathways and development of metabolic organs.[42] Utilization of SAM *via* the DNA methylation reaction leads to the formation of S-adenosyl homocysteine (SAH), which is an inhibitor of DNA methyltransferases.[60] Of note, SAH levels are high in obesity in rodents and humans,[61,62] and can therefore alter DNA methylation in the developing fetus in response to maternal obesity. Similarly, an imbalance in methyl donor vitamins such as folate and B12 has been shown to alter DNA methylation in somatic tissues and male germline, and can impair metabolic homeostasis.[63] The effect of dietary methyl donor supplementation on epigenetic regulation is best highlighted by studies on the obese *Agouti* mice; feeding pregnant *Agouti* females a methyl donor-rich diet leads to hypermethylation of the IAP (Intracisternal A-particle) region in the offspring, and protects against obesity.[64,65] Besides nutritional quality and accessibility, developmental exposure to synthetic chemicals and nutritional supplements has also presented itself as a major driver of epigenetic aberrations that can predispose to diabetes. In the past few decades, the exposure to chemicals such as bisphenol A (BPA) and phthalates *via* food and water has become increasingly prevalent. Several studies show that exposure to BPA or phthalates during development and growth leads to increased risk for obesity and diabetes in rodents and humans, corresponding to epigenetic changes in multiple metabolic tissues (reviewed in ref. 42).

2.3.1 Gametes, Early Embryonic Development, and Transmission of Disease Risk

The recognition of early development as a particularly sensitive time window for future diabetes risk opens up a new avenue for diabetes prevention. Epigenetic changes during this time period can severely impact the differentiation, functional maturation, and growth of metabolic organs and predispose to poor metabolic fitness and diabetes. The sperm and oocytes are especially vulnerable cell types, given that any epigenetic changes in the gametes can potentially be transmitted to the next generation. The epigenome undergoes extensive modulation throughout development, but the early embryonic development is particularly marked by large- scale epigenetic changes. The mammalian genome undergoes two waves of global DNA demethylation and re-methylation, the first occurs immediately after fertilization and constitutes the pre-implantation reprogramming, and the second occurs in the germline during gametogenesis (reviewed in ref. 66–68). The demethylation upon fertilization serves to create a clean epigenetic template that reflects the developmental multipotency, and can subsequently be used to build stage-specific DNA methylation patterns to guide differentiation.

Using *in vitro* fertilization to model epigenetic inheritance exclusively *via* either affected male or female gametes, Huypens and colleagues demonstrated that diet-induced obesity in the parent increases the risk for obesity and diabetes in the offspring in a sex- and parent of origin-specific manner[69]. Recent work has shown that environmental factors, such as diet, exercise, synthetic chemical exposure, and trauma, can remodel small RNAs and methylome in the sperm with evident phenotypic consequences in the offspring.[57,70,71] Similarly, dietary imbalances and other environmental exposures in female mice can alter the oocyte epigenome, leading to metabolic impairments in the offspring.[55,72] Such "epimutations" in both maternal and paternal gametes can be transmitted to the next generation, and potentially contribute to phenotypic changes and metabolic defects in the offspring and subsequent generations. Perpetuation of such epigenetic alteration can be multigenerational or transgenerational. The first case involves the parent and parental germline incurring the exposure (in the case of a pregnant female parent, it means that the fetus and fetal germline incur the exposure as well), while the second scenario indicates transmission to a subsequent generation not directly exposed to the environmental challenge.

Imprinted genes represent a particularly vulnerable set of epigenetically regulated loci during development; these are genes with differential allelic regulation of transcription based on parental origin. The epigenetic patterns of imprinted genes are stringently protected through early developmental reprogramming in the gametes to control the gene dosage of key regulators of growth and differentiation[73]. Environmentally induced perturbations in the epigenetic regulation of imprinted genes can therefore not only facilitate disease risk transmission, but also dictate the pattern of its epigenetic inheritance.[74] Collectively, epigenetic factors not only underlie the influence of environmental factors on diabetes risk and pathogenesis, but may also contribute to the intergenerational transmission of disease risk. It remains to be seen whether such epigenetic changes in the gametes can be identified and potentially reversed.

2.4 Epigenome-wide Association Studies (EWAS) for T2D

A large number of genome-wide association studies (GWAS) have identified a large number of genetic variants and loci associated with T2D in humans.[75] However, these variants only constitute a minor fraction of hereditary T2D, suggesting a non-genetic component. Of note, a large majority of T2D-related variants occur in the non-coding portion of the genome, rather than the coding regions.[76] It is likely that such variants mediate functional changes by altering the epigenetic regulation of the neighboring coding regions. For instance, single nucleotide polymorphisms (SNPs) that add or remove CpG sites can alter DNA methylation patterns, while other variants in the non-coding regions may impact the expression of regulatory long non-coding RNAs. Such epigenetic changes can then modify the expression of specific genes or alter regulatory networks to impact the function of different metabolic organs, and potentially contribute to diabetes risk. In addition, changes in the enzymatic activity of epigenetic regulators due to changes in levels of specific metabolites can cause environmentally induced changes in the epigenome, which can independently mediate changes in gene expression and cellular function.

Epigenomic mapping can help explain the functional significance of GWAS variants not directly associated with a coding region. Epigenome-wide association studies (EWAS) can also provide mechanistic insights into the impact of genetic variants on T2D risk, define the disease-relevant gene–environment interactions, and identify the disease-relevant cell types for a given risk variant.[77] Changes in epigenetic status, specifically DNA methylation patterns, of a number of genomic regions have been associated with T2D.[78] However, a large amount of these data are based on the analysis of blood samples rather than specific metabolic tissues, and involve a cross-sectional study design. Thus, while such datasets provide valuable information for designing epigenetic biomarkers for T2D, linking such epigenetic changes to causality is challenging and requires tissue-specific epigenomic profiling. It is important to note that epigenomic changes in the context of disease may either be causal or an effect of the disease process. Integration of epigenomic data with GWAS, and subsequent functional analysis of associated loci can help distinguish these scenarios. Initiatives such as the International Human Epigenome Consortium (IHEC) aim to provide extensive coverage of human epigenomic maps across multiple tissues and cell types, and include efforts to survey data from pedigrees and identical twins to link genetic and epigenetic variants.

Recent advances in such integrative epigenomic analyses have provided important insights into the causal role of genetic variants in T2D risk and pathogenesis. For instance, combining islet-specific epigenomic maps with functional screens of associated target genes, a recent high-throughput screening study aimed at the functional phenotyping of GWAS variants was able to determine the functional relevance of 37 loci associated with T2D risk.[79] Studies such as these have revealed that specific variants associated with T2D risk, such as SNPs for the melatonin receptor *MTNR1B*, are concentrated at islet-specific enhancer regions.[80,81] Similar approaches involving a combination of epigenomic mapping with genetic data have been used to clarify the functional impact and tissue specificity of two of the most well-known T2D risk loci, namely *TCF7L2* and *FTO*.[82–84] These studies show that variants in *TCF7L2* manifest their causal impact primarily on the adipose tissue, and demonstrate that *FTO* variants promote obesity by repressing mitochondrial thermogenesis in adipocyte precursors.[83,84] Studies like these have allowed us to determine the effect of such T2D-related variants on the tissue-specific epigenetic architecture and expression of associated genes, providing mechanistic insights into disease risk, pathogenesis, and complications.

Epigenomics can also identify environmentally induced, systemic, or tissue-specific, disease-relevant changes that can help explain the T2D risk or pathophysiology. A large number of recent studies has led to the identification of tissue-specific epigenetic changes that are causal to the dysregulation of molecular pathways involved in glucose homeostasis. For example, profiling of DNA methylome in islets from diabetic and non-diabetic subjects points to the large-scale epigenetic dysregulation of pathways involved in insulin secretion, islet survival, and islet adaptation.[85–87] Similarly, longitudinal epigenomic profiling has been instrumental in unraveling the molecular basis of why early and rigorous glycemic control after diagnosis is the key to the prevention of diabetes complications.[88,89] The significance of longitudinal epigenomic mapping is further highlighted by a recent set of studies by Snyder and colleagues, called the Integrative Personal Omics Profiling (IPOP) initiative. By integrating longitudinal personal epigenomic, transcriptomic, and metabolic datasets, this study showed that molecular signatures related to inflammation and cardiomyopathy are up-regulated in insulin-resistant individuals as a function of body weight gain.[90] Altogether, these studies point to the power of tissue-specific epigenomic profiling in elucidating the epigenetic basis of T2D pathogenesis.

2.4.1 Methods for Epigenomic Profiling

EWAS work has relied on methods of epigenomic profiling to establish single base-pair resolution maps of DNA methylation patterns, profile histone modifications, and establish chromatin architecture maps.[91] For DNA methylation, this has involved increasingly sophisticated means to map the methylome. These range from affinity-based DNA pull-down techniques such as MeDIP-seq (**Me**thylated **DNA** **I**mmuno**p**recipitation using anti-methyl-CpG antibody), and using specific reagents such as methylation-sensitive enzymes or bisulfite treatment, to distinguish between methylated and unmethylated CpG residues. Bisulfite treatment is common to most of the popular methods, and bisulfite-treated DNA can either be hybridized to arrays (Illumina's Infinium arrays for human methylome), or directly sequenced (**w**hole **g**enome **b**isulfite **s**equencing: WGBS), to determine methylation patterns in a given sample. In combination with methyl-sensitive enzyme digestion, bisulfite sequencing can be used to generate samples for **r**educed **r**epresentation **b**isulfite **s**equencing (RRBS), Although RRBS provides less coverage than WGBS, it is significantly more cost-effective and hence quite popular (Table 2.1). Chromatin immunoprecipitation-sequencing (ChIP-seq)-based profiling of specific histone modifications and chromatin modifier recruitment can be highly informative about chromatin features such as promoter

Table 2.1 Methods for establishing DNA methylation profiles for EWAS include either affinity-based enrichment of methylated regions or distinguishing the methylation status of a CpG residue by either methyl-sensitive restriction enzyme-based assays or bisulfite treatment.

Method	Description	Class
Whole-genome bisulfite sequencing (WGBS)	Whole-genome bisulfite sequencing (WGBS) involves bisulfite treatment followed by sequencing, and is the most comprehensive method to determine genome-wide methylation at the single-cytosine level.	Bisulfite treatment
Reduced representation bisulfite sequencing (RRBS)	Reduced representation bisulfite sequencing (RRBS) analyzes the genome-wide methylation profiles on a single nucleotide level. It combines methylation-sensitive restriction enzymes and bisulfite sequencing to enrich for areas of the genome with a high CpG content.	Bisulfite treatment, combined with methylation-sensitive restriction enzyme treatment
Methylation-sensitive restriction enzyme sequencing (MRE-seq)	This is a restriction enzyme-based method which incorporates parallel digestions with different methylation-sensitive restriction endonucleases, followed by deep sequencing. Methylation status of single CpG residues is inferred from the pattern of restriction digestion.	Based on methylation-sensitive restriction enzyme treatment
Methylated DNA immunoprecipitation (MeDIP-seq)	MeDIP-seq involves isolation of methylated DNA fragments *via* an antibody raised against 5-methylcytosine (5mC), and coupling with next-generation short read sequencing.	Affinity enrichment
Methylated DNA-binding proteins (MBD-seq)	MBD-seq method is based on capture of methylated DNA using a methyl-CpG binding domain-based (MBD) protein.	Affinity enrichment
DNA methyl arrays	These methods couple bisulfite treatment with hybridization to gene arrays to determine the pattern of DNA methylation. *E.g.*, Illumina BeadChip and EPIC arrays.	Array based, coupled with bisulfite treatment

activity and enhancer engagement, while methods such as ATAC-seq (Assay for Transposase-Accessible Chromatin sequencing) are frequently used to determine chromatin accessibility. When combined with transcriptome profiling, these EWAS methods provide valuable insights into the epigenetic regulation of gene expression.[91]

EWAS can involve a variety of study designs such as case *vs.* control, longitudinal studies, disease-discordant monozygotic twins, and family cohorts. The first method usually allows highly powered sampling, while the latter three can be highly informative about causality.[77] The choice of tissue samples is also important. Many tissues such as islets can only be collected from cadaveric donor subjects in most cases. Peripheral blood sampling, on the other hand, is routinely done in a clinical setting, and is therefore ideal for EWAS. However, epigenetic changes often tend to be highly tissue specific, and therefore not all disease-relevant epigenetic associations may be reflected in blood samples.

2.4.2 Epigenetic Mechanisms of Beta-cell Failure

T2D is characterized by the failure of beta-cells to adapt in response to insulin resistance, and involves the progressive decline of beta-cell identity, function, and survival. Epigenetic mechanisms direct all these aspects of beta-cell homeostasis,[13] suggesting that epigenetic changes in response to environmental factors can potentially drive beta-cell failure. A large body of evidence shows that stage-specific patterning of DNA methylation, histone modifications, and chromatin architecture is essential for pancreas lineage specification and endocrine differentiation.[92–96] Epigenetic regulation also plays a pivotal role in regulating the functional beta-cell phenotype. For instance, DNA methylation patterning and regulation *via* miRNAs and lncRNAs are essential for the establishment and maintenance of beta-cell identity and functional maturity in mice and humans.[97–107] Epigenetic mechanisms also regulate beta-cell expansion during postnatal growth, adaptation, and aging.[108–113] These studies suggest that epigenetic changes can potentially alter functional beta-cell mass.

Indeed, epigenetic dysregulation plays an important role in the loss of the functionally mature beta-cell phenotype in diabetes, a phenomenon referred to as de-differentiation.[114] This is exemplified by the epigenetic derepression of Neuropeptide Y (NPY, which inhibits insulin secretion) in diabetic beta-cells, corresponding to impairment of beta-cell identity and function.[115] Recent work on the loss of polycomb repressive complex 2 (PRC2)-dependent epigenetic regulation in T2D islets affirms this idea.[116] The relevance of the epigenetic regulation of functional beta-cell mass to glucose homeostasis is further underscored by imprinting disorders such as the Beckwith–Wiedemann syndrome (BWS) (reviewed in ref. 117). BWS is marked by imprinting defects that lead to the loss of cell-cycle inhibitor CDKN1C ($p57^{Kip2}$), resulting in unchecked expansion of beta-cells and consequent hyperinsulinemia and hypoglycemia.[117] In addition, variants of imprinted loci that regulate insulin secretion, such as *GRB10* and *KCNQ1*, are associated with increased T2D risk.[118,119] Altered imprinting of the *DLK1-MEG3* locus has been reported in human islets from subjects with T2D, and leads to dysregulation of a cluster of miRNAs that regulate genes involved in beta-cell function and survival.[120]

Genome-wide profiling of DNA methylation patterns in human islets from control and T2D donors has revealed that the islet methylation landscape undergoes profound changes in T2D, amounting to the dysregulation of islet transcriptional networks involved in insulin secretion, adaptation, and survival[85–87] (Table 2.2). These studies also demonstrated that histone modifications and DNA methylation play a combinatorial role in regulating islet homeostasis.[87] Functional analysis of some of the T2D-associated differentially methylated regions (DMRs) in human islets supports the relevance of such epigenetic changes to beta-cell failure in T2D. Diabetes risk factors such as obesity, hyperglycemia, and aging have been shown to alter the methylation patterns of T2D-associated islet DMRs in non-diabetic subjects, indicating that epigenetic dysregulation initiates long disease onset.[85] Of note, genes involved in beta-cell replication and function undergo highly specific epigenetic changes with aging,[121] substantiating a causal role for epigenetic dysregulation in mediating T2D risk. Several of the T2D islet DMRs also overlap with DMRs associated with the poor endocrine differentiation of stem cells lacking DNA demethylases,[122] suggesting that dysregulated DNA methylation may contribute to beta-cell de-differentiation. As noted earlier, adverse conditions in fetal life alter the epigenetic program of genes involved in beta-cell replication, function, and survival.[123] Collectively, this discussion supports a direct role for epigenetic dysregulation of the functional beta-cell phenotype in mediating beta-cell failure in T2D.

2.4.3 Epigenetics of Obesity and Insulin Resistance

The postprandial release of insulin facilitates the uptake and storage of nutrients by multiple metabolic tissues such as muscle, liver, and adipose tissue, thus serving to maintain metabolic homeostasis. Insulin signaling promotes glucose uptake and glycogen synthesis by muscle, increased glycogen synthesis and reduced gluconeogenesis in the liver, and increased triglyceride synthesis and reduced lipolysis in adipose tissue. Altered metabolic states such as obesity, chronic over-nutrition, and metabolic disease can dampen the peripheral action of insulin, amounting to insulin resistance.[124] Epigenomic profiling has been instrumental in identification of novel regulatory networks that are altered in the context of insulin resistance and obesity across metabolic tissues. Many such studies have focused on peripheral blood samples, given the ease of sample acquisition and the ability to perform longitudinal studies, and are therefore also highly powered. These studies have identified key genomic regions that are differentially methylated as a function of diabetes, obesity, as well as increased fasting blood glucose and HbA1c.[125,126] Analysis of these data for causal association shows that while a large majority of these DNA methylation differences are a result of obesity, some differential DNA methylation can be causal.

Table 2.2 Epigenome-wide association studies (EWAS) in islets in the context of T2D demonstrate dysregulation of DNA methylation at regions relevant to beta-cell identity, insulin secretion, and survival. In addition, differentially methylated regions include loci associated with T2D risk, such as TCF7L2, FTO, and KCNQ1. Some of the differentially methylated loci in T2D islets have been identified by more than one study, and are marked in bold.

Tissue	Samples	Method	Differentially methylated genes	Reference
Islets	5 diabetic and 11 non-diabetic subjects	Infinium Human Methylation 27K Bead Chip Assay	*GRB10, GUCA2B, MADD, PER2, BCL2, NIBAN, PPP2R4, CHAC1, CASP10, NR4A1*	[86]Volkmar (2012)
	15 diabetic and 34 non-diabetic subjects	Infinium Human Methylation 450K Bead Chip Assay	***TCF7L2***, *FTO, KCNQ1, CDKN1A, PDE7B, SEPT9, SREBF1, TXNIP*	[85,150]Dayeh (2014, 2016)
	6 diabetic and 8 non-diabetic subjects	Whole-genome bisulfite sequencing	***TCF7L2***, *PDX1, ADCY5, NR4A3, PARK2, PID1, SLC2A2, SOCS2*	[87]Volkov (2017)

Epigenome-wide mapping and association studies have also been carried out in peripheral metabolic tissues such as muscle, liver, and adipose tissue, to identify the epigenetic changes underlying metabolic dysfunction in the context of obesity, insulin resistance, and T2D (Table 2.3). Several studies involving DNA methylation profiling of adipose tissue have demonstrated that the differentially methylated regions (DMRs) associated with high BMI and obesity reside in genes involved in lipid and lipoprotein metabolism, nutrient transport, inflammation, and T2D risk. In addition, such obesity-associated changes in adipose DNA methylation appear to be predictive of future risk of T2D.[127–129] Similarly, DNA methylation profiling in adipose samples from patients pre- and post-gastric bypass has led to the identification of key obesity-related DMRs that overlap with known T2D risk loci.[130] These data indicate that the adipose epigenome is altered by changes in adiposity in either direction. Aging, a known risk-factor for T2D, also cause changes in the adipose DNA methylation patterning.[131] The critical role of epigenetic regulation in adipose health is further highlighted by the essential role of DNA methyltransferase Dnmt3a in mediating insulin resistance in adipose tissue.[132] Other studies involving the analysis of multiple histone modifications have identified key genomic regulatory regions associated with insulin resistance in adipocytes, in response to diabetogenic factors such as steroid exposure and inflammation.[133] Samples from monozygotic twins discordant for T2D have been instrumental in determining the contribution of epigenetic factors in diabetes risk and pathogenesis, given the identical genetics in such subjects. Epigenetic profiling in adipose tissue from such subjects has shown that while DNA methylation differences between diabetic and non-diabetic identical twins are relatively small, such methylation differences tend to be located in genes important for adipose metabolic homeostasis,[127,134] pointing to a likely causal role for such changes.

Combinatorial analysis of epigenetic and transcriptomic profiles of skeletal muscle in the context of a recent T2D diagnosis or family history has revealed important differences in the muscle transcriptional regulation and insulin signaling, with some of the differentially regulated regions mapping to known T2D risk loci[28,135] (Table 2.3). Mitochondrial dysfunction and poor glucose uptake are hallmarks of muscle insulin resistance, and the epigenetic profiles of genes regulating mitochondrial function and glucose uptake are altered in T2D,[136,137] as exemplified by *PPARGC1* (key regulator of mitochondrial function) and *GLUT4* (glucose transporter).[138,139] Emerging data from rodent studies suggest that the muscle-specific epigenetic changes in response to obesity can also be transmitted to subsequent generations.[140] Muscle mass and insulin sensitivity decline with age, contributing to increased T2D risk.[78] Human skeletal muscle undergoes age-related changes in DNA methylation and expression of several genes involved in oxidative phosphorylation,[141–143] suggesting that age-induced DNA methylation changes can directly impact muscle function. Physical activity and exercise have long been known to improve muscle insulin sensitivity and metabolic health. Long-term exercise

Table 2.3 EWAS in insulin-sensing peripheral tissues has led to the identification of DNA methylation-dependent dysregulation of genes involved pathways related to insulin action and tissue function in the context of obesity and exercise. The majority of these differentially methylated regions are tissue specific, while some others can be dysregulated in more than one tissue. This is exemplified by some sites identified in both muscle and liver (references). Differentially methylated sites identified by multiple studies in a given tissue are indicated in bold.

Tissue	Samples	Method	Differentially methylated genes	Reference
Adipose Tissue	5 monozygotic twin pairs discordant for T2D	Infinium Human Methylation 27K Bead Chip array (Illumina)	***KCNQ1***, ***IRS1***, *ADCY5*, *CAV1*, *CIDEC*, *CDKN2A*, *CDKN2B*, *DUSP9*, *HNF4A*, *IDE*, *MTNR1B*, *TSPAN8*, *WFS1*	[134]Ribel-Madsen (2012)
	28 non-diabetic and 28 diabetic subjects	Infinium Human Methylation 450K Bead Chip array (Illumina)	***KCNQ1***, ***IRS1***, ***TCF7L2***, *PPARG*, *ABCG1*	[127]Nilsson (2014) [150]Dayeh (2016)
	22 obese subjects- pre (14) and post (8) Roux-en-Y gastric bypass (RYGB) surgery and 7 non-obese subjects	Bisulfite Pyrosequencing	27 genetic T2D risk loci including ***TCF7L2***	[130]Multhaup (2015)
	96 male subjects (23–80 years), 94 females (21–37 years)	Infinium Human Methylation 450K Bead Chip	*ELOVL2*, *FTO*, *FHL2*, *PENK*, *KLF1*, *SST*, *GLRA1*, *TP73*, *GATA4*, *THRB*, *DLX5*, *NEFM*, *TMEM17*, *ATP8A2*, *FOXE3*	[131]Ronn (2015)
	36 morbidly obese Caucasian patients with different degrees of insulin sensitivity	Infinium Human Methylation 450K Bead Chip array (Illumina)	*COL9A1*, *COL11A2*, *CD44*, *MUC4*, *ADAM2*, *IGF2BP1*, *GATA4*, *TET1*, *ZNF714*, *ADCY9*, *TBX5*, *HDACM*	[128]Crujeiras (2016)
Liver	35 diabetic and 60 non-diabetic subjects	Infinium Human Methylation 450K Bead Chip array (Illumina)	*GRB10*, *ABCC3*, *MOGAT1*, and *PRDM16*	[145]Nilsson (2015)
	11 non-obese, 13 obese non-diabetic and 11 obese T2D subjects	Illumina Infinium Human Methylation 450K Bead Chip array (Illumina), Bisulfite pyrosequencing	*GCK*, *PFKL*, *ACACA*, *ACLY*, *FASN* and *PRKCE*	[144]Kirchner (2016)

	Liver biopsies from 95 individuals obtained during a Roux-en-Y gastric bypass surgery	Illumina Infinium Human Methylation 450K Bead Chip array (Illumina)	*ATP8A2, DLX5, ELOVL2, FOXE3, FHL2, GATA4, GLRA1, KLF14, NEFM, PENK, SST, THRB, TMEM1, TP73* **Sites similar to those in muscle in Ronn, 2015.*	[149]Bysani (2017)
Muscle	14 young and 14 elderly subjects	Bisulfite sequencing	*NDUFB6*	[143]Ling (2007)
	10 young and 10 elderly non-diabetic twins	Bisulfite sequencing	*COX7A1*	[141]Ronn (2008)
	17 normal glucose-tolerant subjects, 8 impaired glucose-tolerant subjects, and 17 subjects with T2D	MeDIP Assay	*PPARGC1A*	[138]Barres (2009)
	11 monozygotic twin pairs discordant for T2D	Infinium Human Methylation 27K Bead Chip array (Illumina)	*CDKN2A, DUSP9, HNF4A, HHEX, KCNQ1, KLF11, PPARGC1A, SLC30A8*	[134]Ribel-Madsen (2012)
	15 subjects with and 13 subjects without family history of T2D, with 6 months of supervised exercise	MeDIP-Chip analysis	*MAPK1, MYO18B, HOXC6, MEF2A, RUNX1, NDUFC2, THADA, ADIPOR1, BDKRB2, TRIB1*	[28]Nitert (2012)
	17 subjects with T2D, sampled at baseline and post exercise training	Infinium Human Methylation 450K Bead Chip array (Illumina)	*GLUT4, NRF2*	[139]Rowlands (2014)
	17 monozygotic twin pairs discordant for T2D	Infinium Human Methylation 450K Bead Chip array (Illumina)	*PHOSPHO1*	[150]Dayeh (2016)

intervention has been shown to modify the skeletal muscle DNA methylation profiles in individuals with and without a family history of T2D.[28] Such exercise-induced changes in DNA methylation and histone modifications are associated with regions involved in muscle function and T2D pathogenesis.[134,137] Of interest, acute exercise can induce favorable epigenetic changes in the skeletal muscle of sedentary individuals, as demonstrated by DNA methylation and transcriptional profiling of skeletal muscle biopsies before and after acute bursts of exercise.[29] It therefore appears plausible that exercise intervention can partly restore the "healthy" skeletal muscle epigenetic signatures in diabetes, and improve insulin sensitivity and metabolic health.

Similar to the adipose tissue and skeletal muscle, the liver epigenome also undergoes profound changes in response to obesity and hyperglycemia, as demonstrated by multiple epigenetic profiling studies (Table 2.3). Obesity and T2D can alter the liver methylome and gene-expression profiles to mediate increased glycolysis and lipolysis, which can contribute to the development of insulin resistance. Several of these DMRs map to known loci associated with several T2D risks.[144,145] The relevance of liver DNA methylation to T2D risk becomes even more evident with aging. Similar to adipose and skeletal muscle, the liver also experiences age-dependent reprogramming of DNA methylation. It is well known that the cellular DNA methylation landscape changes with age, and is predictive of the functional aging of a cell.[146] Notably, obesity leads to increased DNA methylation age in the liver, which can have implications for higher risk for hepatic insulin resistance with age.[147] Aging is associated with increased expression of genes encoding *de novo* DNA methyltransferases (*DNMT3A* and *DNMT3B*) in the human liver, which can contribute to age-related changes in liver DNA methylation.[148] In support of this, EWAS studies have shown that aging leads to altered blood DNA methylation profiles in humans, corresponding to functionally relevant changes in hepatic gene expression.[149]

All of these studies collectively emphasize that epigenetic dysregulation is an underlying feature in insulin resistance and T2D, and impacts all metabolic tissues. Altered epigenome, in turn, leads to aberrant gene expression with functional consequences. Another common point that emerges from these studies is the impact of age-related DNA methylation changes on T2D risk. As discussed earlier, aging is a major factor predisposing to insulin resistance and T2D risk. Age-related changes in DNA methylation with consequent changes in gene expression that impact tissue function across all insulin-sensitive tissues likely contribute to increased disease risk. The studies discussed here also highlight the interaction between genetics and epigenetics, and imply that epigenetic changes can attribute functionality to disease-associated genetic variants. As the epigenetic program and gene expression are highly tissue specific, the genome–epigenome interaction may differentially impact specific tissues. That said, common T2D-related epigenetic changes do exist across tissues. For example, increased DNA methylation at the cholesterol transport gene *ABCG1* is associated with higher T2D risk, and is seen in both blood and adipose tissue.[150] While many of the epigenetic changes evident in T2D may be secondary to disease pathogenesis, emerging evidence strongly suggests the existence of epigenetic changes that can be causal. Longitudinal studies focused on following epigenetic changes through disease onset and progression will be necessary to identify the causal epigenetic underpinnings of diabetes pathogenesis.

2.4.4 Epigenetic Dysregulation as a Mediator of Diabetes Complications

Many patients with T2D develop severe and life-threatening secondary health problems due to persistent hyperglycemia. The most prevalent complications include microvascular issues such as nephropathy, retinopathy, and neuropathy, as well as macrovascular diseases like atherosclerosis, hypertension, and cardiovascular disease.[151] Long-standing metabolic dysregulation due to hyperglycemia is the major precipitating factor for diabetes complications; and can cause vascular damage through multiple pathways, including oxidative stress, accumulation of advanced glycation end-products (AGEs), changes in extracellular matrix (ECM), and the activation of pro-fibrotic and pro-inflammatory pathways such as transforming growth factor-beta (TGF-beta), NF-κB, and angiotensin-II (AngII),[152] which promote the expression of genes involved in growth, inflammation, apoptosis, and fibrosis. This leads to endothelial dysfunction, macrophage infiltration, and inflammation, growth, and fibrosis in vascular smooth muscle and renal cells, ultimately resulting in dysfunction of different organs.[152] This suggests that the cellular environment plays a critical role in the pathogenesis of diabetes complications, and points to the involvement of epigenetic dysregulation.

Epigenetic mapping has been instrumental in elucidating the causative mechanisms of diabetic complications (reviewed in ref. 151,153). For instance, genome-wide DNA methylation profiling of renal tubuli from humans with chronic kidney disease including diabetic nephropathy and control subjects showed significant differences in DNA methylation in genes with a known role in fibrosis. Similarly, analysis of human peripheral blood samples revealed specific DNA methylation differences that are predictive of renal function decline in diabetic nephropathy.[154,155] Of note, changes in both DNA methylation and several histone modifications have been implicated in driving the TGF-β-dependent activation of genes associated with renal fibrosis (reviewed in ref. 151,153). In line with this, promoters of pro-fibrotic, redox-regulatory, and pro-inflammatory genes associated with diabetic nephropathy are marked by activating histone modifications *in vivo*, in rodent models of diabetes.[151] A hyperglycemic milieu has also been shown to disrupt the DNA methylation patterns of key genes involved in vascular complications and neuropathy, in primary endothelial and Schwann cells, respectively.[156,157] Altered histone modifications also contribute to the dysregulation of genes involved in redox homeostasis, cell–ECM interaction, and inflammation in retinal endothelium, in both *in vitro* and *in vivo* models of diabetic retinopathy.[158] Monocytes and macrophages are key cellular players involved in the pathogenesis of vascular complications, and hyperglycemia leads to large-scale changes in the histone modification profiles in human monocytes *in vitro* and *in vivo.*[159] Many such epigenetic changes occur in genomic regions associated with inflammation and oxidative stress, and likely contribute to vascular dysfunction.[159] Similarly, alterations in histone modification profiles drive the upregulation of inflammatory gene expression in vascular smooth muscle cells (VSMCs), in response to hyperglycemia *in vitro*, and in mouse models of T2D.[160] In addition to DNA methylation and histone modification, non-coding RNAs (miRNAs and LncRNAs) also serve as key epigenetic mediators of diabetes complications (reviewed in ref. 153,161–163). For instance, the *miR-216/miR-217* cluster, the *miR-379* mega-cluster, and its host lncRNA, *lnc-MGC* are all involved in promoting gene expression changes that cause cellular hypertrophy and fibrosis, in the context of diabetic nephropathy.[164,165] Similarly, lncRNAs *Dnm3os* and *Giver* are involved in driving inflammation and VSMC dysfunction, respectively.[166,167] Collectively, these studies underscore the mechanistic contribution of multiple layers of epigenetic dysregulation in diabetes complications.

The strongest evidence for the involvement of epigenetic mechanisms in diabetes complications comes from studies on the phenomenon of metabolic memory, which alludes to cells retaining memory of a prior hyperglycemic exposure, even after restoration of normoglycemia. Metabolic memory is highly clinically relevant as it underlies the continued progression of diabetes complications even upon achieving rigorous glycemic control after long-standing hyperglycemia. This phenomenon has been reported in experimental models of diabetes, as well as in epidemiological studies such as the Diabetes Control and Complications Trial (DCCT), and the subsequent long-term follow-up Epidemiology of Diabetes Interventions and Complications (EDIC) study on T1D.[151,168] This study demonstrated that while all the subjects maintained similar intensive glycemic control during the follow-up EDIC phase, subjects with conventional glycemic control during the DCCT phase were at a higher risk for diabetes complications compared to those who maintained intensive glycemic control throughout.[168] Epidemiological studies on T2D similarly point to a "legacy effect" showing persistent beneficial effects associated with intensive glycemic control, long after such an intervention is over.[169] These studies suggest that prolonged exposure to a hyperglycemic environment likely leads to the persistence of epigenetic changes associated with tissue dysfunction, thus establishing a metabolic memory. In line with this hypothesis, profiling of DNA methylation and multiple histone modifications in peripheral blood cells from the DCCT/EDIC study revealed significant differences in epigenetic marks at several loci associated with inflammation, between conventional *vs.* intensive glycemic control.[89,151] Longitudinal profiling of DNA methylation in blood samples collected from the same patients in the DCCT *vs.* EDIC phases of this study showed the persistence of DNA methylation patterns at key loci associated with diabetes complications, demonstrating an epigenetic basis for metabolic memory.[88] In summary, changes in the cellular environment can foster epigenetic changes, which may persist even after the initial metabolic assault has ceased. Studies focusing on the mechanistic understanding of such epigenetic persistence are required to overcome the issue of metabolic memory in diabetes management and therapy.

2.5 Extracellular Vesicles, the Epigenome, and Diabetes Pathogenesis

The pathogenesis of T2D involves the functional interaction of multiple cell types and organs. Extracellular vesicles (EVs), membrane-bound small vesicles released by multiple cell types, are a key mode of cell–cell communication, and have recently been implicated in the pathogenesis of multiple diseases, including T2D. EVs contain unique molecular cargos, the composition of which is reflective of the physiologic and environmental conditions of the source cells.[170] These cargos include a variety of biomolecules including proteins, mRNAs, miRNAs, and lncRNAs, which can modulate the epigenome and therefore the phenotype of the target cells. The involvement of EVs in multiple aspects of T2D pathogenesis and its vascular complications is now well recognized. One of the earliest reports showed a significant increase in platelet-derived EVs in patients with T2D, and suggested their contribution to atherosclerosis in these subjects,[171] an observation supported by multiple subsequent studies reporting increased numbers of total and platelet-derived EVs in patients with T2D.[172] Most of the work on the intercellular transmission of epigenetic information *via* EVs in T2D has focused on the miRNA components of these vesicles. Emerging evidence shows that the miRNA signatures of EVs are altered in the T2D, and these disease-specific signatures play an active role in various aspects of diabetes pathogenesis and its various complications.

Two major studies published in 2017 firmly established the role of circulating EV-dependent epigenetic regulation in metabolic homeostasis; both focusing on adipose tissue. One study showed that mice harboring an adipose-specific loss of the miRNA processing enzyme Dicer had a marked reduction in circulating EV miRNAs, a phenomenon also witnessed in humans with lipodystrophy. Transplantation of wild-type adipose into these KO animals was able to restore a variety of EV miRNAs that are known to be associated with improved glucose tolerance. Notably, serum EVs from the wild-type mice were able to mimic this "rescue." This study also demonstrated that adipose tissues serves as a source of circulating EV miRNAs that can modulate gene expression in other tissues.[173] Another study showed that adipose-resident macrophages from obese mice release EVs harboring *miR-155*, that cause insulin resistance in lean recipient mice. On the other hand, macrophage-derived EVs from lean mice were able to improve insulin sensitivity in obese recipients.[174] Subsequent work showed that circulating EVs from mice fed with a high-fat diet contain key miRNA species associated with metabolic dysregulation, and can induce glucose intolerance and dysregulation of lipid metabolism in lean animals. The effect of these circulating EVs was mimicked by loading these miRNAs into EVs from lean mice, further cementing the importance of EV-dependent epigenetic dysregulation in metabolic disease.[175]

Other insulin-sensitive tissues also produce EVs with altered miRNA cargos in response to diabetogenic cues. EVs produced by hepatocytes in response to a lipotoxic exposure are enriched in *miR-128-3p*, which can drive a pro-fibrotic response in the liver stellate cells.[176] Similarly, skeletal muscle-derived EVs from mice fed with a palm oil-rich diet can induce myoblast proliferation, and may partly contribute to insulin resistance.[177] These EVs are also targeted to pancreas and induce islet proliferation, mediated in part by *miR-16*.[178] The disease-relevant impact of EVs on beta-cell homeostasis has been clearly demonstrated in a mouse model of T1D; lymphocyte-derived EVs from prediabetic mice contain miRNAs that promote islet inflammation and beta-cell destruction.[179] As inflammation is a known trigger in T2D as well, it is likely that such EV-dependent mechanisms of islet damage operate even in T2D. Human islets have been shown to secrete EVs with unique miRNAs upon exposure to inflammatory cytokines, including *miR-375-3p*, which is also elevated in EV from patients with new-onset diabetes.[180,181] This suggests that EVs likely play an important role in the propagation of beta-cell loss during diabetes pathogenesis. Obesity and aging have been shown to alter the miRNA composition of serum EVs, with consequences for insulin secretion and insulin sensitivity.[182,183] EVs have also emerged as potential mediators of the multiple vascular complications of T2D. EVs derived from endothelial cells exposed to a hyperglycemic milieu and subjects with T2D have reduced *miR-126*, which results in blunted capacity for endothelial repair and may contribute to vascular damage in diabetes.[184] Notably, the abundance of several EV miRNAs associated with angiogenesis is altered in T2D.[185] Collectively, these studies suggest that EV-dependent epigenetic changes play an instrumental role in mediating insulin resistance and diabetes complications.

The data showing improvement of insulin sensitivity upon transfer of EVs derived from lean mice to obese mice points to the therapeutic potential of this approach. However, these data also suggest that the effects of EVs depend on the source and isolation method, as in this study whole-blood EVs derived from lean mice did not appear to replicate the effect of EVs isolated from cultured macrophages of lean animal origin. Several recent studies have identified EV sources that mediate islet-protective and regenerative effects. For instance, EVs harvested from human MSCs ameliorate peripheral insulin resistance and protect against beta-cell destruction in diabetic rats.[186] Notably, human islets release EVs enriched in mRNAs encoding key islet transcription factors and hormones, as well as pro-angiogenic mRNAs and miRNAs; these EVs have been shown to increase the levels of C-peptide in human iPSC-to-beta-cell differentiation cultures, and promote islet survival, increase insulin expression, and enhance islet angiogenesis.[187,188] As discussed earlier, EV contents can now be engineered to include specific miRNAs and other molecules of therapeutic interest, presenting the attractive possibility of targeting beneficial cargos to specific tissues.[189]

2.6 Epigenetic Biomarkers and Therapies

The development of effective blood-based biomarkers is crucial for disease prevention in people with high risk of diabetes, as well as to monitor disease progression and the development of personalized interventional approaches. The key requirements of a suitable peripheral blood biomarker for diabetes would be molecular stability and sensitive detection. In addition, such a biomarker should be able to accurately reflect disease-specific changes, such as beta-cell loss, which begins long before frank onset of disease.[190] Determination of disease-specific DNA methylation patterns in peripheral blood samples has emerged as a popular approach for the development of diabetes biomarkers. Some of these approaches focus on circulating cell-free DNA (cfDNA), which typically originates through active secretion, cell death, or tissue necrosis, and can therefore potentially mirror disease-specific tissue changes. Several studies show that the DNA methylation patterns of cfDNA fragments reflect their tissue of origin.[191,192] This approach has been used to monitor beta-cell loss in type 1 diabetes (T1D), and is based on the principle that beta-cell death releases cfDNA fragments harboring beta-cell-specific DNA methylation signatures, which can then be assayed in peripheral blood samples. The detection of beta-cell-specific DNA methylation signatures of the human *INSULIN* (*INS*) has frequently been used as a candidate biomarker for recent beta-cell death in T1D progression[193–196] (Table 2.4).

Plenty of data shows that DNA methylation-based blood biomarker assays are also highly relevant for T2D[131,197] (Table 2.4). A large number of highly powered DNA methylation mapping studies comparing blood samples from diabetic and non-diabetic subjects have led to the identification of candidate regions with consistent T2D-associated DNA methylation differences across cohorts from multiple ethnic and geographical backgrounds[150,198] (and reviewed in ref. 199). Such differentially methylated regions include loci associated with increased T2D risk (such as *FTO* and *TCF7L2*), and also regions with conserved T2D-associated DNA methylation differences across multiple tissues, making them ideally suited for T2D biomarker development.[200–202] In addition to DNA methylation patterns in peripheral blood and cfDNA, circulating miRNAs represent another highly stable molecular entity with profiles that are reflective of disease conditions, such as T2D.[163] For instance, the serum levels of miR-192 and miR-193b increase in subjects with pre-T2D,[203] while miR-155 levels are reduced in blood samples of T2D patients.[204] Several factors, however, can impact the outcome of such biomarker assays. Given the temporal and cell-type specificity of epigenetic marks, such approaches may be sensitive to the age of subjects and the rate of disease progression. Harnessing available data on metabolic tissue DNA methylation and miRNA profiles, coupled with longitudinal epigenetic profiling of blood samples during T2D progression will enable the development of highly robust and clinically useful biomarkers.

Therapeutic approaches involving the targeting of epigenetic mechanisms such as DNA methylation and histone modifications are currently being used with some degree of success for diseases such as cancer, and emerging as potential candidates for diabetes. Small-molecule inhibitors of the bromodomain (BRD) family of epigenetic regulators have shown therapeutic promise for a wide variety of diseases such as cancer and neurological diseases. Similarly, small-molecule inhibitors of DNA methyltransferases, 5-aza-deoxycytidine (5azadC; Decitabine) and 5-aza-cytidine (5azaC; Azacitidine; Vidaza), are clinically approved for use as anti-tumor agents.[205] Proof-of-principle work now shows that BRD inhibition promotes islet regeneration and prevents islet inflammation to protect against diabetes development in mice.[206] Several such epigenetic inhibitors are being tested in improving the efficiency of *in vitro* stem cell and induced pluripotent stem cell (iPSC) differentiation protocols for beta-cell replacement approaches. For example, BRD inhibition, 5-aza-dC , and histone deacetylase inhibitor BIX-01294 have been shown to promote pancreatic endocrine lineage in various differentiation models.[207–209] All of these molecules, however, target a whole class of epigenetic regulators, and may therefore result in undesirable systemic and off-target effects,[205] calling their therapeutic suitability into question. The development of more selective and context-specific epigenetic inhibitors is required to overcome these challenges.

Non-coding RNAs, such as miRNAs and lncRNAs, represent an epigenetic mechanism with a high degree of innate specificity, and are emerging as popular therapeutic candidates for epigenetic targeting in T2D and its multiple complications.[153,162] For example, inhibition of *miR-192* and *lnc-MGC* using modified antisense oligonucleotides has been shown to attenuate early diabetic nephropathy in mice.[162,165] Recent advances in gene-editing using methods such as CRISPR/Cas9 and TALEN further permit the epigenetic engineering of specific loci, and offer an avenue to either rectify pathogenic epigenetic changes or introduce therapeutically beneficial modifications. Such methods can target epigenetic regulators to specific genomic sequences to modify their epigenetic marks.[210] In one recent example, targeted DNA demethylation of the imprinted cell-cycle inhibitor gene *CDKN1C*, achieved by TALEN-dependent targeting of TET1 enzyme, was used to drive the replication of adult human beta-cells which typically do not proliferate.[211] Similarly, a CRISPR/Cas9 epigenetic remodeling system was recently used to recruit specific regulatory factors to modulate histone marks and activate the expression of beta-cell lineage factor Pdx1, to facilitate the reprogramming of liver cells to a beta-cell-like phenotype.[212] CRISPR/Cas9-based epigenetic engineering has also been used in VSMCs to delete specific enhancers regulated by AngII and block the angiotensin-dependent gene expression associated with a hypertensive phenotype.[213] Such approaches have also been tested *in vivo* using mouse models. A recent study used a Cas9-Dnmt3a construct to drive the DNA methylation and repression of

Table 2.4 DNA methylation analysis of peripheral blood samples can be used for the development of potential biomarkers for diabetes. This table presents such examples, including differentially methylated sites that have been used by multiple groups (indicated in bold).

Tissue	Samples	Method	Differentially methylated genes	Reference
Blood	5 diabetic and 6 non-diabetic subjects	Bisulfite sequencing	***INS***	[193]Akirav (2011)
	11 T1D progressors and 10 non-progressors	Bisulfite sequencing	***INS***	[195]Herold (2015)
	Follow-up of 25 372 participants; 1608 subjects of Indian descent and 7088 subjects of European descent developed T2D	Illumina Human Methylation 450 Bead Chip Array	***ABCG1, PHOSPHO1, TXNIP***	[198]Chambers (2015)
	2770 participants, non-diabetic at baseline, followed prospectively, to detect progression toward T2D	Infinium Human Methylation 450 Bead Chips (Illumina), Bisulfite pyrosequencing	***ABCG1, PHOSPHO1, TXNIP***	[150]Dayeh (2016)
	11 pairs of T2D-discordant MZTW	Infinium Human Methylation 450 Bead Chip Array	*ELOVL5*	[200]Hwang (2018)

alpha-cell fate regulator *Arx* in mouse pancreatic progenitors, leading to enhanced beta-cell formation during development.[214] Epigenome editing has also been used to tailor histone modifications. These proof-of-principle successes highlight the therapeutic potential of epigenomic editing; for islet expansion, cellular reprogramming, and iPSC differentiation to generate beta-cell for replacement therapies, as well as to correct causal epigenetic changes associated with diabetes and disease complications, with minimal off-target effects.

2.7 Conclusions

The epigenome acts as a regulatory tool to modify the cellular phenotype in response to the changing environment, by mediating a context-specific interpretation of the genetic code. It is likely that changes in environmental factors such as diet, sleep, activity, temperature, *etc.* initially cause epigenetic changes to facilitate adaptation. Persistence of an adverse environmental exposure may ultimately drive epigenetic dysregulation and cause adaptive failure. The recent advances in EWAS of T2D strongly affirm this hypothesis, and suggest that epigenetic dysregulation can play a causal role in diabetes pathogenesis and complications. Understanding gene–environment interactions by combining EWAS and GWAS has led to the identification and validation of putative causal variants in T2D. Epigenetic studies have also clarified our understanding of the mechanisms underlying the developmental origins of disease, inheritance of T2D risk, and the phenomenon of metabolic memory, and point to epigenetic dysregulation as the basis of dysfunction in multiple metabolic organs. Emerging work suggests that intercellular communication through EVs may contribute to such systemic epigenetic dysregulation in T2D. Elucidation of the epigenetic underpinnings of T2D pathogenesis now offers the possibility of developing robust biomarkers for disease risk and progression, and designing novel therapeutic approaches involving targeted epigenetic engineering. The work discussed here also suggests the possibility of reversing adverse epigenetic changes *via* personalized lifestyle interventions towards mitigation of T2D risk and better disease management. A clearer understanding of the mechanisms that dictate the epigenetic plasticity of a given genomic region, along with the identification of acute *versus* chronic epigenetic changes in response to increased metabolic load, will be required to realize these therapeutic possibilities.

Acknowledgements

We are grateful to members of the Dhawan Lab for helpful discussions. Work in Dhawan laboratory is supported by grants from the National Institutes of Health (R01DK120523), the Wanek Family Project to Cure Type 1 Diabetes at City of Hope, and Human Islet Research Network (NIH) UC4 DK104162 (to SD).

References

1. P. Z. Zimmet, D. J. Magliano, W. H. Herman and J. E. Shaw, *Lan. Diabetes Endocrinol.*, 2014, **2**, 56–64.
2. J. B. Meigs, L. A. Cupples and P. W. Wilson, *Diabetes*, 2000, **49**, 2201–2207.
3. P. Almgren, M. Lehtovirta, B. Isomaa, L. Sarelin, M. R. Taskinen, V. Lyssenko, T. Tuomi, L. Groop and G. Botnia Study, *Diabetologia*, 2011, **54**, 2811–2819.
4. L. C. Groop and T. Tuomi, *Ann. Med.*, 1997, **29**, 37–53.
5. J. Kaprio, J. Tuomilehto, M. Koskenvuo, K. Romanov, A. Reunanen, J. Eriksson, J. Stengard and Y. A. Kesaniemi, *Diabetologia*, 1992, **35**, 1060–1067.
6. P. M. Visscher, M. A. Brown, M. I. McCarthy and J. Yang, *Am. J. Hum. Genet.*, 2012, **90**, 7–24.
7. P. M. Visscher, N. R. Wray, Q. Zhang, P. Sklar, M. I. McCarthy, M. A. Brown and J. Yang, *Am. J. Hum. Genet.*, 2017, **101**, 5–22.
8. T. A. Manolio, F. S. Collins, N. J. Cox, D. B. Goldstein, L. A. Hindorff, D. J. Hunter, M. I. McCarthy, E. M. Ramos, L. R. Cardon, A. Chakravarti, J. H. Cho, A. E. Guttmacher, A. Kong, L. Kruglyak, E. Mardis, C. N. Rotimi, M. Slatkin, D. Valle, A. S. Whittemore, M. Boehnke, A. G. Clark, E. E. Eichler, G. Gibson, J. L. Haines, T. F. Mackay, S. A. McCarroll and P. M. Visscher, *Nature*, 2009, **461**, 747–753.
9. E. D. Rosen, K. H. Kaestner, R. Natarajan, M. E. Patti, R. Sallari, M. Sander and K. Susztak, *Diabetes*, 2018, **67**, 1923–1931.
10. C. Ling and L. Groop, *Diabetes*, 2009, **58**, 2718–2725.
11. A. Bird, *Nature*, 2007, **447**, 396–398.
12. C. D. Allis and T. Jenuwein, *Nat. Rev. Genet.*, 2016, **17**, 487–500.
13. L. Arnes and L. Sussel, *Trends Genet.*, 2015, **31**, 290–299.
14. P. Sassone-Corsi, *Science*, 2013, **339**, 148–150.
15. S. T. Keating and A. El-Osta, *Circ. Res.*, 2015, **116**, 715–736.
16. C. Lu and C. B. Thompson, *Cell Metab.*, 2012, **16**, 9–17.
17. C. N. Hales and D. J. Barker, *Br. Med. Bull.*, 2001, **60**, 5–20.
18. J. W. Rich-Edwards, G. A. Colditz, M. J. Stampfer, W. C. Willett, M. W. Gillman, C. H. Hennekens, F. E. Speizer and J. E. Manson, *Ann. Intern. Med.*, 1999, **130**, 278–284.
19. J. C. Jimenez-Chillaron, M. Hernandez-Valencia, A. Lightner, R. R. Faucette, C. Reamer, R. Przybyla, S. Ruest, K. Barry, J. P. Otis and M. E. Patti, *Diabetologia*, 2006, **49**, 1974–1984.
20. S. Dhawan and R. Natarajan, *Curr. Diabetes Rep.*, 2019, **19**, 47.
21. H. Zhang and T. I. Pollin, *Curr. Diabetes Rep.*, 2018, **18**, 121.
22. M. C. Cornelis and F. B. Hu, *Annu. Rev. Nutr.*, 2012, **32**, 245–259.
23. I. Sluijs, J. W. Beulens, A. D. van der, A. M. Spijkerman, D. E. Grobbee and Y. T. van der Schouw, *Diabetes Care*, 2010, **33**, 43–48.
24. M. van Nielen, E. J. Feskens, M. Mensink, I. Sluijs, E. Molina, P. Amiano, E. Ardanaz, B. Balkau, J. W. Beulens, H. Boeing, F. Clavel-Chapelon, G. Fagherazzi, P. W. Franks, J. Halkjaer, J. M. Huerta, V. Katzke, T. J. Key,

K. T. Khaw, V. Krogh, T. Kuhn, V. V. Menendez, P. Nilsson, K. Overvad, D. Palli, S. Panico, O. Rolandsson, I. Romieu, C. Sacerdote, M. J. Sanchez, M. B. Schulze, A. M. Spijkerman, A. Tjonneland, R. Tumino, A. D. van der, A. M. Wurtz, R. Zamora-Ros, C. Langenberg, S. J. Sharp, N. G. Forouhi, E. Riboli and N. J. Wareham and InterAct Consortium, *Diabetes Care*, 2014, **37**, 1854–1862.

25. S. C. Jacobsen, L. Gillberg, J. Bork-Jensen, R. Ribel-Madsen, E. Lara, V. Calvanese, C. Ling, A. F. Fernandez, M. F. Fraga, P. Poulsen, C. Brons and A. Vaag, *Diabetologia*, 2014, **57**, 1154–1158.
26. L. Gillberg, A. Perfilyev, C. Brons, M. Thomasen, L. G. Grunnet, P. Volkov, F. Rosqvist, D. Iggman, I. Dahlman, U. Riserus, T. Ronn, E. Nilsson, A. Vaag and C. Ling, *Diabetologia*, 2016, **59**, 799–812.
27. M. R. Keleher, R. Zaidi, L. Hicks, S. Shah, X. Xing, D. Li, T. Wang and J. M. Cheverud, *BMC Genomics*, 2018, **19**, 888.
28. M. D. Nitert, T. Dayeh, P. Volkov, T. Elgzyri, E. Hall, E. Nilsson, B. T. Yang, S. Lang, H. Parikh, Y. Wessman, H. Weishaupt, J. Attema, M. Abels, N. Wierup, P. Almgren, P. A. Jansson, T. Ronn, O. Hansson, K. F. Eriksson, L. Groop and C. Ling, *Diabetes*, 2012, **61**, 3322–3332.
29. R. Barres, J. Yan, B. Egan, J. T. Treebak, M. Rasmussen, T. Fritz, K. Caidahl, A. Krook, D. J. O'Gorman and J. R. Zierath, *Cell Metab.*, 2012, **15**, 405–411.
30. O. Fabre, L. R. Ingerslev, C. Garde, I. Donkin, D. Simar and R. Barres, *Epigenomics*, 2018, **10**, 1033–1050.
31. H. C. Roh, L. T. Y. Tsai, M. Shao, D. Tenen, Y. Shen, M. Kumari, A. Lyubetskaya, C. Jacobs, B. Dawes, R. K. Gupta and E. D. Rosen, *Cell Metab.*, 2018, **27**, 1121–1137.
32. W. Sun, H. Dong, A. S. Becker, D. H. Dapito, S. Modica, G. Grandl, L. Opitz, V. Efthymiou, L. G. Straub, G. Sarker, M. Balaz, L. Balazova, A. Perdikari, E. Kiehlmann, S. Bacanovic, C. Zellweger, D. Peleg-Raibstein, P. Pelczar, W. Reik, I. A. Burger, F. von Meyenn and C. Wolfrum, *Nat. Med.*, 2018, **24**, 1372–1383.
33. C. S. Colwell and A. V. Matveyenko, *Diabetes*, 2014, **63**, 1826–1828.
34. P. Sassone-Corsi, in *A Time for Metabolism and Hormones*, ed. P. Sassone-Corsi and Y. Christen, Cham (CH), 2016, pp. 1–11.
35. G. Benegiamo, S. A. Brown and S. Panda, *Adv. Exp. Med. Biol.*, 2016, **907**, 107–122.
36. A. Chaix, T. Lin, H. D. Le, M. W. Chang and S. Panda, *Cell Metab.*, 2019, **29**, 303–319.
37. P. Zimmet, Z. Shi, A. El-Osta and L. Ji, *Nat. Rev. Endocrinol.*, 2018, **14**, 738–746.
38. S. M. Clee and A. D. Attie, *Endocr. Rev.*, 2007, **28**, 48–83.
39. K. Almind and C. R. Kahn, *Diabetes*, 2004, **53**, 3274–3285.
40. Y. Kido, *Diabetol Int*, 2017, **8**, 7–13.
41. C. C. Jensen, M. Cnop, R. L. Hull, W. Y. Fujimoto, S. E. Kahn and American Diabetes Association GENNID Study Group, *Diabetes*, 2002, **51**, 2170–2178.

42. A. Bansal and R. A. Simmons, *Am. J. Physiol.: Endocrinol. Metab.*, 2018, **315**, E15–E28.
43. B. T. Heijmans, E. W. Tobi, A. D. Stein, H. Putter, G. J. Blauw, E. S. Susser, P. E. Slagboom and L. H. Lumey, *Proc. Natl. Acad. Sci. U. S. A.*, 2008, **105**, 17046–17049.
44. L. C. Schulz, *Proc. Natl. Acad. Sci. U. S. A.*, 2010, **107**, 16757–16758.
45. P. D. Gluckman, M. A. Hanson and H. G. Spencer, *Trends Ecol Evol*, 2005, **20**, 527–533.
46. F. Y. Ideraabdullah and S. H. Zeisel, *Physiol. Rev.*, 2018, **98**, 667–695.
47. F. A. Van Assche, F. De Prins, L. Aerts and M. Verjans, *Br. J. Obstet. Gynaecol.*, 1977, **84**, 751–753.
48. U. Nicolini, C. Hubinont, J. Santolaya, N. M. Fisk and C. H. Rodeck, *Horm. Metab. Res.*, 1990, **22**, 426–430.
49. T. D. Clausen, E. R. Mathiesen, T. Hansen, O. Pedersen, D. M. Jensen, J. Lauenborg and P. Damm, *Diabetes Care*, 2008, **31**, 340–346.
50. T. Holder, C. Giannini, N. Santoro, B. Pierpont, M. Shaw, E. Duran, S. Caprio and R. Weiss, *Diabetologia*, 2014, **57**, 2413–2420.
51. C. C. Li, P. E. Young, C. A. Maloney, S. A. Eaton, M. J. Cowley, M. E. Buckland, T. Preiss, D. C. Henstridge, G. J. Cooney, M. A. Febbraio, D. I. Martin, J. E. Cropley and C. M. Suter, *Epigenetics*, 2013, **8**, 602–611.
52. Z. Vucetic, J. Kimmel, K. Totoki, E. Hollenbeck and T. M. Reyes, *Endocrinology*, 2010, **151**, 4756–4764.
53. B. R. Carone, L. Fauquier, N. Habib, J. M. Shea, C. E. Hart, R. Li, C. Bock, C. Li, H. Gu, P. D. Zamore, A. Meissner, Z. Weng, H. A. Hofmann, N. Friedman and O. J. Rando, *Cell*, 2010, **143**, 1084–1096.
54. S. F. Ng, R. C. Lin, D. R. Laybutt, R. Barres, J. A. Owens and M. J. Morris, *Nature*, 2010, **467**, 963–966.
55. V. M. Sales, A. C. Ferguson-Smith and M. E. Patti, *Cell Metab.*, 2017, **25**, 559–571.
56. D. Martinez, T. Pentinat, S. Ribo, C. Daviaud, V. W. Bloks, J. Cebria, N. Villalmanzo, S. G. Kalko, M. Ramon-Krauel, R. Diaz, T. Plosch, J. Tost and J. C. Jimenez-Chillaron, *Cell Metab.*, 2014, **19**, 941–951.
57. T. de Castro Barbosa, L. R. Ingerslev, P. S. Alm, S. Versteyhe, J. Massart, M. Rasmussen, I. Donkin, R. Sjogren, J. M. Mudry, L. Vetterli, S. Gupta, A. Krook, J. R. Zierath and R. Barres, *Mol. Metab.*, 2016, **5**, 184–197.
58. I. Donkin, S. Versteyhe, L. R. Ingerslev, K. Qian, M. Mechta, L. Nordkap, B. Mortensen, E. V. Appel, N. Jorgensen, V. B. Kristiansen, T. Hansen, C. T. Workman, J. R. Zierath and R. Barres, *Cell Metab.*, 2016, **23**, 369–378.
59. G. S. Ducker and J. D. Rabinowitz, *Cell Metab.*, 2017, **25**, 27–42.
60. C. D. Davis and E. O. Uthus, *Exp. Biol. Med.*, 2004, **229**, 988–995.
61. B. Vanderkruk and B. G. Hoffman, *FEBS J.*, 2020, **288**, 3683–3693.
62. J. Pereda, S. Perez, J. Escobar, A. Arduini, M. Asensi, G. Serviddio, L. Sabater, L. Aparisi and J. Sastre, *PLoS One*, 2012, **7**, e44383.
63. D. Kaspar, S. Hastreiter, M. Irmler, M. Hrabe de Angelis and J. Beckers, *Mamm. Genome*, 2020, **31**, 119–133.

64. G. L. Wolff, R. L. Kodell, S. R. Moore and C. A. Cooney, *FASEB J.*, 1998, **12**, 949–957.
65. J. E. Cropley, C. M. Suter, K. B. Beckman and D. I. Martin, *Proc. Natl. Acad. Sci. U. S. A.*, 2006, **103**, 17308–17312.
66. M. A. Surani, K. Hayashi and P. Hajkova, *Cell*, 2007, **128**, 747–762.
67. H. Cedar and Y. Bergman, *Annu. Rev. Biochem.*, 2012, **81**, 97–117.
68. D. M. Messerschmidt, B. B. Knowles and D. Solter, *Genes Dev.*, 2014, **28**, 812–828.
69. P. Huypens, S. Sass, M. Wu, D. Dyckhoff, M. Tschop, F. Theis, S. Marschall, M. Hrabe, de Angelis and J. Beckers, *Nat. Genet.*, 2016, **48**, 497–499.
70. K. Gapp, A. Jawaid, P. Sarkies, J. Bohacek, P. Pelczar, J. Prados, L. Farinelli, E. Miska and I. M. Mansuy, *Nat. Neurosci.*, 2014, **17**, 667–669.
71. L. R. Ingerslev, I. Donkin, O. Fabre, S. Versteyhe, M. Mechta, P. Pattamaprapanont, B. Mortensen, N. T. Krarup and R. Barres, *Clin. Epigenetics*, 2018, **10**, 12.
72. H. J. Clarke and K. F. Vieux, *Semin. Cell Dev. Biol.*, 2015, **43**, 106–116.
73. D. Lucifero, C. Mertineit, H. J. Clarke, T. H. Bestor and J. M. Trasler, *Genomics*, 2002, **79**, 530–538.
74. A. Kong, V. Steinthorsdottir, G. Masson, G. Thorleifsson, P. Sulem, S. Besenbacher, A. Jonasdottir, A. Sigurdsson, K. T. Kristinsson, M. L. Frigge, A. Gylfason, P. I. Olason, S. A. Gudjonsson, S. Sverrisson, S. N. Stacey, B. Sigurgeirsson, K. R. Benediktsdottir, H. Sigurdsson, T. Jonsson, R. Benediktsson, J. H. Olafsson, O. T. Johannsson, A. B. Hreidarsson, G. Sigurdsson, A. C. Ferguson-Smith, D. F. Gudbjartsson, U. Thorsteinsdottir and K. Stefansson, *Nature*, 2009, **462**, 868–874.
75. K. K. Mattis and A. L. Gloyn, *J. Mol. Biol.*, 2020, **432**, 1551–1578.
76. H. Giral, U. Landmesser and A. Kratzer, *Front. Cardiovasc. Med.*, 2018, **5**, 181.
77. V. K. Rakyan, T. A. Down, D. J. Balding and S. Beck, *Nat. Rev. Genet.*, 2011, **12**, 529–541.
78. C. Davegardh, S. Garcia-Calzon, K. Bacos and C. Ling, *Mol. Metab.*, 2018, **14**, 12–25.
79. S. K. Thomsen, A. Ceroni, M. van de Bunt, C. Burrows, A. Barrett, R. Scharfmann, D. Ebner, M. I. McCarthy and A. L. Gloyn, *Diabetes*, 2016, **65**, 3805–3811.
80. M. Thurner, M. van de Bunt, J. M. Torres, A. Mahajan, V. Nylander, A. J. Bennett, K. J. Gaulton, A. Barrett, C. Burrows, C. G. Bell, R. Lowe, S. Beck, V. K. Rakyan, A. L. Gloyn and M. I. McCarthy, *eLife*, 2018, **7**, e31977.
81. M. van de Bunt, J. E. Manning Fox, X. Dai, A. Barrett, C. Grey, L. Li, A. J. Bennett, P. R. Johnson, R. V. Rajotte, K. J. Gaulton, E. T. Dermitzakis, P. E. MacDonald, M. I. McCarthy and A. L. Gloyn, *PLoS Genet.*, 2015, **11**, e1005694.
82. C. G. Bell, S. Finer, C. M. Lindgren, G. A. Wilson, V. K. Rakyan, A. E. Teschendorff, P. Akan, E. Stupka, T. A. Down, I. Prokopenko, I. M. Morison, J. Mill, R. Pidsley, C. International Type 2 Diabetes 1q,

P. Deloukas, T. M. Frayling, A. T. Hattersley, M. I. McCarthy, S. Beck and G. A. Hitman, *PLoS One*, 2010, **5**, e14040.

83. K. A. Bailey, D. Savic, M. Zielinski, S. Y. Park, L. J. Wang, P. Witkowski, M. Brady, M. Hara, G. I. Bell and M. A. Nobrega, *Hum. Mol. Genet.*, 2015, **24**, 1646–1654.
84. M. Claussnitzer, S. N. Dankel, K. H. Kim, G. Quon, W. Meuleman, C. Haugen, V. Glunk, I. S. Sousa, J. L. Beaudry, V. Puviindran, N. A. Abdennur, J. Liu, P. A. Svensson, Y. H. Hsu, D. J. Drucker, G. Mellgren, C. C. Hui, H. Hauner and M. Kellis, *N. Engl. J. Med.*, 2015, **373**, 895–907.
85. T. Dayeh, P. Volkov, S. Salo, E. Hall, E. Nilsson, A. H. Olsson, C. L. Kirkpatrick, C. B. Wollheim, L. Eliasson, T. Ronn, K. Bacos and C. Ling, *PLoS Genet.*, 2014, **10**, e1004160.
86. M. Volkmar, S. Dedeurwaerder, D. A. Cunha, M. N. Ndlovu, M. Defrance, R. Deplus, E. Calonne, U. Volkmar, M. Igoillo-Esteve, N. Naamane, S. Del Guerra, M. Masini, M. Bugliani, P. Marchetti, M. Cnop, D. L. Eizirik and F. Fuks, *EMBO J.*, 2012, **31**, 1405–1426.
87. P. Volkov, K. Bacos, J. K. Ofori, J. L. Esguerra, L. Eliasson, T. Ronn and C. Ling, *Diabetes*, 2017, **66**, 1074–1085.
88. Z. Chen, F. Miao, A. D. Paterson, J. M. Lachin, L. Zhang, D. E. Schones, X. Wu, J. Wang, J. D. Tompkins, S. Genuth, B. H. Braffett, A. D. Riggs, D. E. R. Group and R. Natarajan, *Proc. Natl. Acad. Sci. U. S. A.*, 2016, **113**, E3002–3011.
89. F. Miao, Z. Chen, S. Genuth, A. Paterson, L. Zhang, X. Wu, S. M. Li, P. Cleary, A. Riggs, D. M. Harlan, G. Lorenzi, O. Kolterman, W. Sun, J. M. Lachin, R. Natarajan and D. E. R. Group, *Diabetes*, 2014, **63**, 1748–1762.
90. R. Chen, G. I. Mias, J. Li-Pook-Than, L. Jiang, H. Y. Lam, R. Chen, E. Miriami, K. J. Karczewski, M. Hariharan, F. E. Dewey, Y. Cheng, M. J. Clark, H. Im, L. Habegger, S. Balasubramanian, M. O'Huallachain, J. T. Dudley, S. Hillenmeyer, R. Haraksingh, D. Sharon, G. Euskirchen, P. Lacroute, K. Bettinger, A. P. Boyle, M. Kasowski, F. Grubert, S. Seki, M. Garcia, M. Whirl-Carrillo, M. Gallardo, M. A. Blasco, P. L. Greenberg, P. Snyder, T. E. Klein, R. B. Altman, A. J. Butte, E. A. Ashley, M. Gerstein, K. C. Nadeau, H. Tang and M. Snyder, *Cell*, 2012, **148**, 1293–1307.
91. C. Ling and T. Ronn, *Cell Metab.*, 2019, **29**, 1028–1044.
92. C. R. Xu, P. A. Cole, D. J. Meyers, J. Kormish, S. Dent and K. S. Zaret, *Science*, 2011, **332**, 963–966.
93. B. McKenna, M. Guo, A. Reynolds, M. Hara and R. Stein, *Cell Rep.*, 2015, **10**, 2032–2042.
94. S. Georgia, M. Kanji and A. Bhushan, *Genes Dev.*, 2013, **27**, 372–377.
95. O. Lenoir, K. Flosseau, F. X. Ma, B. Blondeau, A. Mai, R. Bassel-Duby, P. Ravassard, E. N. Olson, C. Haumaitre and R. Scharfmann, *Diabetes*, 2011, **60**, 2861–2871.
96. A. Wang, F. Yue, Y. Li, R. Xie, T. Harper, N. A. Patel, K. Muth, J. Palmer, Y. Qiu, J. Wang, D. K. Lam, J. C. Raum, D. A. Stoffers, B. Ren and M. Sander, *Cell Stem Cell*, 2015, **16**, 386–399.

97. J. B. Papizan, R. A. Singer, S. I. Tschen, S. Dhawan, J. M. Friel, S. B. Hipkens, M. A. Magnuson, A. Bhushan and L. Sussel, *Genes Dev.*, 2011, **25**, 2291–2305.
98. S. Dhawan, S. Georgia, S. I. Tschen, G. Fan and A. Bhushan, *Dev. Cell*, 2011, **20**, 419–429.
99. S. Dhawan, S. I. Tschen, C. Zeng, T. Guo, M. Hebrok, A. Matveyenko and A. Bhushan, *J. Clin. Invest.*, 2015, **125**, 2851–2860.
100. D. Neiman, J. Moss, M. Hecht, J. Magenheim, S. Piyanzin, A. M. J. Shapiro, E. J. P. de Koning, A. Razin, H. Cedar, R. Shemer and Y. Dor, *Proc. Natl. Acad. Sci. U. S. A.*, 2017, **114**, 13525–13530.
101. A. Martinez-Sanchez, G. A. Rutter and M. Latreille, *Front Genet*, 2016, **7**, 226.
102. R. A. Singer and L. Sussel, *Diabetes*, 2018, **67**, 1461–1470.
103. M. P. LaPierre and M. Stoffel, *Mol Metab*, 2017, **6**, 1010–1023.
104. M. S. Kanji, M. G. Martin and A. Bhushan, *Diabetes*, 2013, **62**, 1602–1611.
105. F. C. Lynn, P. Skewes-Cox, Y. Kosaka, M. T. McManus, B. D. Harfe and M. S. German, *Diabetes*, 2007, **56**, 2938–2945.
106. C. Jacovetti, S. J. Matkovich, A. Rodriguez-Trejo, C. Guay and R. Regazzi, *Nat. Commun.*, 2015, **6**, 8084.
107. L. Arnes, I. Akerman, D. A. Balderes, J. Ferrer and L. Sussel, *Genes Dev.*, 2016, **30**, 502–507.
108. S. K. Karnik, C. M. Hughes, X. Gu, O. Rozenblatt-Rosen, G. W. McLean, Y. Xiong, M. Meyerson and S. K. Kim, *Proc. Natl. Acad. Sci. U. S. A.*, 2005, **102**, 14659–14664.
109. S. Dhawan, S. I. Tschen and A. Bhushan, *Genes Dev.*, 2009, **23**, 906–911.
110. H. Chen, X. Gu, I. H. Su, R. Bottino, J. L. Contreras, A. Tarakhovsky and S. K. Kim, *Genes Dev.*, 2009, **23**, 975–985.
111. C. Sanchez-Parra, C. Jacovetti, O. Dumortier, K. Lee, M. L. Peyot, C. Guay, M. Prentki, D. R. Laybutt, E. Van Obberghen and R. Regazzi, *Diabetes*, 2018, **67**, 2254–2267.
112. H. Chen, X. Gu, Y. Liu, J. Wang, S. E. Wirt, R. Bottino, H. Schorle, J. Sage and S. K. Kim, *Nature*, 2011, **478**, 349–355.
113. S. Dhawan, E. Dirice, R. N. Kulkarni and A. Bhushan, *Diabetes*, 2016, **65**, 1208–1218.
114. C. Talchai, S. Xuan, H. V. Lin, L. Sussel and D. Accili, *Cell*, 2012, **150**, 1223–1234.
115. P. Rodnoi, M. Rajkumar, A. S. M. Moin, S. K. Georgia, A. E. Butler and S. Dhawan, *JCI Insight*, 2017, **2**.
116. T. T. Lu, S. Heyne, E. Dror, E. Casas, L. Leonhardt, T. Boenke, C. H. Yang, Sagar, L. Arrigoni, K. Dalgaard, R. Teperino, L. Enders, M. Selvaraj, M. Ruf, S. J. Raja, H. Xie, U. Boenisch, S. H. Orkin, F. C. Lynn, B. G. Hoffman, D. Grun, T. Vavouri, A. M. Lempradl and J. A. Pospisilik, *Cell Metab.*, 2018, **27**, 1294–1308 e1297.
117. M. L. Golson and K. H. Kaestner, *Mol Metab*, 2017, **6**, 1066–1076.
118. I. Prokopenko, W. Poon, R. Magi, B. R. Prasad, S. A. Salehi, P. Almgren, P. Osmark, N. Bouatia-Naji, N. Wierup, T. Fall, A. Stancakova, A. Barker,

V. Lagou, C. Osmond, W. Xie, J. Lahti, A. U. Jackson, Y. C. Cheng, J. Liu, J. R. O'Connell, P. A. Blomstedt, J. Fadista, S. Alkayyali, T. Dayeh, E. Ahlqvist, J. Taneera, C. Lecoeur, A. Kumar, O. Hansson, K. Hansson, B. F. Voight, H. M. Kang, C. Levy-Marchal, V. Vatin, A. Palotie, A. C. Syvanen, A. Mari, M. N. Weedon, R. J. Loos, K. K. Ong, P. Nilsson, B. Isomaa, T. Tuomi, N. J. Wareham, M. Stumvoll, E. Widen, T. A. Lakka, C. Langenberg, A. Tonjes, R. Rauramaa, J. Kuusisto, T. M. Frayling, P. Froguel, M. Walker, J. G. Eriksson, C. Ling, P. Kovacs, E. Ingelsson, M. I. McCarthy, A. R. Shuldiner, K. D. Silver, M. Laakso, L. Groop and V. Lyssenko, *PLoS Genet.*, 2014, **10**, e1004235.

119. M. E. Travers, D. J. Mackay, M. Dekker Nitert, A. P. Morris, C. M. Lindgren, A. Berry, P. R. Johnson, N. Hanley, L. C. Groop, M. I. McCarthy and A. L. Gloyn, *Diabetes*, 2013, **62**, 987–992.
120. V. Kameswaran, N. C. Bramswig, L. B. McKenna, M. Penn, J. Schug, N. J. Hand, Y. Chen, I. Choi, A. Vourekas, K. J. Won, C. Liu, K. Vivek, A. Naji, J. R. Friedman and K. H. Kaestner, *Cell Metab.*, 2014, **19**, 135–145.
121. D. Avrahami, C. Li, J. Zhang, J. Schug, R. Avrahami, S. Rao, M. B. Stadler, L. Burger, D. Schubeler, B. Glaser and K. H. Kaestner, *Cell Metab.*, 2015, **22**, 619–632.
122. J. Li, X. Wu, M. Lee, J. Ke, Q. Lan, J. Li, Y. Huang, D.-Q. Sun and R. Xie, *Biorxiv*, 2020, **5**, 2.107532.
123. R. F. Thompson, M. J. Fazzari, H. Niu, N. Barzilai, R. A. Simmons and J. M. Greally, *J. Biol. Chem.*, 2010, **285**, 15111–15118.
124. V. T. Samuel and G. I. Shulman, *J. Clin. Invest.*, 2016, **126**, 12–22.
125. A. P. Feinberg, R. A. Irizarry, D. Fradin, M. J. Aryee, P. Murakami, T. Aspelund, G. Eiriksdottir, T. B. Harris, L. Launer, V. Gudnason and M. D. Fallin, *Sci. Transl. Med.*, 2010, **2**, 49ra67.
126. E. Walaszczyk, M. Luijten, A. M. W. Spijkerman, M. J. Bonder, H. L. Lutgers, H. Snieder, B. H. R. Wolffenbuttel and J. V. van Vliet-Ostaptchouk, *Diabetologia*, 2018, **61**, 354–368.
127. E. Nilsson, P. A. Jansson, A. Perfilyev, P. Volkov, M. Pedersen, M. K. Svensson, P. Poulsen, R. Ribel-Madsen, N. L. Pedersen, P. Almgren, J. Fadista, T. Ronn, B. Klarlund Pedersen, C. Scheele, A. Vaag and C. Ling, *Diabetes*, 2014, **63**, 2962–2976.
128. A. B. Crujeiras, A. Diaz-Lagares, J. M. Moreno-Navarrete, J. Sandoval, D. Hervas, A. Gomez, W. Ricart, F. F. Casanueva, M. Esteller and J. M. Fernandez-Real, *Transl. Res.*, 2016, **178**, 13–24 e15.
129. L. D. Orozco, C. Farrell, C. Hale, L. Rubbi, A. Rinaldi, M. Civelek, C. Pan, L. Lam, D. Montoya, C. Edillor, M. Seldin, M. Boehnke, K. L. Mohlke, S. Jacobsen, J. Kuusisto, M. Laakso, A. J. Lusis and M. Pellegrini, *Hum. Mol. Genet.*, 2018, **27**, 1830–1846.
130. M. L. Multhaup, M. M. Seldin, A. E. Jaffe, X. Lei, H. Kirchner, P. Mondal, Y. Li, V. Rodriguez, A. Drong, M. Hussain, C. Lindgren, M. McCarthy, E. Naslund, J. R. Zierath, G. W. Wong and A. P. Feinberg, *Cell Metab.*, 2015, **21**, 138–149.

131. T. Ronn, P. Volkov, L. Gillberg, M. Kokosar, A. Perfilyev, A. L. Jacobsen, S. W. Jorgensen, C. Brons, P. A. Jansson, K. F. Eriksson, O. Pedersen, T. Hansen, L. Groop, E. Stener-Victorin, A. Vaag, E. Nilsson and C. Ling, *Hum. Mol. Genet.*, 2015, **24**, 3792–3813.
132. D. You, E. Nilsson, D. E. Tenen, A. Lyubetskaya, J. C. Lo, R. Jiang, J. Deng, B. A. Dawes, A. Vaag, C. Ling, E. D. Rosen and S. Kang, *eLife*, 2017, **6**, e30766.
133. S. Kang, L. T. Tsai, Y. Zhou, A. Evertts, S. Xu, M. J. Griffin, R. Issner, H. J. Whitton, B. A. Garcia, C. B. Epstein, T. S. Mikkelsen and E. D. Rosen, *Nat. Cell Biol.*, 2015, **17**, 44–56.
134. R. Ribel-Madsen, M. F. Fraga, S. Jacobsen, J. Bork-Jensen, E. Lara, V. Calvanese, A. F. Fernandez, M. Friedrichsen, B. F. Vind, K. Hojlund, H. Beck-Nielsen, M. Esteller, A. Vaag and P. Poulsen, *PLoS One*, 2012, **7**, e51302.
135. L. J. Scott, M. R. Erdos, J. R. Huyghe, R. P. Welch, A. T. Beck, B. N. Wolford, P. S. Chines, J. P. Didion, N. Narisu, H. M. Stringham, D. L. Taylor, A. U. Jackson, S. Vadlamudi, L. L. Bonnycastle, L. Kinnunen, J. Saramies, J. Sundvall, R. D. Albanus, A. Kiseleva, J. Hensley, G. E. Crawford, H. Jiang, X. Wen, R. M. Watanabe, T. A. Lakka, K. L. Mohlke, M. Laakso, J. Tuomilehto, H. A. Koistinen, M. Boehnke, F. S. Collins and S. C. Parker, *Nat. Commun.*, 2016, **7**, 11764.
136. J. M. Dos Santos, M. L. Moreli, S. Tewari and S. A. Benite-Ribeiro, *Metabolism*, 2015, **64**, 1619–1628.
137. S. L. McGee, E. Fairlie, A. P. Garnham and M. Hargreaves, *J. Physiol.*, 2009, **587**, 5951–5958.
138. R. Barres, M. E. Osler, J. Yan, A. Rune, T. Fritz, K. Caidahl, A. Krook and J. R. Zierath, *Cell Metab.*, 2009, **10**, 189–198.
139. D. S. Rowlands, R. A. Page, W. R. Sukala, M. Giri, S. D. Ghimbovschi, I. Hayat, B. S. Cheema, I. Lys, M. Leikis, P. W. Sheard, S. J. Wakefield, B. Breier, Y. Hathout, K. Brown, R. Marathi, F. E. Orkunoglu-Suer, J. M. Devaney, B. Leiken, G. Many, J. Krebs, W. G. Hopkins and E. P. Hoffman, *Physiol. Genomics*, 2014, **46**, 747–765.
140. P. S. Alm, T. de Castro Barbosa, R. Barres, A. Krook and J. R. Zierath, *Mol. Metab.*, 2017, **6**, 621–630.
141. T. Ronn, P. Poulsen, O. Hansson, J. Holmkvist, P. Almgren, P. Nilsson, T. Tuomi, B. Isomaa, L. Groop, A. Vaag and C. Ling, *Diabetologia*, 2008, **51**, 1159–1168.
142. A. Zykovich, A. Hubbard, J. M. Flynn, M. Tarnopolsky, M. F. Fraga, C. Kerksick, D. Ogborn, L. MacNeil, S. D. Mooney and S. Melov, *Aging Cell*, 2014, **13**, 360–366.
143. C. Ling, P. Poulsen, S. Simonsson, T. Ronn, J. Holmkvist, P. Almgren, P. Hagert, E. Nilsson, A. G. Mabey, P. Nilsson, A. Vaag and L. Groop, *J. Clin. Invest.*, 2007, **117**, 3427–3435.
144. H. Kirchner, I. Sinha, H. Gao, M. A. Ruby, M. Schonke, J. M. Lindvall, R. Barres, A. Krook, E. Naslund, K. Dahlman-Wright and J. R. Zierath, *Mol. Metab.*, 2016, **5**, 171–183.

145. E. Nilsson, A. Matte, A. Perfilyev, V. D. de Mello, P. Kakela, J. Pihlajamaki and C. Ling, *J. Clin. Endocrinol. Metab.*, 2015, **100**, E1491–1501.
146. C. G. Bell, R. Lowe, P. D. Adams, A. A. Baccarelli, S. Beck, J. T. Bell, B. C. Christensen, V. N. Gladyshev, B. T. Heijmans, S. Horvath, T. Ideker, J. J. Issa, K. T. Kelsey, R. E. Marioni, W. Reik, C. L. Relton, L. C. Schalkwyk, A. E. Teschendorff, W. Wagner, K. Zhang and V. K. Rakyan, *Genome Biol.*, 2019, **20**, 249.
147. S. Horvath, W. Erhart, M. Brosch, O. Ammerpohl, W. von Schonfels, M. Ahrens, N. Heits, J. T. Bell, P. C. Tsai, T. D. Spector, P. Deloukas, R. Siebert, B. Sipos, T. Becker, C. Rocken, C. Schafmayer and J. Hampe, *Proc. Natl. Acad. Sci. U. S. A.*, 2014, **111**, 15538–15543.
148. Y. Xiao, B. Word, A. Starlard-Davenport, A. Haefele, B. D. Lyn-Cook and G. Hammons, *Cell Biol. Toxicol.*, 2008, **24**, 265–272.
149. M. Bysani, A. Perfilyev, V. D. de Mello, T. Ronn, E. Nilsson, J. Pihlajamaki and C. Ling, *Epigenomics*, 2017, **9**, 105–122.
150. T. Dayeh, T. Tuomi, P. Almgren, A. Perfilyev, P. A. Jansson, V. D. de Mello, J. Pihlajamaki, A. Vaag, L. Groop, E. Nilsson and C. Ling, *Epigenetics*, 2016, **11**, 482–488.
151. M. A. Reddy, E. Zhang and R. Natarajan, *Diabetologia*, 2015, **58**, 443–455.
152. J. M. Forbes and M. E. Cooper, *Physiol. Rev.*, 2013, **93**, 137–188.
153. M. Kato and R. Natarajan, *Nat. Rev. Nephrol.*, 2014, **10**, 517–530.
154. A. Y. Chu, A. Tin, P. Schlosser, Y. A. Ko, C. Qiu, C. Yao, R. Joehanes, M. E. Grams, L. Liang, C. A. Gluck, C. Liu, J. Coresh, S. J. Hwang, D. Levy, E. Boerwinkle, J. S. Pankow, Q. Yang, M. Fornage, C. S. Fox, K. Susztak and A. Kottgen, *Nat. Commun.*, 2017, **8**, 1286.
155. C. Qiu, R. L. Hanson, G. Fufaa, S. Kobes, C. Gluck, J. Huang, Y. Chen, D. Raj, R. G. Nelson, W. C. Knowler and K. Susztak, *Kidney Int.*, 2018, **93**, 1417–1431.
156. L. Pirola, A. Balcerczyk, R. W. Tothill, I. Haviv, A. Kaspi, S. Lunke, M. Ziemann, T. Karagiannis, S. Tonna, A. Kowalczyk, B. Beresford-Smith, G. Macintyre, M. Kelong, Z. Hongyu, J. Zhu and A. El-Osta, *Genome Res.*, 2011, **21**, 1601–1615.
157. E. S. Kim, F. Isoda, I. Kurland and C. V. Mobbs, *Endocrinology*, 2013, **154**, 3054–3066.
158. R. A. Kowluru and M. Mishra, *Free Radic. Biol. Med.*, 2017, **103**, 155–164.
159. F. Miao, X. Wu, L. Zhang, Y. C. Yuan, A. D. Riggs and R. Natarajan, *J. Biol. Chem.*, 2007, **282**, 13854–13863.
160. L. M. Villeneuve, M. A. Reddy, L. L. Lanting, M. Wang, L. Meng and R. Natarajan, *Proc. Natl. Acad. Sci. U. S. A.*, 2008, **105**, 9047–9052.
161. A. Leung, V. Amaram and R. Natarajan, *Vascul. Pharmacol.*, 2018, **114**, 139–144.
162. A. Leung and R. Natarajan, *Antioxid Redox Signal*, 2018, **29**, 1064–1073.
163. M. Kato and R. Natarajan, *Ann. N. Y. Acad. Sci.*, 2015, **1353**, 72–88.
164. M. Kato, S. Putta, M. Wang, H. Yuan, L. Lanting, I. Nair, A. Gunn, Y. Nakagawa, H. Shimano, I. Todorov, J. J. Rossi and R. Natarajan, *Nat. Cell Biol.*, 2009, **11**, 881–889.

165. M. Kato, M. Wang, Z. Chen, K. Bhatt, H. J. Oh, L. Lanting, S. Deshpande, Y. Jia, J. Y. Lai, C. L. O'Connor, Y. Wu, J. B. Hodgin, R. G. Nelson, M. Bitzer and R. Natarajan, *Nat. Commun.*, 2016, **7**, 12864.
166. S. Das, M. A. Reddy, P. Senapati, K. Stapleton, L. Lanting, M. Wang, V. Amaram, R. Ganguly, L. Zhang, S. Devaraj, D. E. Schones and R. Natarajan, *Arterioscler., Thromb., Vasc. Biol.*, 2018, **38**, 1806–1820.
167. S. Das, E. Zhang, P. Senapati, V. Amaram, M. A. Reddy, K. Stapleton, A. Leung, L. Lanting, M. Wang, Z. Chen, M. Kato, H. J. Oh, Q. Guo, X. Zhang, B. Zhang, H. Zhang, Q. Zhao, W. Wang, Y. Wu and R. Natarajan, *Circ. Res.*, 2018, **123**, 1298–1312.
168. Writing Team for the Diabetes Control and Complications Trial/Epidemiology of Diabetes Interventions and Complications Research Group, JAMA, 2003, 290, 2159–2167.
169. J. Chalmers and M. E. Cooper, *N. Engl. J. Med.*, 2008, **359**, 1618–1620.
170. H. Valadi, K. Ekstrom, A. Bossios, M. Sjostrand, J. J. Lee and J. O. Lotvall, *Nat. Cell Biol.*, 2007, **9**, 654–U672.
171. S. Nomura, M. Suzuki, K. Katsura, G. L. Xie, Y. Miyazaki, T. Miyake, H. Kido, H. Kagawa and S. Fukuhara, *Atherosclerosis*, 1995, **116**, 235–240.
172. S. Y. Li, J. Wei, C. H. Zhang, X. D. Li, W. T. Meng, X. M. Mo, Q. Y. Zhang, Q. L. Liu, K. Y. Ren, R. Du, H. M. Tian and J. W. Li, *Cell. Physiol. Biochem.*, 2016, **39**, 2439–2450.
173. T. Thomou, M. A. Mori, J. M. Dreyfuss, M. Konishi, M. Sakaguchi, C. Wolfrum, T. N. Rao, J. N. Winnay, R. Garcia-Martin, S. K. Grinspoon, P. Gorden and C. R. Kahn, *Nature*, 2017, **545**, 252.
174. W. Ying, M. Riopel, G. Bandyopadhyay, Y. Dong, A. Birmingham, J. B. Seo, J. M. Ofrecio, J. Wollam, A. Hernandez-Carretero, W. Fu, P. Li and J. M. Olefsky, *Cell*, 2017, **171**, 372–384 e312.
175. C. Castano, S. Kalko, A. Novials and M. Parrizas, *Proc. Natl. Acad. Sci. U. S. A.*, 2018, **115**, 12158–12163.
176. D. Povero, N. Panera, A. Eguchi, C. D. Johnson, B. G. Papouchado, L. de Araujo Horcel, E. M. Pinatel, A. Alisi, V. Nobili and A. E. Feldstein, *Cell Mol. Gastroenterol. Hepatol.*, 2015, **1**, 646–663.
177. H. Aswad, A. Forterre, O. P. Wiklander, G. Vial, E. Danty-Berger, A. Jalabert, A. Lamaziere, E. Meugnier, S. Pesenti, C. Ott, K. Chikh, S. El-Andaloussi, H. Vidal, E. Lefai, J. Rieusset and S. Rome, *Diabetologia*, 2014, **57**, 2155–2164.
178. A. Jalabert, G. Vial, C. Guay, O. P. Wiklander, J. Z. Nordin, H. Aswad, A. Forterre, E. Meugnier, S. Pesenti, R. Regazzi, E. Danty-Berger, S. Ducreux, H. Vidal, S. El-Andaloussi, J. Rieusset and S. Rome, *Diabetologia*, 2016, **59**, 1049–1058.
179. C. Guay, J. K. Kruit, S. Rome, V. Menoud, N. L. Mulder, A. Jurdzinski, F. Mancarella, G. Sebastiani, A. Donda, B. J. Gonzalez, C. Jandus, K. Bouzakri, M. Pinget, C. Boitard, P. Romero, F. Dotta and R. Regazzi, *Cell Metab.*, 2019, **29**, 348–361.

180. P. B. Saravanan, S. Vasu, G. Yoshimatsu, C. M. Darden, X. Wang, J. Gu, M. C. Lawrence and B. Naziruddin, *Diabetologia*, 2019, **62**, 1901–1914.
181. Q. Fu, H. Jiang, Z. Wang, X. Wang, H. Chen, Z. Shen, L. Xiao, X. Guo and T. Yang, *Aging*, 2018, **10**, 3986–3999.
182. H. Xu, X. Du, J. Xu, Y. Zhang, Y. Tian, G. Liu, X. Wang, M. Ma, W. Du, Y. Liu, L. Dai, W. Huang, N. Tong, Y. Wei and X. Fu, *PLoS Biol.*, 2020, **18**, e3000603.
183. T. Su, Y. Xiao, Y. Xiao, Q. Guo, C. Li, Y. Huang, Q. Deng, J. Wen, F. Zhou and X. H. Luo, *ACS Nano*, 2019, **13**, 2450–2462.
184. F. Jansen, X. Yang, M. Hoelscher, A. Cattelan, T. Schmitz, S. Proebsting, D. Wenzel, S. Vosen, B. S. Franklin, B. K. Fleischmann, G. Nickenig and N. Werner, *Circulation*, 2013, **128**, 2026–2038.
185. F. Jansen, H. Wang, D. Przybilla, B. S. Franklin, A. Dolf, P. Pfeifer, T. Schmitz, A. Flender, E. Endl, G. Nickenig and N. Werner, *Cardiovasc. Diabetol.*, 2016, **15**, 49.
186. Y. Sun, H. Shi, S. Yin, C. Ji, X. Zhang, B. Zhang, P. Wu, Y. Shi, F. Mao, Y. Yan, W. Xu and H. Qian, *ACS Nano*, 2018, **12**, 7613–7628.
187. D. Ribeiro, E. M. Andersson, N. Heath, A. Persson-Kry, R. Collins, R. Hicks, N. Dekker and A. Forslow, *PLoS One*, 2017, **12**, e0187665.
188. F. Figliolini, V. Cantaluppi, M. De Lena, S. Beltramo, R. Romagnoli, M. Salizzoni, R. Melzi, R. Nano, L. Piemonti, C. Tetta, L. Biancone and G. Camussi, *PLoS One*, 2014, **9**, e102521.
189. P. Garcia-Manrique, M. Matos, G. Gutierrez, C. Pazos and M. C. Blanco-Lopez, *J Extracell Vesicles*, 2018, **7**, 1422676.
190. J. J. Meier, T. G. Breuer, R. C. Bonadonna, A. Tannapfel, W. Uhl, W. E. Schmidt, H. Schrader and B. A. Menge, *Diabetologia*, 2012, **55**, 1346–1354.
191. M. W. Snyder, M. Kircher, A. J. Hill, R. M. Daza and J. Shendure, *Cell*, 2016, **164**, 57–68.
192. J. Moss, J. Magenheim, D. Neiman, H. Zemmour, N. Loyfer, A. Korach, Y. Samet, M. Maoz, H. Druid, P. Arner, K. Y. Fu, E. Kiss, K. L. Spalding, G. Landesberg, A. Zick, A. Grinshpun, A. M. J. Shapiro, M. Grompe, A. D. Wittenberg, B. Glaser, R. Shemer, T. Kaplan and Y. Dor, *Nat. Commun.*, 2018, **9**, 5068.
193. E. M. Akirav, J. Lebastchi, E. M. Galvan, O. Henegariu, M. Akirav, V. Ablamunits, P. M. Lizardi and K. C. Herold, *Proc. Natl. Acad. Sci. U. S. A.*, 2011, **108**, 19018–19023.
194. M. I. Husseiny, A. Kaye, E. Zebadua, F. Kandeel and K. Ferreri, *PLoS One*, 2014, **9**, e94591.
195. K. C. Herold, S. Usmani-Brown, T. Ghazi, J. Lebastchi, C. A. Beam, M. D. Bellin, M. Ledizet, J. M. Sosenko, J. P. Krischer, J. P. Palmer and G. Type 1 Diabetes TrialNet Study, *J. Clin. Invest.*, 2015, **125**, 1163–1173.
196. M. M. Fisher, R. A. Watkins, J. Blum, C. Evans-Molina, N. Chalasani, L. A. DiMeglio, K. J. Mather, S. A. Tersey and R. G. Mirmira, *Diabetes*, 2015, **64**, 3867–3872.

197. K. Bacos, L. Gillberg, P. Volkov, A. H. Olsson, T. Hansen, O. Pedersen, A. P. Gjesing, H. Eiberg, T. Tuomi, P. Almgren, L. Groop, L. Eliasson, A. Vaag, T. Dayeh and C. Ling, *Nat. Commun.*, 2016, **7**, 11089.
198. J. C. Chambers, M. Loh, B. Lehne, A. Drong, J. Kriebel, V. Motta, S. Wahl, H. R. Elliott, F. Rota, W. R. Scott, W. Zhang, S. T. Tan, G. Campanella, M. Chadeau-Hyam, L. Yengo, R. C. Richmond, M. Adamowicz-Brice, U. Afzal, K. Bozaoglu, Z. Y. Mok, H. K. Ng, F. Pattou, H. Prokisch, M. A. Rozario, L. Tarantini, J. Abbott, M. Ala-Korpela, B. Albetti, O. Ammerpohl, P. A. Bertazzi, C. Blancher, R. Caiazzo, J. Danesh, T. R. Gaunt, S. de Lusignan, C. Gieger, T. Illig, S. Jha, S. Jones, J. Jowett, A. J. Kangas, A. Kasturiratne, N. Kato, N. Kotea, S. Kowlessur, J. Pitkaniemi, P. Punjabi, D. Saleheen, C. Schafmayer, P. Soininen, E. S. Tai, B. Thorand, J. Tuomilehto, A. R. Wickremasinghe, S. A. Kyrtopoulos, T. J. Aitman, C. Herder, J. Hampe, S. Cauchi, C. L. Relton, P. Froguel, R. Soong, P. Vineis, M. R. Jarvelin, J. Scott, H. Grallert, V. Bollati, P. Elliott, M. I. McCarthy and J. S. Kooner, *Lancet Diabetes Endocrinol.*, 2015, **3**, 526–534.
199. T. Willmer, R. Johnson, J. Louw and C. Pheiffer, *Front. Endocrinol.*, 2018, **9**, 744.
200. J. Y. Hwang, H. J. Lee, M. J. Go, H. B. Jang, N. H. Choi, J. B. Bae, J. E. Castillo-Fernandez, J. T. Bell, T. D. Spector, H. J. Lee and B. J. Kim, *Sci. Rep.*, 2018, **8**, 14862.
201. S. D. van Otterdijk, A. M. Binder, K. Szarc Vel Szic, J. Schwald and K. B. Michels, *PLoS One*, 2017, **12**, e0180955.
202. S. Canivell, E. G. Ruano, A. Siso-Almirall, B. Kostov, L. Gonzalez-de Paz, E. Fernandez-Rebollo, F. A. Hanzu, M. Parrizas, A. Novials and R. Gomis, *PLoS One*, 2014, **9**, e99310.
203. M. Parrizas and A. Novials, *Best Pract. Res., Clin. Endocrinol. Metab.*, 2016, **30**, 591–601.
204. X. Lin, Y. Qin, J. Jia, T. Lin, X. Lin, L. Chen, H. Zeng, Y. Han, L. Wu, S. Huang, M. Wang, S. Huang, R. Xie, L. Liang, Y. Liu, R. Liu, T. Zhang, J. Li, S. Wang, P. Sun, W. Huang, K. Yao, K. Xu, T. Du and D. Xiao, *PLoS Genet.*, 2016, **12**, e1006308.
205. A. Ganesan, P. B. Arimondo, M. G. Rots, C. Jeronimo and M. Berdasco, *Clin Epigenetics*, 2019, **11**, 174.
206. W. Fu, J. Farache, S. M. Clardy, K. Hattori, P. Mander, K. Lee, I. Rioja, R. Weissleder, R. K. Prinjha, C. Benoist and D. Mathis, *eLife*, 2014, **3**, e04631.
207. L. Huijbregts, M. B. Kjaer Petersen, C. Berthault, M. Hansson, V. Aiello, L. Rachdi, A. Grapin-Botton, C. Honore and R. Scharfmann, *Diabetes*, 2019, **68**, 761–773.
208. G. S. Manzar, E. M. Kim and N. Zavazava, *J. Biol. Chem.*, 2017, **292**, 14066–14079.
209. E. Akinci, A. Banga, K. Tungatt, J. Segal, D. Eberhard, J. R. Dutton and J. M. Slack, *PLoS One*, 2013, **8**, e82424.

210. X. S. Liu, H. Wu, X. Ji, Y. Stelzer, X. Wu, S. Czauderna, J. Shu, D. Dadon, R. A. Young and R. Jaenisch, *Cell*, 2016, **167**, 233–247.
211. K. Ou, M. Yu, N. G. Moss, Y. J. Wang, A. W. Wang, S. C. Nguyen, C. Jiang, E. Feleke, V. Kameswaran, E. F. Joyce, A. Naji, B. Glaser, D. Avrahami and K. H. Kaestner, *J. Clin. Invest.*, 2019, **129**, 209–214.
212. H. K. Liao, F. Hatanaka, T. Araoka, P. Reddy, M. Z. Wu, Y. Sui, T. Yamauchi, M. Sakurai, D. D. O'Keefe, E. Nunez-Delicado, P. Guillen, J. M. Campistol, C. J. Wu, L. F. Lu, C. R. Esteban and J. C. Izpisua Belmonte, *Cell*, 2017, **171**, 1495–1507.
213. S. Das, P. Senapati, Z. Chen, M. A. Reddy, R. Ganguly, L. Lanting, V. Mandi, A. Bansal, A. Leung, S. Zhang, Y. Jia, X. Wu, D. E. Schones and R. Natarajan, *Nat. Commun.*, 2017, **8**, 1467.
214. J. Liu, A. Banerjee, C. A. Herring, J. Attalla, R. Hu, Y. Xu, Q. Shao, A. J. Simmons, P. K. Dadi, S. Wang, D. A. Jacobson, B. Liu, E. Hodges, K. S. Lau and G. Gu, *Dev. Cell*, 2019, **48**, 49–63.

CHAPTER 3

Optimizing the Current Type 2 Diabetes Antidiabetics with Nanotechnologies: Where Do We Stand?

AMAR ABDERRAHMANI,*[a] SABINE SZUNERITS,[a] STEPHANE DALLE[b] AND RABAH BOUKHERROUB[a]

[a] Univ. Lille, CNRS, Centrale Lille, Univ. Polytechnique Hauts-de-France, UMR 8520-IEMN, F-59000 Lille, France; [b] Institut de Génomique Fonctionnelle, INSERM, CNRS UMR5203, Université de Montpellier, F-34000 Montpellier, France

*Email: amar.abderrahmani@univ-lille.fr

3.1 Introduction

Diabetes is defined by the World Health Organization (WHO) as a fasting plasma glucose level above or equal to 126 mg dL^{-1} (or 7.0 mmol L^{-1}) or plasma glucose above or equal to 200 mg dL^{-1} (11.1 mmol L^{-1}) after a 2-hour oral glucose tolerance test. In the absence of treatment capable of lowering the plasma glucose level, several diseases can develop, including heart, peripheral arterial and cerebrovascular diseases. Those with diabetes are also at increased risk of retinopathy, erectile dysfunction, non-alcoholic fatty liver disease (NAFLD), nephropathy and neuropathy. They are also at increased risk of some infectious diseases, such as tuberculosis and COVID-19.[1]

Nanoscience & Nanotechnology Series No. 59
Nanotechnology for Diabetes Management
Edited by Amar Abderrahmani, Sabine Szunerits, Rabah Boukerroub and Abdelfattah El Ouaamari

Published by the Royal Society of Chemistry, www.rsc.org

Hyperglycemia results from heterogeneous etiopathology. This heterogeneity has led to the classification of diabetes into several types (Table 3.1), with type 2 diabetes (T2D) as the most commonly diagnosed.[2] T2D accounts for between 90% and 95% of diabetes diagnosed worldwide, with the highest proportions in low- and middle-income countries.[3] The disease is most common in adults, but an increasing number of children and adolescents are also affected.[4] T2D often remains undiagnosed for many years, because the hyperglycemia is not severe enough to provoke noticeable symptoms of diabetes.[5] Hyperglycemia develops when pancreatic β-cells from islets of Langerhans no longer adapt the production of insulin plasma concentration to the increased insulin body requirements.[5] In the history of this disease, the need for insulin is progressively increased as the consequence of the decrease of insulin sensitivity of liver, muscle and adipose tissue.[6] The overall loss of insulin sensitivity is referred to as systemic insulin resistance.[6] For years, insulin resistance is compensated for by an elevated insulin production resulting from increased β-cell function (BCF) and number, also referred to as beta-cell mass (BCM).[6] However, in the history of the disease, adaptation of BCF and BCM to insulin resistance fails, leading to T2D onset. Improvement of insulin sensitivity, insulin secretion and inhibition of glucose absorption and reabsorption constitute the main therapeutic strategies of T2D. Several drugs, which can be classified into orally delivered and subcutaneously injected medications, are currently prescribed in this regard. Oral drugs are the insulin sensitizer biguanide, sodium–glucose cotransporter-2 (SGLT-2) inhibitors and insulin secretagogues including sulfonylurea, glinides and gliptins. The insulin secretagogue glucagon like peptide-1 receptor agonists (GLP-1RA) and insulin are currently subcutaneously injectable drugs. Oral antidiabetics with lifestyle changes are the first lines of T2D therapy. Although they are mostly well-tolerated, some patients experience some adverse effects, mainly due to gastrointestinal factors. For some patients, these effects result from current dosage and/or dosing frequency caused by the low aqueous solubility and stability of most oral antidiabetics. These concerns can contribute to treatment interruption and/or long-term loss of compliance to the treatment, which finally leads to poor diabetes management. WHO reported that only around 50% of patients from developed countries adhere to their long-term therapy,[7] with a marked effect 6 months after starting treatment.[8] After the failure of oral antidiabetics, the only alternative is injection of GLP-1RA and ultimately insulin. Unfortunately, this painful and invasive mode of administration, associated with needle phobia, can lead some patients to dramatically reduce their adherence or compliance with the treatment. To achieve long-term glycemic control and, thereby, to avoid the risks of developing critical complications, it is key to improve the efficacy of all current antidiabetics.

This chapter deals with the role of nanotechnologies for improving the gastrointestinal permeability, half-life and bioavailability of the current oral antidiabetics, and for finding painless and non-invasive alternative routes for administering the currently injectable drugs.

Table 3.1 Types of diabetes.

Type 1 diabetes	β-Cell destruction (mostly immune-mediated) and absolute insulin deficiency; onset most common in childhood and early adulthood
Type 2 diabetes	The most common type, various degrees of β-cell dysfunction and insulin resistance; commonly associated with overweight and obesity, aging populations, increasing and unplanned urbanization, dietary changes such as increased consumption of highly processed foods and sugars/sweetened beverages, reduced physical activity, unhealthy lifestyle and behavioral patterns, fetal malnutrition, and increasing fetal exposure to hyperglycemia during pregnancy
Hybrid forms of diabetes	
Slowly evolving, immune-mediated diabetes of adults	Similar to slowly evolving type 1 in adults but more often has features of the metabolic syndrome and a single GAD autoantibody
Ketosis-prone type 2 diabetes	Present with ketosis and insulin deficiency but later does not require insulin; common episodes of ketosis, not immune-mediated
Other specific types	
Familial forms of diabetes Monogenic defects of β-cell function	Caused by specific gene mutations, has several clinical manifestations requiring different treatments, some occurring in the neonatal period, others at early adulthood
Monogenic defects in insulin action	Caused by specific gene mutations; has features of severe insulin resistance without obesity; diabetes develops when β-cells do not compensate for insulin resistance
Diseases of the exocrine pancreas	Various conditions that affect the pancreas can result in hyperglycemia (trauma, tumor, inflammation, *etc.*)
Endocrine disorders	Occurs in diseases with excess secretion of hormones that are insulin antagonists
Drug- or chemical-induced	Some medicines and chemicals impair insulin secretion or action, some can destroy β-cells
Infection-related diabetes	Some viruses have been associated with direct β-cell destruction
Uncommon specific forms of immune-mediated diabetes	Associated with rare immune-mediated diseases
Other genetic syndromes sometimes associated with diabetes	Many genetic disorders and chromosomal abnormalities increase the risk of diabetes
Unclassified diabetes	Diabetes that does not clearly fit into other categories. This category should be used temporarily when there is not a clear diagnostic category, especially close to the time of diagnosis
Hyperglycemia first detected during pregnancy	
Diabetes mellitus in pregnancy	Type 1 or type 2 diabetes first diagnosed during pregnancy
Gestational diabetes mellitus	Hyperglycemia below diagnostic thresholds for diabetes in pregnancy

3.2 Biguanides

Biguanides originate from compounds identified in the *Galega Officinalis*, also known as French lilac, Italian fitch or goat's rue.[9] Biguanides are a class of oral antidiabetic that includes metformin, phenformin (phenylethyl biguanide) and buformin (1-butylbiguanide). Because of the high incidence of hyperlactatemia and metabolic acidosis, phenformin and buformin were removed from the market in many countries in the late 1970s.[10–12] Only metformin is now widely prescribed as a first-line drug for reducing hyperglycemia by treating insulin resistance in patients with T2D. To achieve its antihyperglycemic effect, metformin suppresses liver glucose production and adipocyte lipolysis, enhances insulin-stimulated glucose disposal in muscle and improves insulin receptor activity.[13,14] Metformin is also suggested to have anticancer activity in various cancer types including hepatocellular carcinoma, the ultimate stage of NAFLD.[15–17] Following oral administration, metformin is mainly absorbed from the upper small intestine. The half-life ($t_{1/2}$) of metformin is in the range of 0.9–2.6 h. Its absolute bioavailability is about 50–60%. The relatively low bioavailability, upon oral administration, leads to repeated delivery of high doses of metformin (500 mg two or three times daily, or 850 mg once or twice daily with or after meals) for an effective treatment. As a result, patient compliance is reduced and/or the incidence of side effects, such as diarrhea and nausea, increases. To improve the bioavailability of metformin orally delivered, the medicine was either encapsulated in or conjugated to sugar-derived nanoparticles including alginate or chitosan (Table 3.2). Alginate has been approved by the Food and Drug Administration (FDA) in food, pharma and medicine (wounds, bone) applications since 1970.[18] Several clinical trials have been undertaken to test Algisyl-LVRTM, a commercially available alginate hydrogel-based medicine, for cardiac regeneration.[19] The use of chitosan in food and medicine is also approved by the FDA.[20,21] So far, chitosan and its derivatives have been used in the food industry, agriculture, pharmacy, medicine, cosmetology, textile and paper industries, and chemistry.[22] Most preclinical studies show that the formulation of metformin with chitosan nanoparticles can reduce the dose while preserving the activity of the compound (Table 3.2). These data could soon promote and accelerate preclinical and clinical studies

Table 3.2 Nanoparticle-based systems for oral delivery of metformin.

Nanoparticles	*In vitro* and preclinical studies
Encapsulation of metformin in nanoparticles for oral delivery	
Alginate	Threefold reduction in the concentration of metformin to administrate in mice[89]
Alginate/chitosan complex	Prolonged metformin activity for 12 h in rats[90]
Hyaluronic acid polymer	Increased permeability in intestinal membrane in *ex vivo* experiments[91]
Chitosan/eudragit complex	Higher pharmacokinetic and prolonged metformin activity[92]
Conjugation of metformin to nanoparticles for oral delivery	
Carbon nanotubes	Prolonged metformin delivery and activity in diabetic rats[93]

releasing the potential and toxicity of an oral nanoparticle-based delivery system of metformin.

In addition to oral delivery, the transdermal administration of metformin has been investigated to bypass the gastrointestinal (GI) tract, and thereby its adverse side effects. Several transdermal delivery systems have been set up. One is a graphene-hybrid device array with sweat-based glucose and pH monitoring that is interconnected with a sweat-control layer coupled to bioresorbable temperature-responsive microneedles.[23] Polymer-based microneedles are coated with the phase-change material (PCM) in which metformin can be loaded. High glucose concentration recordings trigger the embedded heaters to dissolve PCM and, as a result, bioresorbable microneedles release metformin.[23] Metformin delivered through this transdermal delivery system efficiently reduces hyperglycemia of diabetic *db*/*db* mice.[23] Patches with hydrogel-based microneedles have also been investigated for transdermal delivery of metformin hydrochloride (HCl) in a sustained manner.[24,25] In one study, the patch was developed from an aqueous blend of poly(methylvinylether-*co*-maleic acid) and polyethylene glycol (PEG) to which a metformin drug reservoir was attached.[24] In rats, metformin released from the delivery system was monitored in the plasma at 1 h post-microneedle treatment.[24] In addition, unlike with the earlier needles, hydrogel-based microneedles can be withdrawn and remain intact after being applied to the skin. Another hydrogel-based microneedle patch for transdermal release of metformin by near-infrared light has been developed.[25] The patch was prepared by a simple mixture of carboxyl-enriched reduced graphene oxide and metformin HCl.[25] The delivery of metformin was achieved by weakening the gelation forces through photothermal heating.[25] Photothermal activation resulted in a stepwise dissolution of the gels and the release of active metformin in an intermittent on/off cycle with near-infrared light.[25] *In vivo* and *in vitro* studies confirmed that metformin was present in plasma 1 h after patch activation and that heat stimulation did not alter the metformin activity in liver.[25]

3.3 Sulfonylureas

The antidiabetic effect of sulfonylureas was discovered in 1942 by Janbon and Loubatière.[26] While the sulfonamides were investigated to combat *Salmonella* in typhoid fever, hypoglycemia episodes were observed in animals and humans.[26] The first antidiabetic medicine of the sulfonylurea class was carbutamide (1-butyl-3-sulfonylurea). However, this medicine was withdrawn from the market because of its adverse effects on bone marrow. So far, several generations of sulfonylurea have been developed and improved based on their absorption, metabolism and dosing (Table 3.3). They all act by binding to the receptor for sulfonylureas on β-pancreatic cells, leading to the closure of the ATP-dependent potassium (K^+) channel. As a consequence of this closure, the flow of K^+ is abolished, whereas the calcium influx is stimulated due to the opening of the voltage-dependent calcium channels. Finally, the increased influx of calcium into β-cells causes insulin secretion. Sulfonylureas

Table 3.3 Representative sulfonylureas.

Generation	Sulfonylurea	Dose (mg)	Duration of action $t_{1/2}$ (h)
1st	Tolbumatide	500–2000	4–7
2nd	Glibenclamide	2.5–15	2–15
	Gliclazide	40–320	10
	Glipizide	2.5–20	2–5
	Gliquidone	15–180	3–4
3rd	Glimepiride	1–6	1–6

are prescribed to non-obese patients as monotherapy or bitherapy with metformin, when the latter is not efficient to achieve adequate glycemic control. Sulfonylureas lower blood glucose concentrations by about 20% and glycosylated hemoglobin (HbA1C) by 1–2%.[27] They are effective only when residual pancreatic β-cells are present. However, most sulfonylureas are poorly soluble in water. This poor aqueous solubility impinges the dissolution of the drugs and, therefore, its bioavailability.[28] In addition, the low oral bioavailability results in wasting of a large portion of an oral dose and adds to the cost of drug therapy. Another concern resulting from excessive concentration of sulfonylureas is hypoglycemia. All sulfonylureas can cause hypoglycemia, but the risk is greater with glibenclamide.[29,30] The persistence of hypoglycemia for many hours can lead to patients being admitted to hospital. Other infrequent side effects caused by the intake of high concentrations of sulfonylureas include weight gain, nausea, skin reactions such as erythema multiforme, exfoliative dermatitis and also, more rarely, photosensitivity.[31] The increase in the aqueous solubility of sulfonylureas is one of key objectives for reducing the concentration of sulfonylureas while preserving their therapeutic efficiency. To reach this objective, several studies have attempted to engineer new oral nanoparticle-based sulfonylureas formulations. These nanoparticles include niosomes, polymeric micelles, polymeric nanocrystals, and solid lipid nanoparticles. All these formulations can improve the solubility and intestine permeability of sulfonylureas (Table 3.4). Next, investigations will be required to compare the most suitable formulation for the toxicity tests and clinical trials. These studies, with the increased numbers of nanoparticles approved by drug agencies, will enable the release in the next years of new performant formulations of sulfonylureas with limited side effects.

Changing the route of delivery may also be an alternative for increasing the bioavailability of sulfonylureas. To this end, some studies have emerged showing the possibility of the inhalation of glibenclamide.[32,33] Indeed, delivery of drugs into the lungs offers several advantages. Due to its high vascular nature and large surface area, the pulmonary system enables rapid diffusion of drugs in the systemic circulation while avoiding or circumventing gastrointestinal permeability and the first-pass effect. Reports have revealed a 50–60% reduction of hyperglycemia from diabetic rats that have inhaled glibenclamide formulated into a niosome-based inhaler.[32,33] Again, toxicity studies followed by clinical trials are required to confirm the efficiency and relevance of this delivery system in the clinical situation.

Table 3.4 Nanoparticle-based formulations of sulfonylureas for oral delivery.

Medicines	Nanoparticles	Materials
Glibenclamide	Polymers	Hydroxypropyl-β-cyclodextrin and poly(methylvinyl ether-*co*-maleic anhydride)[94,95]
	Solid lipid nanoparticles	Glyceryl palmitostearate, lecithin and polyethylene glycol[96]
Gliclazide	Niosomes	Maltodextrine[97]
	Polymeric micelles	Methoxypoly(ethylene glycol), poly(ε-caprolactone)/(mPEG/PCL)[98]
	Polymeric nanocrystals	Poly lactic-*co*-glycolic acid[99]
	Coating with nanocrystals	*p*-Tolylsulfonylisocyanate[100]
Glipizide	Polymeric	Alginate[101]
	Polymeric	Methacrylic polymer and polyvinyl alcohol[102]
Glimepiride	Polymeric micelles	Cyclodextrin[103]
	Cellulosic polymers	Hydroxypropyl methylcellulose and polyvinylpyrrolidone[104]
	Niosomes	Nonionic surfactants (Tween)[105]
Gliquidone		Polyvinylpyrrolidone[106]

3.4 Glinides

Glinides are a class of oral drugs including repaglinide, nateglinide and mitiglinide. These stimulate insulin secretion independently of glucose. Compared with sulfonylureas, glinides display faster insulinotropic activity through binding to its distinct sites on the pancreatic β-cell membrane.[35] Clinical trials have proven that, as a monotherapy, glinides could decrease HbA1c levels effectively in patients with T2D,[36] as well as in combination therapy.[37] Repaglinide combined with lifestyle interventions can significantly reduce HbA1c levels.[39] However, glinides have very low aqueous solubility, poor oral bioavailability and a short half-life (~1 h).[34] This concern leads to enhanced drug dosage, which increases the associated side effects. To improve its bioavailability and half-life, attempts have been made to encapsulate glinides into nanoparticles. Lipid-based formulations (incorporation of the active lipophilic component into inert lipid vehicles), including micro- or nanoemulsions, have been used for improving the oral bioavailability of repaglinide.[35] Because of its high solubilization capacity, toxicological safety, long shelf-life and rapid onset of action, the nanoemulsion has been particularly investigated in this respect.[35] In diabetic rats induced by streptozotocin, the repaglinide nanoemulsion formulation triggers a higher antihyperglycemic effect than that of the tablet repaglinide.[35] Other studies have reported several formulations of repaglinide using niosomes,[36] poly(methyl methacrylate) (PMMA),[37] nanocrystals[38] and solid lipid nanoparticles.[39–41] The antidiabetic effect of repaglinide has been shown when it is encapsulated into niosomes.[36] In diabetic rats, the antihyperglycemic effect of the repaglinide niosome formulation is two-fold longer than that of free repaglinide.[36] The important reduction of the plasma glucose level is achieved using the repaglinide nanocrystal formulation[38] and solid lipid nanoparticles.[40] In diabetic rats, the hypoglycemic activity of repaglinide nanocrystals is more prominent than that obtained with oral repaglinide.[38] Although the effect of the PMMA-based formulation on glucose metabolism has not been tested, it is safe and *in vitro* it prolongs the release of the antidiabetic.[37]

As for metformin, attempts have also been made to investigate the transdermal route for delivering repaglinide. To this end, repaglinide has been prepared into a mixture of biodegradable nanoparticles including chitosan, poly(DL-lactic acid) (PLA) and poly(ε-caprolactone) (PCL) by the solvent evaporation method.[42] The formulation has been incorporated into a transdermal patch for *ex vivo* and *in vivo* studies. The transdermal patch containing repaglinide nanoparticles showed a 75-fold increase in efficiency compared with conventional oral administration of the drug. In diabetic rats, the release of repaglinide *via* the transdermal route was slower but longer than the oral administration of a repaglinide (2 mg) patch.[42] When compared to oral repaglinide, the transdermal release of repaglinide leads to a gradual reduction of plasma glucose levels with a maximal effect reached after 36 h post-patch application.[42]

3.5 GLP-1 Receptor Agonists

The glucagon-like peptide-1 (GLP-1) receptor agonists (GLP-1RA) are the last discovered insulin secretagogues for T2D treatment. GLP-1RA mimic the native GLP-1, a peptide secreted by neuroendocrine L cells localized in the ilium and colon.[43] Unlike the glinides and sulfonylureas, GLP-1RA only stimulate insulin secretion in response to the elevation of glucose. The effect of GLP-1RA is achieved by a series of intracellular mechanisms in which cAMP production and protein kinase A (PKA) activity play a central role.[43] cAMP is produced from the transformation of ATP into cAMP. As the ATP level is stimulated by glucose, GLP1-RA require glucose for producing a high cAMP level, and thereby increasing insulin secretion. As a result of this glucose-dependent mechanism effect of GLP-1RA, the risk of hypoglycemia upon the intake of these antidiabetics is almost absent. In addition to stimulating insulin secretion, GLP-1RA offer other benefits for patients with T2D. They inhibit gastric emptying, appetite and food intake, weight gain and cardiovascular risk.[44] In addition, GLP-1RA have a longer half-life than native GLP-1 (Table 3.5). Indeed, the half-life of native GLP-1 is below 2 min.[45] The very low half-life of GLP-1 results from proteolysis induced by dipeptidyl peptidase-4 (DPP-4).[43] For this reason, long-acting GLP-1 analogs resistant to DPP-4 degradation are currently being developed. So far, seven GLP-1RA have been released for treating T2D (Table 3.5). These consist of short-acting GLP-1RA, including exenatide and lixisenatide, and long-acting ones such as liraglutide, taspoglutide, semaglutide, albiglutide and dulaglutide.[46] Liraglutide is derived from the native GLP-1 sequence in which a 16-carbon fatty acid has been conjugated to the ε-amino group of lysine26.

Table 3.5 Schematic description of the various GLP-1 receptor agonists currently available.[a]

GLP-1RA	Sequence (amino acids)	% of Amino acids similarity to native GLP-1	Half-life	Clearance ($l\ h^{-1}$)
Native human GLP-1	HAEGTFTSDVSSYLEGQAAKEFIAWLVKGRG		<2 min	
Exenatide	HGEGTFTSDLSKQMEEEAVRLFIEWLKNGGPSSGAPPPS	53	~4 h	9.1
Liraglutide	HAEGTFTSDVSSYLEGQAAKEFIAWLVRGRG E—C16-Fatty Acid Chain	97	~11–13 h	1.2
Semaglutide	HAibEGTFTSDVSSYLEGQAAKEFIAWLVRGRG E—C18-Fatty Acid Chain	94	~7 days	0.05
Lixisenatide	HGEGTFTSDLSKQMEEEAVRLFIEWLKNGGPSSGAPPSKKKKKK	50	~4 h	20–30
Albiglutide	HGEGTFTSDVSSYLEGQAAKEFIAWLVKGRG—Albumin	95	~6–8 h	0.067
Taspoglutide	HA^{m}EGTFTSDVSSYLEGQAAKEFIAWLVKA^{m}RG	93	~10 h	ND
Dulaglutide	HGEGTFTSDVSSYLEEQAAKEFIAWLVKGGG—Ig4-FC HGEGTFTSDVSSYLEEQAAKEFIAWLVKGGG—Ig4-FC	90	~4 days	0.1

[a]Differences of amino acids (Aa) from native GLP-1 with the available GLP-RA are shown in red. GLP-RA with different half-lives have been produced by modifying some Aa from the native GLP-1 or exenatide.

Liraglutide differs from the native GLP-1 primary sequence with lysine34 substitution with arginine. This modification increases albumin binding to 99%, thus protecting liraglutide from DPP-4 degradation and giving it an 11–13 h half-life. Semaglutide looks like liraglutide as it is conjugated to the fatty acid chain to the lysine residue and has a similar primary sequence. Replacement of alanine at position 8 by 2-aminoisobutyric acid (Aib) protects semaglutide against DPP-4 degradation. Lixisenatide is derived from exenatide, a synthetic analog of exendin-4, naturally resistant to DPP-4. Lixisenatide has a six-amino-acid polylysine extension at the C-terminus and has a half-life of around 4 h but has been formulated for once a day injection. Albiglutide is a genetic fusion of a GLP-1 concatamer to human albumin. Taspoglutide is also based on the native GLP-1 that is resistant to DPP-4 degradation thanks to methylation of Ala8 and Ala35. Dulaglutide results from the fusion of GLP-1 *via* a linker to the Fc domain of human immunoglobulin (IgG4). All of these are delivered by subcutaneous injection. This mode of administration is uncomfortable, leading to some patients developing needle phobia, which can interrupt the treatment. Therefore, an oral route is preferable, as it can increase patients' compliance to treatment. However, this mode of administration is challenging as it compromises the stability of the peptides in the gastrointestinal tract and their absorption efficiency. In addition, using the oral administration, some side effects including nausea, headache and nasopharyngitis can develop.[47] To protect against proteolysis and to permit intestinal permeation, several GLP-1RA formulations using chemical permeation enhancers (CPEs) and nanoparticles have been engineered. Because the *N*-acetylated amino-acid derivative of salicylic acid (SNAC)[48] increases the systemic bioavailability of heparin, ibandronate and vitamin B12 in humans,[49–51] it has been chosen for increasing absorption of the GLP-1RA semaglutide. The formulation of semaglutide into tablets with SNAC eases the transient opening of the epithelial tight junctions in the intestine and the paracellular transport.[52] In addition, the formulation prevents the gastric degradation of semaglutide by neutralizing the low pH microenvironment surrounding the tablet.[53,54] Regarding the hypoglycemic effect, oral and subcutaneous semaglutide are similar.[55] Oral semaglutide is as efficient as injected semaglutide for stimulating insulin secretion as it becomes free when the oral formulation reaches the bloodstream.[55] Indeed, after intestinal absorption, semaglutide and SNAC dissociate, liberating semaglutide, which becomes identical to injected semaglutide in terms of structure.[55] Clinical trials confirmed the safety and noninferiority of the oral semaglutide when compared to subcutaneous injection of several GLP-1RA including semaglutide, liraglutide and dulaglutide.[56] These results have led the FDA and the European Medical Agency to grant marketing authorization for oral semaglutide (sold as Rybelsus®). Oral semaglutide is expected to allow many patients to continue with oral treatment and, therefore, to prevent escalation of therapy using either an injectable GLP-1RA or insulin. In addition, oral semaglutide is safe for patients with cardiovascular disease or high cardiovascular risk.[57]

The cardiovascular risks of oral semaglutide, placebo and other GLP-1RA are similar.[57,58] These clinical trials could soon make oral semaglutide among the first line of oral antidiabetic treatment and, therefore, could change the paradigm of T2D management. However, as the SNAC and enhancers increase intestinal permeability, many substances including toxins and pathogens can penetrate into the systemic circulation. This concern alerts us to the possible damage to the bio-membrane, which can lead to local gastrointestinal inflammation, infection and systemic inflammation.[55] For this reason, important efforts are still required to discover alternatives for increasing the bioavailability and proteolytic resistance of oral GLP-1RA. Several alternatives might result from the use of nanoparticle-based carriers capable of delivering DNA into nanoparticles (Table 3.6).

3.6 Dipeptidyl Peptidase-4 (DDP-4) Inhibitors

DPP-4 inhibitors were released in 2000 with clinical evidence for lowering glycemia in patients with T2D.[59] The first DPP-4 inhibitor to be used in a clinical study in type 2 diabetes was NVP DPP728.[60] DPP-4 inhibitors were introduced as the oral antidiabetic class of gliptin into the clinical market in 2006 and are now used as monotherapy or in combination with other oral antidiabetics in many countries.[61] DPP-4 inhibitors function as a competitive inhibitor of the enzyme DPP-4. These inhibitors prevent the inactivation of enzyme substrates including GLP-1. This results in the increase of the half-life of GLP-1 active forms after meals and, therefore, insulin secretion.[62] Sitagliptin, linagliptin, vildagliptin, saxagliptin and alogliptin constitute the available DPP-4 inhibitors. Most gliptins have relatively good oral bioavailability.[63] DPP-4 inhibitors are safe and display a very limited risk of hypoglycemia. However, they can lead to some adverse effects, with gastrointestinal disorders and nasopharyngitis being the most frequently reported ones.[62,64] Some cases of pancreatitis and pancreatic cancer have also been diagnosed after DPP-4 inhibitors treatment, although this is still debated.[65] Generally, gliptin is prescribed as a single daily oral administration with the exception for vildagliptin, which is given twice per day because of its shorter half-life. To increase the half-life of vildagliptin and to reduce the adverse effects caused by its gastrointestinal passage, some efforts have been made to improve the formulation of gliptin using nanoparticles such as chitosan, eudragit and poly(DL-lactic-acid-*co*-glycolic acid) (PLGA) (Table 3.7). Most studies report an improvement in intestinal permeability and release of the antidiabetic, which may promote more preclinical studies to investigate the long-term effects of these new formulations from a clinical perspective.

3.7 Sodium–Glucose Co-transporter 2 Inhibitors (SGLT2i)

SGLT2i emerged as the last antidiabetic oral class of gliflozin 5–10 years ago.[66] The first SGLT2i was phlorizin, a molecule that has was discovered in the root,

Table 3.6 Nanoparticle-based oral formulations of native GLP-1 and GLP-1RA.

Nanoparticles	GLP-RA	*In vitro* and preclinical studies
Liposomes	Native GLP-1	In rats, 1.7-fold higher insulin secretion, 3.6-fold higher elevation of serum GLP-1 level with liposomal formulation of GLP-1 as compared to GLP-1 injection[107]
Poly(DL-lactic-acid-*co*-glycolic acid) PLGA and porous silicon (PSi) nanoparticles coated with chitosan (CS)	Native GLP-1	The CS-coated PSi and PLGA nanoparticles have the highest GLP-1 permeation across the intestinal cell models when compared to non-coated nanoparticles[108]
CS-coated PSi or PLGA encapsulated into hydroxypropylmethylcellulose acetylsuccinate polymer	Native GLP-1	Encapsulated formulation of GLP-1 has higher cell permeation than non-encapsulated CS-coated PSi and PLGA nanoparticles[109]
Encapsulation into alginate coated with CS	Liraglutide	High stability and controlled release of liraglutide in intestinal cell model over time[110]
Encapsulation into PLGA	Liraglutide	High cell permeation and high protection against enzymatic degradation[111]
Encapsulation into propyl-amine cyclodextrin	Liraglutide	Effective protection against enzymatic degradation and similar delivery of liraglutide in blood glucose levels to subcutaneous administration of non-encapsulated liraglutide[112]
Encapsulation with deoxycholic acid and low-molecular-weight protamine enveloped with polysialic acid and glycolic acid	Exenatide	Accumulation of Exenatide in the distal ileum, increases plasma concentration of Exenatide, and reduces the frequency of medication administration[113]
Complexation with cholic acid (CA) and phosphatidylcholine (PC)	Exenatide	Reduction of plasma glucose level and HbA1c during and after 7 weeks of once-daily oral administration in db/db mice. Improved β-cell mass after 7 weeks of treatment[114]
Non-viral gene delivery based on GLP-1 cDNA conjugated with protamine sulfate-conjugated taurocholic acid and calcium phosphate	Native cDNA GLP-1	A single oral dose leads to normoglycemia for up to 7 days in three different diabetes mice models and 14 days in diabetic monkeys[115]
Complexation with polyethyleneimine coated with a mixture of 1,2-dipalmitoyl-*sn*-glycero-3-phosphocholine and 1,2-dimyristoyl-*rac*-glycero-3-methoxy poly(ethylene glycol)-2000	Native cDNA GLP-1	After a single oral dose, sustained expression of GLP-1 was detected in the liver, lung and small intestine 12–36 h post-administration. Blood glucose is reduced to the normal range in 6 h following oral dose and maintained within the normal range (5–10 mmol L^{-1}) for at least 18 h in *db/db* diabetic mice[116]

Table 3.7 Oral formulation of DPP-4 inhibitors (iDDP-4) using nanoparticles.

Nanoparticles	iDDP-4	*In vitro* and preclinical studies
Encapsulation into nanosphere made by cationic Eudragit polymer coated with DNA	Vildagliptin	Nanospheres effectively bypass the acidic pH of the stomach and enhance glycemic control in *db*/*db* mice without any risk of pancreatitis or pancreatic cancer[117]
Microencapsulation into Eudragit polymer	Vildagliptin	Reduced plasma glucose level in diabetic rats induced by streptozotocin[118]
Mixture with zinc oxide nanoparticles (ZnO)	Vildagliptin	Vildapliptin with ZnO leads to higher β-cell function, glucose tolerance, weight loss and insulin levels than free vildapliptin in diabetic rats fed a high-fat diet induced by streptozotocin[119]
Solid lipid nanoparticles using palmitic acid, poloxamer 188 and Tween 80	Linagliptin	High bioavailability and permeability with hypoglycemic effect in rats[120]
Chitosan conjugated poly(dl-lactic-acid-*co*-glycolic acid) (PLGA) nanoparticles conjugated with cell-penetrating peptides	NVP DPP 728 together with native GLP-1	The combination of both GLP-1 and iDPP-4 resulted in strong and prolonged hypoglycemic effects in non-obese diabetic rats induced by streptozotocin and nicotinamide[121]
Functionalized porous silicon coated with chitosan	NVP DPP 728 together with native GLP-1	High delivery of GLP1 and iDPP-4 and improved oral bioavailability[122]

Table 3.8 SGLT2i currently prescribed in the United States and Europe.

SGLT2i	Dose (mg)	Duration of action $t_{1/2}$(h)
Canagliflozin	100–300	13.1
Dapagliflozin	5–10	13
Empagliflozin	10–25	12.4
Ertugliflozin	5–15	17

bark, leaves, shoots and fruit of the apple tree 150 years ago.[67] Phlorizin increases urinary glucose excretion in healthy humans.[67] The poor water solubility, specificity and oral bioavailability of phlorizin have led to the search for other more suitable mimetics for clinical application.[66] In Europe, there are four members of the class of gliflozin used for treating T2D[68] (Table 3.8). However, there are also marginal members in this class, as those that are only approved by the European Medicines Agency for T1D, or those that are exclusively found in Japan.[68] The SGLT2i inhibit the reabsorption of glucose *via* inhibition of the sodium–glucose co-transporter 2 in the proximal convoluted renal tubule.[69] As a consequence, SGLT2i increase glycosuria and natriuresis, leading to a decrease in HbA1c, body weight, blood pressure and albuminuria.[70] Beyond improving glycemic control, SGLT2i reduce the risk of cardiovascular disease events caused by atherosclerosis.[71] The cardiovascular benefits of SGLT2i result from several mechanisms including reduction of blood pressure, enhancement of cardiac energy metabolism, reduction of inflammation and improvements to vascular function.[71] Several mechanism(s) responsible for the benefits to cardiovascular risk of SGLT2 inhibition have been suggested.[72] Beside the cardioprotective effect, SGLT2i reduce the risk of kidney failure and the rate of all-cause mortality.[71] However, these benefits differ according to the type of SGLT2i.[72] Serious adverse events have been described that threaten the tolerability of this type of drug, including genital and urinary tract infections, diabetic ketoacidosis, bone fractures, foot and leg amputations and cancer.[73] Although great progresses have been made in SGLT2i since their discovery, they remain poorly water soluble. an improvement in the aqueous solubility using nanoparticles might reduce the dose for oral delivery, the onset of adverse effects and also improve the bioavailability of SGLT2i patients with T2D. To this end, very few studies have been published so far. One study has investigated a nanoparticle-based formulation of dapagliflozin using lipids and surfactants.[74] This report showed that the best *in vitro* permeability and sustained release of the drug were obtained when dapagliflozin was mixed with labrasol and tween 20 as the lipid and surfactant, respectively.[74] This study may pave the way for other preclinical studies in the future.

3.8 Basal Insulin

Insulin therapy can be prescribed in T2D when oral monotherapy does not achieve or maintain the HbA1C target over 3 months and if there is evidence of ongoing catabolism (weight loss), ketosis or features of hyperglycemia including polyuria/polydipsia and/or severely elevated blood glucose.[75] In this condition, 0.1–0.2 U kg^{-1} of basal insulin is typically injected. If the insulin treatment requires concentrations greater than 0.5 U kg^{-1}, insulin is associated with metformin, DPP-4 inhibitors or SGLT2i. Sometimes, bolus insulin needs to be administered in addition to basal insulin. Rapid-acting insulin analogs are used as bolus formulations due to their prompt onset of action. However, the invasive mode for administering insulin *via* needles leads to numerous drawbacks, resulting in poor patient compliance and unsatisfactory glycemic control. Therefore, there is a significant need for alternative and non-invasive strategies to deliver insulin. In addition, these strategies could include the real-time release of insulin in an appropriate dose–response manner over a prolonged period of time. From this perspective, an insulin pump (continuous subcutaneous insulin infusion) may be used. Unfortunately, patients and physicians are reluctant to intensify therapy due to the fear of hypoglycemia, regimen complexity, and increased multiple daily injections. Other options, such as inhaled insulin, have been investigated. Inhaled insulin (II, Technosphere insulin-inhalation system, Afrezza) is now available for prandial use.[76] Pulmonary administration of II improves glycemia in insulin-naïve T2D individuals.[77] In addition, it induces less hypoglycemia than subcutaneously

Table 3.9 Clinical trials for oral formulation of insulin with nanoparticles.

Nanoparticles	Clinical trials identifiers	Phase	Results
Nanoparticles with a calcium phosphate core and pegylated salts of fatty acids, coated with carbomer and cellulose acetate phthalate silica-based nanoparticles	ChiCTR-TRC-12001872	I	Magnitude, and duration of the glucose response induced by oral insulin were similar to those of subcutaneous injection of insulin[123]
Silica-based nanoparticles	NCT01973920	II	Not provided
Liver-targeted liposomes	NCT00814294	III	Not provided

injected insulin.[77] However, the dosing range is limited. The use of inhaled insulin requires pulmonary function testing prior to and after starting therapy. It is contraindicated in subjects with asthma or other lung diseases. Other alternative routes of insulin delivery including the nasal, buccal and transdermal routes have been continuously under investigation. The current progress in the development of painless and on-demand transdermal delivery of insulin using reduced graphene oxide[78–81] may lead soon to clinical trials. Intensive research carried out over the last 30 years has enabled new nanoparticle-based formulations of insulin for oral delivery to be released.[82] Many attempts have been done using polymeric- (*e.g.*, PLGA), chitosan-, lipid- (*e.g.*, liposomes) and silica-based nanoparticles.[82] A formulation of insulin with cationic liposomes containing DOTAP with bovine serum albumin has shown an impressive capacity to diffuse within the mucosal barrier and to lead to high bioavailability of oral insulin.[83] High intestinal permeation and bioavailability have also been achieved with insulin that is encapsulated in silica-coated nanoparticles (SiNPs).[84,85] Clinical trials testing the efficacy and toxicity of oral insulin delivered *via* liposomes and SiNPs (Table 3.9) have been completed and should enable conclusions to be drawn about their effectiveness.

3.9 Conclusion

With the exception of GLP-1RA and insulin, most current antidiabetic drugs come in the form of tablets. Unfortunately, their low aqueous solubility impinges on the intestinal permeation and bioavailability of the drug, thereby resulting in an increase in the drug dosage and the onset of undesirable side effects. These concerns may lead to the stopping of the treatment or, with due to negligence and forgetfulness with the chronic intake of drugs, to the reduction of patient adherence to the treatment. Indeed, between 15–20% of patients discontinue oral antidiabetic therapy after 6 months of treatment.[7] This poor patient compliance leads to the progression of diabetes, with major risks of disabilities and premature death, mainly because of cardiovascular disease. Indeed, it has been estimated that less than 50% of patients truly achieve the glycemic goals and that a majority of these patients die prematurely of cardiovascular disease.[86] The improvements in the formulation of the current antidiabetics listed in this chapter, with the application of suitable measures (*e.g.*, communication and education), would contribute to improving drug efficiency and the adherence of patients with an intolerance to the antidiabetics. In this respect, the use of safe nanoparticles would be a key solution, as exemplified for oral insulin. The new nanoparticle-based formulations would increase the therapeutic options, allowing clinicians to choose and adapt different formulations of drugs to individual patients, creating a cornerstone for the future implementation of personalized diabetes medicine. This achievement is crucial for achieving better long-term glycemic control and for stopping the progression of diabetes toward other diseases such as cardiovascular diseases and the risk of infection, as currently exemplified by the COVID-19 pandemic. The potential socioeconomic benefits of reducing diabetes progression are huge. Indeed, diabetes costs to the health systems in most countries worldwide are substantial, reaching 10% of total spending on adults in 2019.[3] Several studies have shown that better long-term glycemic control benefits the healthcare system by reducing hospitalizations risk of a first cardiovascular event, and mortality by at least of 50%, and leading to a saving of approximately 1500 euros/year/patient.[87,88]

Acknowledgements

The Centre National de la Recherche Scientifique, the University of Lille, I-SITE-ULNE, the Hauts-de-France region and the CPER "Photonics for Society" are gratefully acknowledged for their financial support.

References

1. S. Lim, J. H. Bae, H.-S. Kwon and M. A. Nauck, *Nat. Rev. Endocrinol.*, 2021, **17**, 11–30.
2. S. Chatterjee and M. J. Davies, *Nat. Rev. Endocrinol.*, 2018, **14**, 386–387.
3. R. Williams, S. Karuranga, B. Malanda, P. Saeedi, A. Basit, S. Besançon, C. Bommer, A. Esteghamati, K. Ogurtsova, P. Zhang and S. Colagiuri, *Diabetes Res. Clin. Pract.*, 2020, **162**, 108072.
4. T. Barrett, M. Y. Jalaludin, S. Turan, M. Hafez and N. Shehadeh, *Pediatr. Diabetes*, 2020, **21**, 158–172.
5. J. Cantley and F. M. Ashcroft, *BMC Biol.*, 2015, **13**, 33.
6. R. A. DeFronzo, E. Ferrannini, L. Groop, R. R. Henry, W. H. Herman, J. J. Holst, F. B. Hu, C. R. Kahn, I. Raz, G. I. Shulman, D. C. Simonson, M. A. Testa and R. Weiss, *Nat. Rev. Dis. Primer*, 2015, **1**, 1–22.
7. L.-E. García-Pérez, M. Alvarez, T. Dilla, V. Gil-Guillén and D. Orozco-Beltrán, *Diabetes Ther. Res. Treat. Educ. Diabetes Relat. Disord.*, 2013, **4**, 175–194.
8. L. Osterberg and T. Blaschke, *N. Engl. J. Med.*, 2005, **353**, 487–497.
9. C. J. Bailey, *Diabetologia*, 2017, **60**, 1566–1576.
10. C. J. Bailey, *Diabetes Care*, 1992, **15**, 755–772.
11. R. I. Misbin, *Ann. Intern. Med.*, 1977, **87**, 591–595.

12. L. F. Verdonck, B. Sangster, A. N. van Heijst, G. de Groot and R. A. Maes, *Diabetologia*, 1981, **20**, 45–46.
13. F. Féry, L. Plat and E. O. Balasse, *Metabolism.*, 1997, **46**, 227–233.
14. R. Giannarelli, M. Aragona, A. Coppelli and S. Del Prato, *Diabetes Metab.*, 2003, **29**, 6S28–35.
15. D. R. Morales and A. D. Morris, *Annu. Rev. Med.*, 2015, **66**, 17–29.
16. Y. Zhang, H. Wang and H. Xiao, *Int. J. Mol. Sci.*, 2021, **22**, 5016.
17. R. J. Dowling, P. J. Goodwin and V. Stambolic, *BMC Med.*, 2011, **9**, 33.
18. J. Sun and H. Tan, *Mater. Basel Switz.*, 2013, **6**, 1285–1309.
19. G. Cattelan, A. Guerrero Gerbolés, R. Foresti, P. P. Pramstaller, A. Rossini, M. Miragoli and C. Caffarra Malvezzi, *Front. Bioeng. Biotechnol.*, 2020, **8**, 414.
20. W. Wang, Q. Meng, Q. Li, J. Liu, M. Zhou, Z. Jin and K. Zhao, *Int. J. Mol. Sci.*, 2020, **21**, E487.
21. A. C. Anselmo and S. Mitragotri, *Bioeng. Transl. Med.*, 2019, **4**, e10143.
22. N. Morin-Crini, E. Lichtfouse, G. Torri and G. Crini, *Environ. Chem. Lett.*, 2019, **17**, 1667–1692.
23. H. Lee, T. K. Choi, Y. B. Lee, H. R. Cho, R. Ghaffari, L. Wang, H. J. Choi, T. D. Chung, N. Lu, T. Hyeon, S. H. Choi and D.-H. Kim, *Nat. Nanotechnol.*, 2016, **11**, 566–572.
24. E. M. Migdadi, A. J. Courtenay, I. A. Tekko, M. T. C. McCrudden, M.-C. Kearney, E. McAlister, H. O. McCarthy and R. F. Donnelly, *J. Controlled Release*, 2018, **285**, 142–151.
25. L. Chengnan, Q. Pagneux, A. Voronova, A. Barras, A. Abderrahmani, V. Plaisance, V. Pawlowski, N. Hennuyer, B. Staels, L. Rosselle, N. Skandrani, M. Li, R. Boukherroub and S. Szunerits, *Nanoscale*, 2019, **11**, 15810–15820.
26. H. Kleinsorge, *Exp. Clin. Endocrinol. Diabetes Off. J. Ger. Soc. Endocrinol. Ger. Diabetes Assoc.*, 1998, **106**, 149–151.
27. L. S. Hermann, B. Scherstén, P. O. Bitzén, T. Kjellström, F. Lindgärde and A. Melander, *Diabetes Care*, 1994, **17**, 1100–1109.
28. N. Seedher and M. Kanojia, *Pharm. Dev. Technol.*, 2009, **14**, 185–192.
29. A. Holstein, A. Plaschke and E. H. Egberts, *Diabetes Metab. Res. Rev.*, 2001, **17**, 467–473.
30. A. S. Gangji, T. Cukierman, H. C. Gerstein, C. H. Goldsmith and C. M. Clase, *Diabetes Care*, 2007, **30**, 389–394.
31. W. Lv, X. Wang, Q. Xu and W. Lu, *Curr. Top. Med. Chem.*, 2020, **20**, 37–56.
32. R. S. Abdel-Rashid, F. I. Abd Allah, A. A. Hassan and F. M. Hashim, *J. Liposome Res.*, 2021, **31**, 291–303.
33. F. M. Hashem, F. I. Abd Allah, R. S. Abdel-Rashid and A. A. A. Hassan, *Drug Dev. Ind. Pharm.*, 2020, **46**, 762–774.
34. C. R. Culy and B. Jarvis, *Drugs*, 2001, **61**, 1625–1660.
35. J. Akhtar, H. H. Siddiqui, S. Fareed, Badruddeen, M. Khalid and M. Aqil, *Drug Deliv*, 2016, **23**, 2026–2034.
36. S. Namdev, K. Gujar, S. Mandlik and P. Jamkar, *Int. J. Pharm. Sci. Nanotechnol.*, 2015, **8**, 2756–2767.

37. U. M. Dhanalekshmi, G. Poovi, N. Kishore and P. N. Reddy, *Int. J. Pharm.*, 2010, **396**, 194–203.
38. R. Gadadare, L. Mandpe and V. Pokharkar, *AAPS PharmSciTech*, 2015, **16**, 787–799.
39. H. A. Ebrahimi, Y. Javadzadeh, M. Hamidi and M. B. Jalali, *Daru, J. Pharm. Sci.*, 2015, **23**, 46.
40. M. K. Rawat, A. Jain and S. Singh, *J. Pharm. Sci.*, 2011, **100**, 2406–2417.
41. M. K. Rawat, A. Jain, A. Mishra, M. S. Muthu and S. Singh, *Ther. Delivery*, 2010, **1**, 63–73.
42. V. Vijayan, K. R. Reddy, S. Sakthivel and C. Swetha, *Colloids Surf., B*, 2013, **111**, 150–155.
43. L. L. Baggio and D. J. Drucker, *Gastroenterology*, 2007, **132**, 2131–2157.
44. J. E. Campbell and D. J. Drucker, *Cell Metab.*, 2013, **17**, 819–837.
45. T. Vilsbøll, H. Agersø, T. Krarup and J. J. Holst, *J. Clin. Endocrinol. Metab.*, 2003, **88**, 220–224.
46. S. Madsbad and J. J. Holst, *Lancet*, 2019, **393**, 1263–1264.
47. T. D. Filippatos, T. V. Panagiotopoulou and M. S. Elisaf, *Rev. Diabet. Stud. RDS*, 2014, **11**, 202–230.
48. C. Twarog, S. Fattah, J. Heade, S. Maher, E. Fattal and D. J. Brayden, *Pharmaceutics*, 2019, **11**, E78.
49. M. C. Castelli, D. F. Wong, K. Friedman and M. G. I. Riley, *Clin. Ther.*, 2011, **33**, 934–945.
50. R. A. Baughman, S. C. Kapoor, R. K. Agarwal, J. Kisicki, F. Catella-Lawson and G. A. FitzGerald, *Circulation*, 1998, **98**, 1610–1615.
51. B. Bittner, C. McIntyre, H. Tian, K. Tang, N. Shah, W. Phuapradit, H. Ahmed, H. Chokshi, M. Infeld, N. Fotaki, H. Ma, A. Portron, P. Jordan and J. Schmidt, *Pharm.*, 2012, **67**, 233–241.
52. M. Davies, T. R. Pieber, M.-L. Hartoft-Nielsen, O. K. H. Hansen, S. Jabbour and J. Rosenstock, *JAMA*, 2017, **318**, 1460–1470.
53. T. A. Bækdal, A. Breitschaft, M. Donsmark, S. J. Maarbjerg, F. L. Søndergaard and J. Borregaard, *Diabetes Ther.*, 2021, **12**, 1915–1927.
54. S. T. Buckley, T. A. Bækdal, A. Vegge, S. J. Maarbjerg, C. Pyke, J. Ahnfelt-Rønne, K. G. Madsen, S. G. Schéele, T. Alanentalo, R. K. Kirk, B. L. Pedersen, R. B. Skyggebjerg, A. J. Benie, H. M. Strauss, P.-O. Wahlund, S. Bjerregaard, E. Farkas, C. Fekete, F. L. Søndergaard, J. Borregaard, M.-L. Hartoft-Nielsen and L. B. Knudsen, *Sci. Transl. Med.*, 2018, **10**, eaar7047.
55. J. D. Bucheit, L. G. Pamulapati, N. Carter, K. Malloy, D. L. Dixon and E. M. Sisson, *Diabetes Technol. Ther.*, 2020, **22**, 10–18.
56. S. Niman, J. Hardy, R. F. Goldfaden, J. Reid, M. Sheikh-Ali, D. Sutton and R. Choksi, *Drugs RD*, 2021, **21**, 133–148.
57. T. K. Thethi, R. Pratley and J. J. Meier, *Diabetes, Obes. Metab.*, 2020, **22**, 1263–1277.
58. M. Husain, A. L. Birkenfeld, M. Donsmark, K. Dungan, F. G. Eliaschewitz, D. R. Franco, O. K. Jeppesen, I. Lingvay, O. Mosenzon, S. D. Pedersen,

C. J. Tack, M. Thomsen, T. Vilsbøll, M. L. Warren and S. C. Bain, *N. Engl. J. Med.*, 2019, **29**, 841–851.
59. B. Ahrén, E. Simonsson, H. Larsson, M. Landin-Olsson, H. Torgeirsson, P.-A. Jansson, M. Sandqvist, P. Båvenholm, S. Efendic, J. W. Eriksson, S. Dickinson and D. Holmes, *Diabetes Care*, 2002, **25**, 869–875.
60. T. E. Hughes, M. D. Mone, M. E. Russell, S. C. Weldon and E. B. Villhauer, *Biochemistry*, 1999, **38**, 11597–11603.
61. B. Gallwitz, *Front. Endocrinol.*, 2019, **10**, 389.
62. B. Ahrén, *J. Diabetes Investig.*, 2021, **12**, 1128–1135.
63. A. J. Scheen, *Diabetes, Obes. Metab.*, 2010, **12**, 648–658.
64. D. Dicker, *Diabetes Care*, 2011, **34**(Suppl 2), S276–278.
65. L. C. Pinto, D. V. Rados, S. S. Barkan, C. B. Leitão and J. L. Gross, *Sci. Rep.*, 2018, **8**, 782.
66. T. Rieg and V. Vallon, *Diabetologia*, 2018, **61**, 2079–2086.
67. H. Chasis, N. Jolliffe and H. W. Smith, *J. Clin. Invest.*, 1933, **12**, 1083–1090.
68. E. Brown, H. J. L. Heerspink, D. J. Cuthbertson and J. P. H. Wilding, *The Lancet*, 2021, **398**, 262–276.
69. E. C. Chao and R. R. Henry, *Nat. Rev. Drug Discov.*, 2010, **9**, 551–559.
70. M. R. Cowie and M. Fisher, *Nat. Rev. Cardiol.*, 2020, **17**, 761–772.
71. T. A. Zelniker, S. D. Wiviott, I. Raz, K. Im, E. L. Goodrich, M. P. Bonaca, O. Mosenzon, E. T. Kato, A. Cahn, R. H. M. Furtado, D. L. Bhatt, L. A. Leiter, D. K. McGuire, J. P. H. Wilding and M. S. Sabatine, *Lancet Lond. Engl.*, 2019, **393**, 31–39.
72. V. Martínez-Vizcaíno, A. Díez-Fernández, C. Álvarez-Bueno, J. Martínez-Alfonso and I. Cavero-Redondo, *J. Clin. Med.*, 2021, **10**, 2713.
73. R. Pelletier, K. Ng, W. Alkabbani, Y. Labib, N. Mourad and J.-M. Gamble, *Ther. Adv. Drug Saf.*, 2021, **12**, 2042098621989134.
74. A. Zafar, *J. Oleo Sci.*, 2020, **69**, 1389–1401.
75. American Diabetes Association, *Diabetes Care*, 2021, **44**, S111–S124.
76. L. Heinemann and C. G. Parkin, *J. Diabetes Res.*, 2018, **2018**, 4568903.
77. J. Rosenstock, D. Franco, V. Korpachev, B. Shumel, Y. Ma, R. Baughman, N. Amin, J. B. McGill and Affinity 2 Study Group, *Diabetes Care*, 2015, **38**, 2274–2281.
78. S. Szunerits, S. Melinte, A. Barras, Q. Pagneux, A. Voronova, A. Abderrahmani and R. Boukherroub, *Chem. Soc. Rev.*, 2021, **50**, 2102–2146.
79. Q. Pagneux, R. Ye, L. Chengnan, A. Barras, N. Hennuyer, B. Staels, D. Caina, J. I. A. Osses, A. Abderrahmani, V. Plaisance, V. Pawlowski, R. Boukherroub, S. Melinte and S. Szunerits, *Nanoscale Horiz.*, 2020, **5**, 663–670.
80. F. Teodorescu, G. Quéniat, C. Foulon, M. Lecoeur, A. Barras, S. Boulahneche, M. S. Medjram, T. Hubert, A. Abderrahmani, R. Boukherroub and S. Szunerits, *J. Controlled Release*, 2017, **245**, 137–146.
81. F. Teodorescu, Y. Oz, G. Quéniat, A. Abderrahmani, C. Foulon, M. Lecoeur, R. Sanyal, A. Sanyal, R. Boukherroub and S. Szunerits, *J. Control. Release*, 2017, **246**, 164–173.

82. M. Durán-Lobato, A. M. López-Estévez, A. S. Cordeiro, T. G. Dacoba, J. Crecente-Campo, D. Torres and M. J. Alonso, *Adv. Drug Deliv. Rev.*, 2021, 113899.
83. A. Wang, T. Yang, W. Fan, Y. Yang, Q. Zhu, S. Guo, C. Zhu, Y. Yuan, T. Zhang and Y. Gan, *Adv. Healthc. Mater.*, 2019, **8**, 1801123.
84. N. G. Lamson, A. Berger, K. C. Fein and K. A. Whitehead, *Nat. Biomed. Eng.*, 2020, **4**, 84–96.
85. D. Hristov, F. McCartney, J. Beirne, E. Mahon, S. Reid, S. Bhattacharjee, G. Penarier, U. Werner, D. Bazile and D. J. Brayden, *ACS Appl. Mater. Interfaces*, 2020, **12**, 1257–1269.
86. C. J. Bailey and M. Kodack, *Int. J. Clin. Pract.*, 2011, **65**, 314–322.
87. E. Y. F. Wan, C. S. C. Fung, F. F. Jiao, E. Y. T. Yu, W. Y. Chin, D. Y. T. Fong, C. K. H. Wong, A. K. C. Chan, K. H. Y. Chan, R. L. P. Kwok and C. L. K. Lam, *Diabetes Care*, 2018, **41**, 49–59.
88. F. F. Jiao, C. S. C. Fung, E. Y. F. Wan, A. K. C. Chan, S. M. McGhee, R. L. P. Kwok and C. L. K. Lam, *Diabetes Care*, 2018, **41**, 250–257.
89. S. Kumar, G. Bhanjana, R. K. Verma, D. Dhingra, N. Dilbaghi and K.-H. Kim, *J. Pharm. Pharmacol.*, 2017, **69**, 143–150.
90. B. Mokhtare, M. Cetin, R. S. Ozakar and H. Bayrakceken, *Trop. J. Pharm. Res.*, 2017, **16**, 287–296.
91. S. Bhujbal and A. K. Dash, *AAPS PharmSciTech*, 2018, **19**, 2543–2553.
92. A. K. Sahu and A. Verma, *J. Pharm. Investig.*, 2016, **46**, 239–252.
93. N. Mirazi, J. Shoaei, A. Khazaei and A. Hosseini, *Eur. J. Drug Metab. Pharmacokinet.*, 2015, **40**, 343–348.
94. D. Lucio, M. C. Martínez-Ohárriz, Z. Gu, Y. He, P. Aranaz, J. L. Vizmanos and J. M. Irache, *Int. J. Pharm.*, 2018, **547**, 97–105.
95. D. Lucio, J. M. Irache, M. Font and M. C. Martínez-Ohárriz, *Int. J. Pharm.*, 2017, **519**, 263–271.
96. L. M. D. Gonçalves, F. Maestrelli, L. Di Cesare Mannelli, C. Ghelardini, A. J. Almeida and P. Mura, *Eur. J. Pharm. Biopharm. Off. J. Arbeitsgemeinschaft Pharm. Verfahrenstechnik EV*, 2016, **102**, 41–50.
97. R. Ku Sahoo, N. Biswas, A. Guha and K. Kuotsu, *Int. J. Biol. Macromol.*, 2014, **69**, 430–434.
98. H. Danafar, H. Jaberizadeh and S. Andalib, *Artif. Cells Nanomedicine Biotechnol.*, 2018, **46**, 1625–1636.
99. B. P. Panda, R. Krishnamoorthy, S. K. Bhattamisra, N. K. H. Shivashekaregowda, L. B. Seng and S. Patnaik, *Sci. Rep.*, 2019, **9**, 17331.
100. F. M. Sroor, S. Y. Abbas, W. M. Basyouni, K. A. M. El-Bayouki, M. F. El-Mansy, H. F. Aly, S. A. Ali, A. F. Arafa and A. A. Haroun, *Bioorganic Chem.*, 2019, **92**, 103290.
101. J. Emami, M. S. S. Boushehri and J. Varshosaz, *Res. Pharm. Sci.*, 2014, **9**, 301–314.
102. P. Saharan, K. Bahmani and S. P. Saharan, *Pharm. Nanotechnol.*, 2019, **7**, 72–85.
103. H. O. Ammar, H. A. Salama, M. Ghorab and A. A. Mahmoud, *Int. J. Pharm.*, 2006, **309**, 129–138.

104. H. Rahim, A. Sadiq, S. Khan, F. Amin, R. Ullah, A. A. Shahat and H. M. Mahmood, *Int. J. Nanomed.*, 2019, **14**, 6287–6296.
105. A. M. Mohsen, M. M. AbouSamra and S. A. ElShebiney, *Drug Dev. Ind. Pharm.*, 2017, **43**, 1254–1264.
106. M. S. Mohamed, W. A. Abdelhafez, G. Zayed and A. M. Samy, *AAPS PharmSciTech*, 2019, **21**, 35.
107. J. Hanato, K. Kuriyama, T. Mizumoto, K. Debari, J. Hatanaka, S. Onoue and S. Yamada, *Int. J. Pharm.*, 2009, **382**, 111–116.
108. F. Araújo, N. Shrestha, M.-A. Shahbazi, P. Fonte, E. M. Mäkilä, J. J. Salonen, J. T. Hirvonen, P. L. Granja, H. A. Santos and B. Sarmento, *Biomaterials*, 2014, **35**, 9199–9207.
109. F. Araújo, N. Shrestha, M.-A. Shahbazi, D. Liu, B. Herranz-Blanco, E. M. Mäkilä, J. J. Salonen, J. T. Hirvonen, P. L. Granja, B. Sarmento and H. A. Santos, *ACS Nano*, 2015, **9**, 8291–8302.
110. F. Shamekhi, E. Tamjid and K. Khajeh, *Int. J. Biol. Macromol.*, 2018, **120**, 460–467.
111. R. Ismail, A. Bocsik, G. Katona, I. Gróf, M. A. Deli and I. Csóka, *Pharmaceutics*, 2019, **11**, E599.
112. E. Presas, S. Tovar, J. Cuñarro, J. P. O'Shea and C. M. O'Driscoll, *J. Pharm. Sci.*, 2021, **110**, 292–300.
113. Y. Han, W. Liu, L. Chen, X. Xin, Q. Wang, X. Zhang, M. Jin, Z. Gao and W. Huang, *Biomaterials*, 2021, **275**, 120944.
114. X. Bao, K. Qian and P. Yao, *J. Nanobiotechnol.*, 2020, **18**, 67.
115. S. M. S. Shahriar, J. M. An, M. N. Hasan, S. S. Surwase, Y.-C. Kim, D. Y. Lee, S. Cho and Y. Lee, *Nano Lett.*, 2021, **21**, 4666–4675.
116. T. Nie, Z. He, Y. Zhou, J. Zhu, K. Chen, L. Liu, K. W. Leong, H.-Q. Mao and Y. Chen, *ACS Appl. Mater. Interfaces*, 2019, **11**, 29593–29603.
117. M. M. F. A. Baig, S. Khan, M. A. Naeem, G. J. Khan and M. T. Ansari, *Biomed. Pharmacother. Biomed. Pharmacother.*, 2018, **97**, 1250–1258.
118. I. Dewan, S. Islam and Md. S. Rana, *J. Drug Deliv.*, 2015, **2015**, 496807.
119. R. M. El-Gharbawy, A. M. Emara and S. E.-S. Abu-Risha, *Biomed. Pharmacother.*, 2016, **84**, 810–820.
120. P. Shah, K. Chavda, B. Vyas and S. Patel, *Drug Deliv. Transl. Res.*, 2021, **11**, 1166–1185.
121. F. Araújo, N. Shrestha, M. J. Gomes, B. Herranz-Blanco, D. Liu, J. J. Hirvonen, P. L. Granja, H. A. Santos and B. Sarmento, *Nanoscale*, 2016, **8**, 10706–10713.
122. N. Shrestha, F. Araújo, M.-A. Shahbazi, E. Mäkilä, M. J. Gomes, M. Airavaara, E. I. Kauppinen, J. Raula, J. Salonen, J. Hirvonen, B. Sarmento and H. A. Santos, *J. Controlled Release*, 2016, **232**, 113–119.
123. J. Li, Y. Wang, L. Han, X. Sun, H. Yu and Y. Yu, *Clin. Ther.*, 2012, **34**, 2333–2338.

CHAPTER 4

Amyloidosis Inhibition and Detection of Human Islet Amyloid Polypeptide with Nanomaterials

ALEKSANDR KAKINEN*[a] AND NICHOLAS ANDRIKOPOULOS[b]

[a] Australian Institute for Bioengineering and Nanotechnology, The University of Queensland, Brisbane Qld 4072, Australia; [b] Monash Institute of Pharmaceutical Sciences, Monash University, 381 Royal Parade, Parkville, VIC 3052, Australia
*Email: a.kakinen@uq.edu.au

4.1 Introduction

Type 2 diabetes (T2D) is a metabolic disorder impacting over 370 million people worldwide. It is thought to arise from genetic and/or acquired defects in both insulin action at insulin-sensitive tissues and insulin secretion from islet β-cells. T2D is associated with progressive β-cell failure and manifests as a decline in insulin secretion and increasing hyperglycemia. Because of the loss of β-cell function, a significant proportion of patients with T2D eventually require insulin therapy. Several mechanisms may contribute to the progressive β-cell failure in T2D including loss of β-cell mass, β-cell exhaustion and the cytotoxic effects of elevated glucose and lipid levels. A growing body of evidence suggests that islet amyloid deposits may play an important role in the loss of β-cells and the progressive decline in insulin

Nanoscience & Nanotechnology Series No. 59
Nanotechnology for Diabetes Management
Edited by Amar Abderrahmani, Sabine Szunerits, Rabah Boukerroub and Abdelfattah El Ouaamari

Published by the Royal Society of Chemistry, www.rsc.org

secretion characteristic of T2D.[1] Islet amyloid polypeptide (IAPP; amylin) is the major component of islet amyloid. IAPP is co-localized with insulin in the islet β-cells and is co-secreted with insulin in response to β-cell stimulation by both glucose and non-glucose secretagogues.[2] The physiological function(s) of IAPP is still not completely understood but proposed biological actions of IAPP include suppression of food intake, gastric emptying, and arginine-stimulated glucagon secretion from pancreatic α-cells. The presence of IAPP amyloids and plaques in the extracellular space of pancreatic β-cells is a characteristic pathologic feature of the pancreas in T2D (Figure 4.1).[3] Islet amyloid is found in up to 90% in postmortem T2D patients[3] and the degree of amyloid deposition correlates with the severity of the disease.[4] Moreover, islet amyloid formation is associated with reduced β-cell mass in diabetic patients.[5]

4.2 IAPP Amyloidosis

Upon its release from β-cells (pH ∼ 5), IAPP experiences a sudden rise in pH and a consequent deprotonation of its residue His-18, which is hypothesized to be the main driving force for IAPP amyloidosis, alongside its disrupted intracellular equilibrium with insulin,[6] zinc[7,8] and C-peptide.[9] IAPP, a 37-residue intrinsically disordered protein (IDP) with the existence of three amyloidogenic regions (8–20, 20–29, 30–37),[10–12] is characterized mainly by its

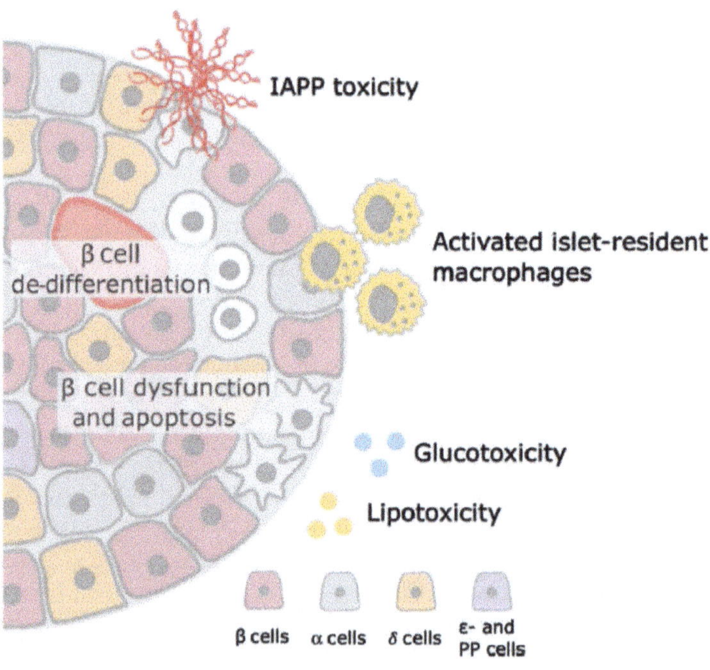

Figure 4.1 Islet amyloid polypeptide (IAPP) toxicity in type 2 diabetes (T2D). Adapted from ref. 18 with permission from Elsevier, Copyright 2020.

hydrophobic (16 residues), slight positively charged (Lys-1, Arg-11, His-18), and aromatic regions (Phe-15, Phe-23, Tyr-37). Moreover, high-resolution NMR structures of IAPP in the presence of SDS micelles at physiological conditions have shown the influence of the amidated IAPP C-terminus on its membrane binding topology.[13] These moieties are the key targets for IAPP amyloidosis inhibition, *via* intermolecular hydrophobic interactions and H-bonding, complemented by π-stacking and electrostatic interactions between the peptide and the nanoparticle inhibitor.[14–16]

Similar to what is known for β-amyloid (Aβ) [Alzheimer's disease (AD)] or α-synuclein (Parkinson's disease), IAPP oligomers are the most toxic species among the aggregation states of the peptide, and β-cells are especially prone to cell membrane damage and the generation of reactive oxygen species (ROS) by IAPP oligomers (see Figure 4.2), contributing to loss of β-cell function and mass in T2D.[17,18]

Notably, Combs *et al.* demonstrated the presence of the amyloid precursor protein (APP) within the pancreas, including pancreatic islets in both mouse and human samples.[19] APP is a single-pass transmembrane protein of uncertain function highly expressed in the CNS and detectable in many other cell types.[20] APP is associated with the production of Aβ, a plaque-forming peptide in AD. Namely, APP is cleaved by proteases such as beta-site amyloid precursor protein cleaving enzyme 1 (BACE1) and gamma secretase to release N-terminal fragments (sAPPα and sAPPβ) and varying size Aβ fragments (Aβ 1–40 and Aβ 1–42). Increased levels of Aβ or Aβ aggregates have also been detected within the pancreas of some transgenic mouse models of AD.[21,22] Whether Aβ itself influences pancreatic IAPP aggregation through cross-seeding or injures islets independently of IAPP is yet to be revealed. However, it has been demonstrated that IAPP and Aβ can accumulate in T2D pancreas and AD brains, and evidence suggests that interaction between the two amyloidogenic proteins can lead to the formation of heterocomplex aggregates with enhanced toxicity in neuronal cells.[23]

4.3 Inhibition of IAPP Amyloidosis

There have been a large number of strategies developed to curb the amyloid aggregation and hence β-cell toxicity elicited by IAPP. The working principle of these approaches is to outcompete the association between IAPP monomers with stronger binding between IAPP and an inhibitor, mediated by H-bonding, hydrophobic, π–π stacking, or electrostatic interactions. Iron oxide nanoparticles (IONPs) decorated with brushed dimethylphosphonate terminated poly(oligo-ethylene glycol) methyl ether acrylate (poly(OEGA)) and poly(methacryloyloxy)ethyl phosphorylcholine (MPC) [poly-(MPC)] moieties (RAFT polymerization utilizing two dimethylphosphonate RAFT agents) elicited non-inhibition and fibril softening in IAPP amyloidosis, respectively, mainly driven by high dipolar interactions between the zwitterionic brushes.[24] Moreover, anionic citrate and cationic branched polyethyleneimine (bPEI)-coated silver NPs (AgNPs) (see Figure 4.3 and Table 4.1) rendered IAPP

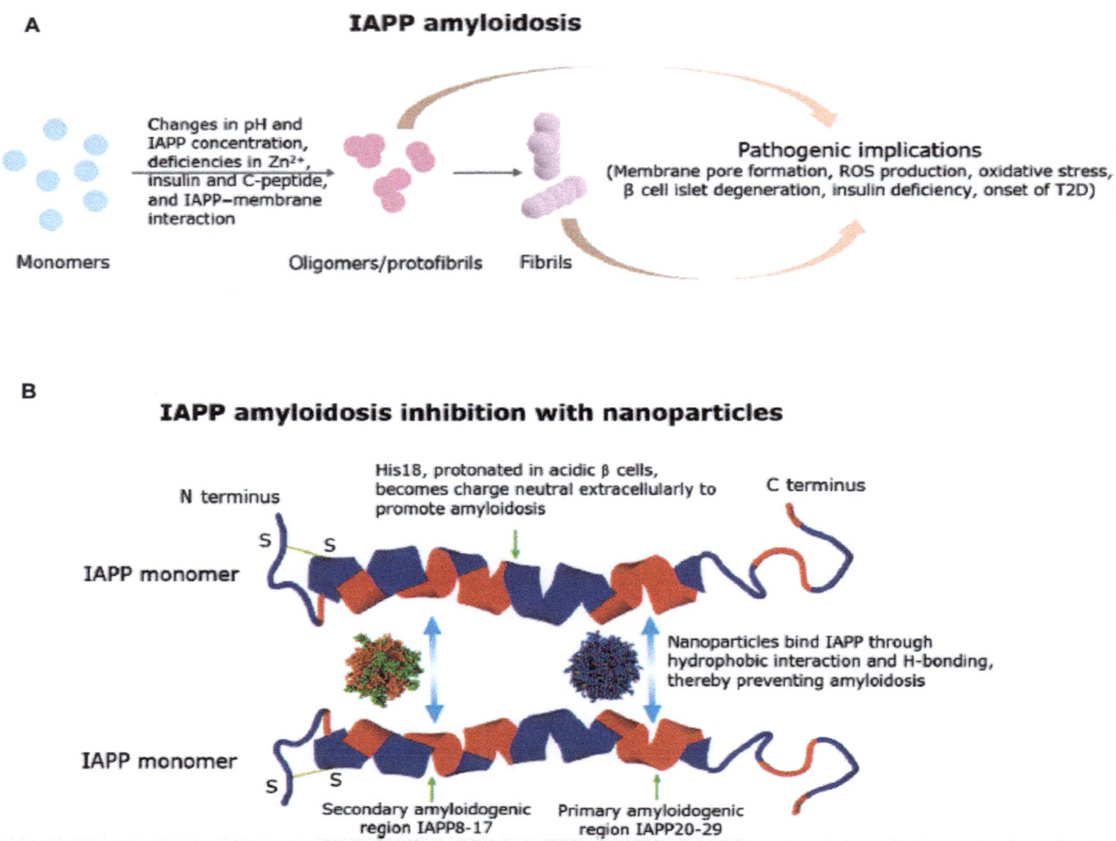

Figure 4.2 Biophysical principle of islet amyloid polypeptide (IAPP) aggregation (amyloidosis), their implications and mitigation with tailor-designed biocompatible nanoparticles. Central to amyloidosis inhibition is the suppression of peptide self-assembly (A and B), triggered by adverse cellular conditions (A, left side), with elevated peptide–nanoparticle interaction (B). ROS, reactive oxygen species; T2D, type 2 diabetes. Reproduced from ref. 18 with permission from Elsevier, Copyright 2020.

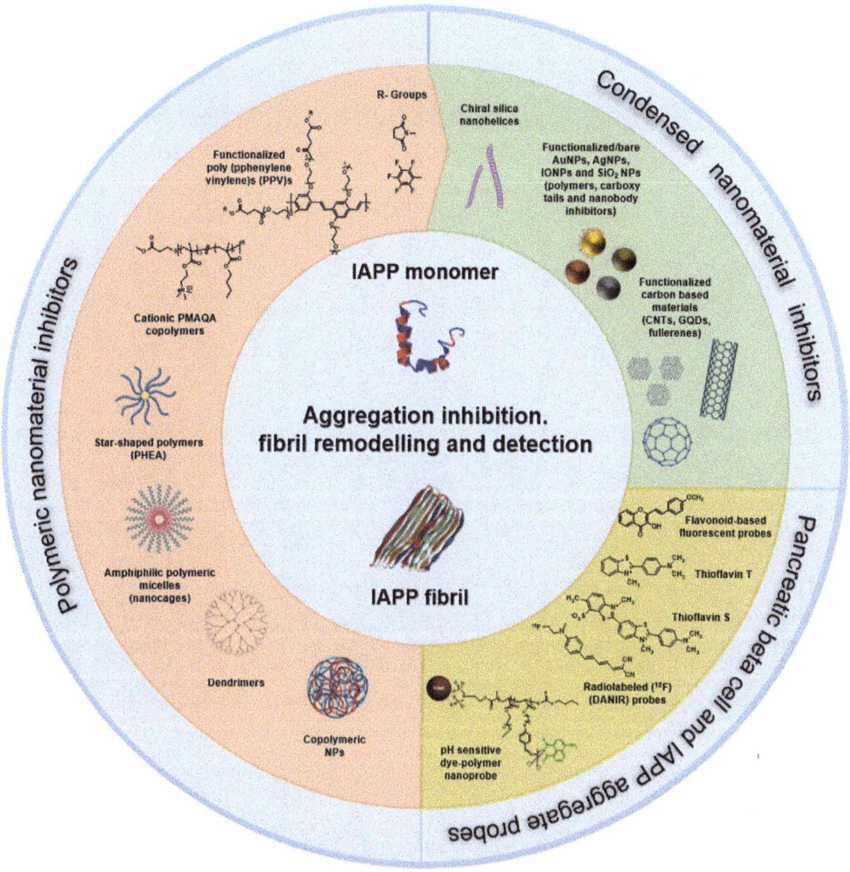

Figure 4.3 Synthesized nanoparticle inhibitors against IAPP amyloidosis and synthesized probes for the detection of IAPP fibrillar aggregates and pancreatic β-cells.

inhibition and fibril shortening, accordingly, mediated by electrostatic interactions between citrate and IAPP as well as hydrophobic interactions and H-bonding between bPEI and IAPP.[24] However, the toxicity of AgNPs limits their applications in amyloidosis.

While highly hydrophilic polymeric moieties are generally considered as ineffective IAPP inhibitors, due to their low ability in establishing hydrophobic interaction with the peptide, PEG-PE micelles can form amphiphilic nanocages against amyloidosis.[25] Apart from the brushed polymeric macromolecular structures, star polymers have found utility in amyloidosis inhibition.[24] Specifically, six-armed cross-linked PHEA [poly(2-hydroxyethylacrylate)] stars, obtained *via* RAFT polymerization using S,S-dibenzyltrithiocarbonate (DBTC) as the RAFT agent, acted as scaffolds for accelerated IAPP fibrillization and cyto-protection. Similarly structured as the star polymers, generation 3 (G3) poly(amidoamine) (PAMAM)

Table 4.1 Nanoparticle inhibitors against IAPP amyloidosis.

Materials		Functional blocks	Synthesis	Effects on IAPP aggregation/ toxicity	Year	Ref.
Polymeric nanomaterials	Co-polymeric NPs	PNiPAM-*b*-PBAM	Free radical polymerization	Inhibition/NA	2008, 2010	28, 57
	Dendrimers	G3 PAMAM-OH	NA (commercially available)	Dose-dependent inhibition/ reduced cytotoxicity	2016	26
	Star polymers	PHEA stars	RAFT followed by bisacrylamide/ AIBN cross-linking	Enhancement/reduced cytotoxicity	2017	24
	Co-polymers	PMAQA	Free radical polymerization	Inhibition/NA	2019	27
	Amphiphilic nanocages	PEG-PE	NA (commercially available)	Inhibition-fibril remodeling/ reduced cytotoxicity	2018	25
	Poly(phenylene vinylene)	PPV-NHS	Heck coupling, substitution	Inhibition-fibril remodeling/ reduced cytotoxicity	2019	29
		PPV-PFP		Enhancement/reduced cytotoxicity		
		PPV-COOH		Enhancement/cytotoxic		
Condensed nanomaterials	Fullerene derivatives	– C_{60} – $C_{60}(OH)_8$	NA	*In silico* inhibition and protofibril disassembly/NA	2020	31
	Silica NPs (SiO_2 NPs)	L/R-SiO_2	Templated polycondensation	Fibril remodeling by R-SiO_2/ reduced cytotoxicity by R-SiO_2 *in vitro* and *in vivo*	2018	34
	Silver NPs (AgNPs)	c-AgNPs	NA (commercially available)	Inhibition/cytotoxic	2017	58
		bPEI-AgNPs		Fibril shortening/reduced cytotoxicity		
	Iron oxide NPs (IONPs)	Bare IONPs bPEG-IONPs	Coprecipitation, sorption mediated coating of AgNPs,	No inhibition/reduced cytotoxicity	2017	58
		bPC-IONPs	RAFT polymerization grafting of bPEG and bPC onto IONPs	Fibril softening/reduced cytotoxicity		
	Graphene quantum dots (GQDs)	FGQDs	Microwave-assisted hydrothermal synthesis from glucose	Inhibition/reduced cytotoxicity	2017	30
	Single-walled carbon nanotubes (SWCNTs)	SWCNT-OH	NA (commercially available)	Inhibition and *in silico* protofibril remodeling/NA	2018	32
	Nanobody-AuNPs	M1-AuNP	*E.coli* expression, ligand exchange	Inhibition/reduced cytotoxicity	2019	36

dendrimers with an ethylenediamine core and amidoethanol surface groups impaired the aggregation of IAPP monomers and suppressed IAPP toxicity *in vivo* and *ex vivo*, owing to the hydrophobic interactions between the dendrimer branches and the amyloidogenic regions of the peptide.[26] On the other hand, the functionalities incorporated in co-polymeric chains can provide binding insights regarding IAPP's amyloidogenic regions. Specifically, hydrophobic butylmethacrylate and cationic methacroylcholine chloride moieties in PMAQA co-polymers inhibited IAPP aggregation by their strong binding with the peptide amyloidogenic core (NFGAIL) through hydrogen bonding. The repulsive force between the cationic PMAQA and IAPP, combined with the enhanced aggregation of the anionic Aβ by PMAQA, indicated the significance of coulombic interactions in amyloidosis inhibition.[27] Furthermore, the ability of amphiphilic co-polymers (see Figure 4.3 and Table 4.1) to adopt nanoparticle shapes through free radical polymerization in the presence of anionic surfactants further enhanced the significance of H-bonding over hydrophobic interactions.[28] Specifically, the higher hydrophobic content in a NiPAM:BAM copolymeric system reduced the number of NH residues available for H-bonding with IAPP, leading to a less efficient inhibitory effect. An ester condensation reaction between carboxy-terminated PEGylated poly(*p*-phenylene vinylene) derivatives (PPV-COOH) and *N*-hydroxysuccinimide (NHS) or pentafluorophenol (PFP) revealed that active ester groups of the produced polymers could form covalent bonds with the amino groups of IAPP to affect the peptide aggregation.[29] This specific inhibitory effect has further promoted the use of perfluorinated compounds as potent inhibitors due to their unique hydrophobicity and lipophilicity. Concerning the fluorinated functionalized nanomaterials (see Figure 4.3 and Table 4.1), microwave hydrothermally synthesized fluorine functionalized graphene quantum dots (FGQDs)[30] displayed an excellent IAPP inhibitory effect, due to the positive mesomeric and negative inductive fluorine effect, which further enhanced H-bonding and electrostatic interactions, accompanied by increased π stacking with IAPP.

Recently, carbon-based nanomaterials have been applied to the field of amyloidosis inhibition, as these nanomaterials can cross biomembranes and their aromatic moieties provide enhanced π stacking with peptides and their fibrillary derivatives. An *in silico* study (see Table 4.1) showed that fullerenes C_{60} and $C_{60}(OH)_8$ both prevented the β-hairpin conformation of IAPP, by weakening inter- and intrapeptide interactions of IAPP dimers.[31] C_{60} preferentially bound to the hydrophobic and aromatic peptide residues and $C_{60}(OH)_8$ had a high propensity for binding with hydrophilic and aromatic residues. Hydroxylated single-walled carbon nanotubes (SWCNT-OH)[32] blocked crucial IAPP peptide–peptide interactions (aromatic F15 and Y37, polar N3, Q10, H18, N31, N35, and charged R11), thus preventing β-sheet and β-hairpin precursor conformations.

In pharmaceutical sciences, the efficacy of a small-molecule drug candidate largely depends on its enantiomeric form, as it can bind differently to the targeted receptors.[33] However, stereoselectivity has not been widely exploited in amyloidosis mitigation. Left- and right-handed (L/R) SiO_2 nanohelices (see Figure 4.3 and Table 4.1),[34] synthesized through sol–gel polycondensation in the presence of self-assembled dicationic surfactants with chiral counterion (L or D tartrate), strongly influenced IAPP fibrillization and toxicity *in vitro* and *in vivo*. Due to steric effects, the R-SiO_2 possessed more binding and nucleation sites per surface area for the primarily left-handed IAPP fibrils, leading to IAPP fibril remodeling and toxicity reduction. In addition, antibodies grafted with specific amyloid peptide segments exploiting the complementarity-determining region (CDR3) have shown effective peptide recognition and suppression of amyloid toxicity mainly through homotypic interactions.[35] AuNPs functionalized with antibody-$IAPP_{22-31}$ nanobodies through citrate ligand exchange led to complete IAPP inhibition and toxicity suppression compared with individual nanobodies.[36]

Since a growing body of knowledge suggests that Aβ is also associated with amyloidosis in T2D, the development of anti-Aβ agents is another important research area central to the development of nanomedicines against T2D. Similar to IAPP, a rational design of nanocomposites against Aβ amyloidosis typically aims at stabilizing the Aβ monomeric state, reducing the oligomeric population, retarding Aβ fibrillization, remodeling Aβ fibrils, or accelerating Aβ elimination *via* H-bonding, hydrophobic, and electrostatic interactions with the peptide.[37,38] However, most studies are focused on eliminating Aβ pathologies associated with AD utilizing *in vitro* and *in vivo* AD models and not T2D.[38–41]

4.4 Nanoprobes for the Detection and Imaging of IAPP Amyloid Deposits

In addition to the mitigation efforts described above, sensing and imaging of IAPP amyloid deposit proteins is another important research area, not as thoroughly discovered as for Aβ plaques and other neurofibrillary tangles, central to the development of anti-T2D theranostics. While amyloid aggregates may be traced with conventional cross-β amyloid histological/non-histological stains such as Thioflavin T (ThT),[42] Thioflavin S (ThS),[43] and Congo Red (CR),[44] there remains much room for the development of new syntheses to exploit the rich physicochemical properties and functionalities entailed by nanomaterials for amyloid inhibition and fluorogenic probing (see Table 4.2).

Watanabe *et al.*[45] constructed malononitrile end group donor-π-acceptor DANIR-OH fluorescent probes for IAPP fibril detection in the pancreas,[46] and as PET imaging agents by functionalizing DANIRs with a radiolabeled [^{18}F] end group. Moreover, dye-conjugated PEG-*b*-PFPHA-coated IONPs[47] enabled multimodal *in vitro*/*ex vivo* detection of IAPP (magnetic resonance imaging/MRI, fluorescence) (see Figure 4.4a). Pansieri *et al.*[48] synthesized two types of multimodal gadolinium (Gd)-based nanoprobes either decorated with a small molecule (Pittsburgh compound B) (AGuIX@PEG@PIB) or with a chimeric protein made of B10 nanobody and *Escherichia coli* alkaline phosphatase (AP) (AGuIX@B10AP@Cy5.5) (see Figure 4.4c). Both functionalized nanoparticles showed a potent specific interaction with IAPP,

Table 4.2 Probes of IAPP amyloidosis.

Materials		Functional groups	Methodology/application	λ_{em} (nm)	Year	Ref.
Radiolabeled ^{18}F molecules	Radiolabeled (^{18}F) (DANIR) probes	– (*E*)-2-(3-(4-((2-fluoroethyl)(methyl)-amino)phenyl)-allylidene)malononitrile (DANIR-F-2b) – 2-((2*E*,4*E*)-5-(4-((2 Fluoroethyl)(methyl)-amino)phenyl)-penta-2,4-dien-1-ylidene)malononitrile (8:DANIR-F 2c)	PET imaging/T2DM patient and healthy pancreatic sections and Tg mice	—	2020	45
Multimodal nanoprobes	Dye-polymer functionalized IONPs	– Iron oxide nanoparticles (IONPs) – PEG-*b*-PFPA (polyethylene glycol methyl ether acrylate) block (poly 6-(4-formylphenoxy) acrylate) – PEG-*b*-PFPHA (polyethylene glycol methyl ether acrylate) block (poly 6-(4-formylphenoxy) hexyl acrylate) – NaP-NH_2 dye	Multimodal sensing (pH, fluorescence)/cell lines (BTC-6, CHO, SH-SY5Y, HEK, and Raw 264.7) and mouse islets (C57BL)	540	2020	47
	Pittsburgh compound B or nanobody (B10AP) functionalized gadolinium-based NPs	AGuIX@PEG@PIB – Gd-based NPs – DOTAGA (Gd^{3+}) chelates grafted into a polysiloxane network – (APTES+TEOS) – PEG – Pittsburgh compound B	Specific interaction with amyloid deposits ($A\beta_{1-42}$, $IAPP_{1-37}$ and Val30MetTTR fibrils) localized in pathological tissue. K_D values: 10 μM range (AGuIX@PIB), 10 nM (AGuIX@Cya5.5@B10AP).	405	2017	59
		AGuIX@B10AP@Cy5.5 – Gd-based NPs – DOTAGA (Gd^{3+}) chelates grafted into a polysiloxane network (APTES+TEOS) – Cyanine 5.5 (Cy5.5) – B10AP chimeric protein (B10 nanobody and *E. coli* alkaline phosphatase (AP)		670/613		
Small molecules	Flavonoid-based fluorescent probes	– 3-HF-4′-OMe – 3-HF-ene-4′-OMe – 3-HF-diene-4′-Ome	Fluorescence sensing/MIN6 cells	535	2020	49

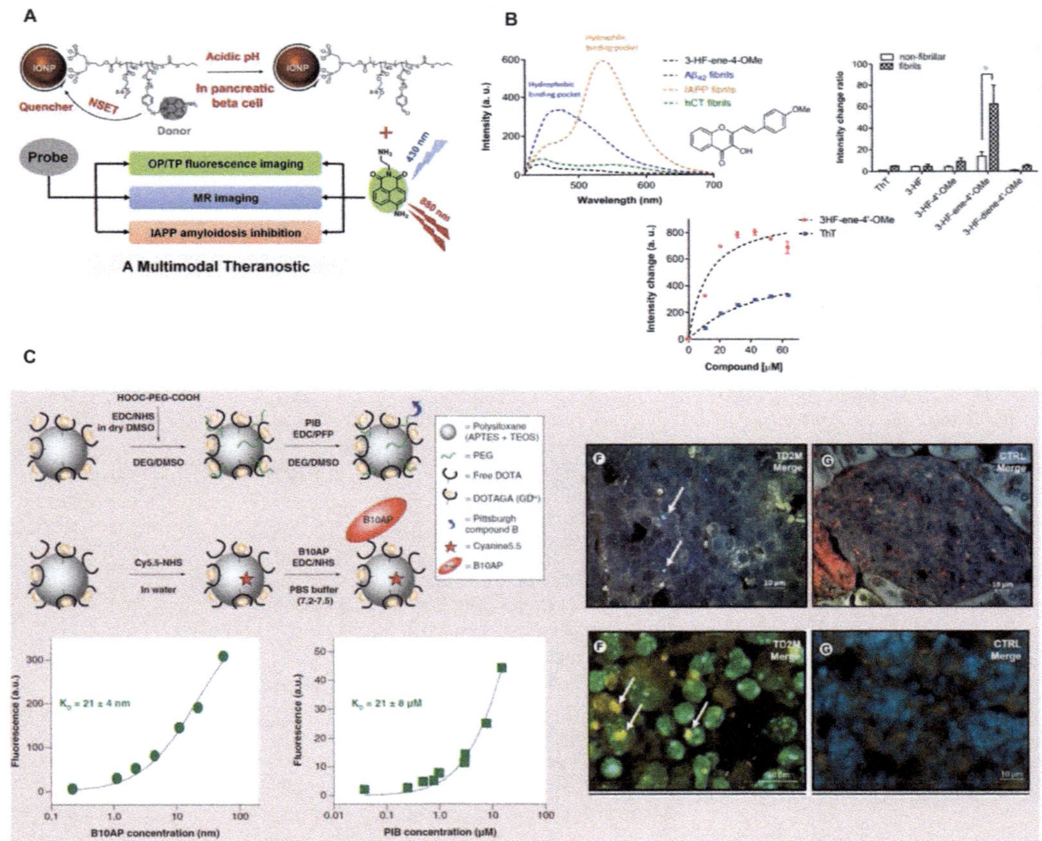

Figure 4.4 Recent advances on strategies for constructing IAPP amyloid nanoprobes. (a) Novel multimodal theragnostic system for imaging and inhibition of β-cell degeneration by IAPP amyloidosis. Reproduced from ref. 47 with permission from American Chemical Society, Copyright 2020. (b) 3-Hydroxyflavone stilbene derivatives (3-HF-ene-4′-OMe most sensitive dye) for distinguishing non-fibrillar and fibrillar IAPP and Aβ. Reproduced from ref. 49 with permission from American Chemical Society, Copyright 2020. (c) Multimodal imaging Gd-nanoparticles functionalized with Pittsburgh compound B or a nanobody for amyloid plaques targeting. Reproduced from ref. 48 with permission from Future Medicine LTD, Copyright 2017.

amyloid-β, and Val30MetTTR amyloid deposits localized in pathological tissues representing a useful tool for the early diagnosis of T2D, Alzheimer's disease, and the familial amyloidotic polyneuropathy. In addition to the use of flavonoid derivatives for amyloid inhibition, Chen *et al.*[49] designed a kinetic amyloid probe consisting of 3-hydroxyflavone (3-HF) and stilb(ene)$_n$ ($n = 0$–2) moieties. The exceptional efficiency of 3-HF-ene-4′-OMe for tracking IAPP fibrils in inhibitory conditions (*e.g.*, with resveratrol) and in β-cells appeared promising for the design of an *in vivo* IAPP probe (see Figure 4.4b).

It is worth noting that the development of IAPP detection methods is also important for other diseases such as AD. There is evidence that T2D, obesity, and other prediabetic states of insulin resistance could promote AD development.[50] IAPP deposition has been also described in the brains of AD patients.[51] The IAPP deposits detected in brain blood vessels and the parenchyma of subjects with AD, with or without T2D, support the hypothesis that IAPP oligomers could circulate and reach different organs. Notably, it was recently reported that IAPP accelerates Aβ aggregation and brain inoculation of pancreatic IAPP aggregates into AD transgenic mice and resulted in a more severe AD phenotype, indicating that IAPP might influence Aβ aggregation.[52] With the emergence of novel methods for the detection of protein aggregates in biological fluids, the level of circulating IAPP aggregates could be used, in the future, as a biomarker for AD.[53]

4.5 Summary and Outlook

Much progress has been made in IAPP amyloidosis inhibition and detection over the past decade, yet many challenges remain as the majority of the literature in this emerging field has been dealing with simple *in vitro* IAPP-nanoparticle systems without the validation of *in vivo* data. Accordingly, a range of factors pertinent to the T2D amyloid pathology has to be considered for the design of nanoparticles to develop into viable anti-IAPP amyloidosis nanomedicines and sensors (see Figure 4.5).

Although it is non-trivial to predict the exact outcome resulting from IAPP-nanoparticle binding, as the current design of nanoparticle inhibitors is still guided by semiempirical principles and performed through trial and error, it is clear that such inhibitors should possess the central ability of initiating strong interactions with the IAPP, typically *via* H-bonding, hydrophobic, and electrostatic interactions with the amyloidogenic regions and C-terminus of the IAPP. The nanoparticles as well as their surface functionalities should also be biocompatible and, preferably, biodegradable. Moreover, the nanoparticle inhibitors and probes should reach the pancreas, and hence should possess the capacity of systemic circulation without significant post-synthesis modifications by the protein corona[54–56] to trigger opsonization.

Lastly, the design of nanoparticle inhibitors and, especially, probes, should acquire a certain degree of responsiveness to *in vivo* conditions as biological systems are extremely sophisticated in terms of their anatomy such as low pH value of ~5 in the pancreatic β-cells.

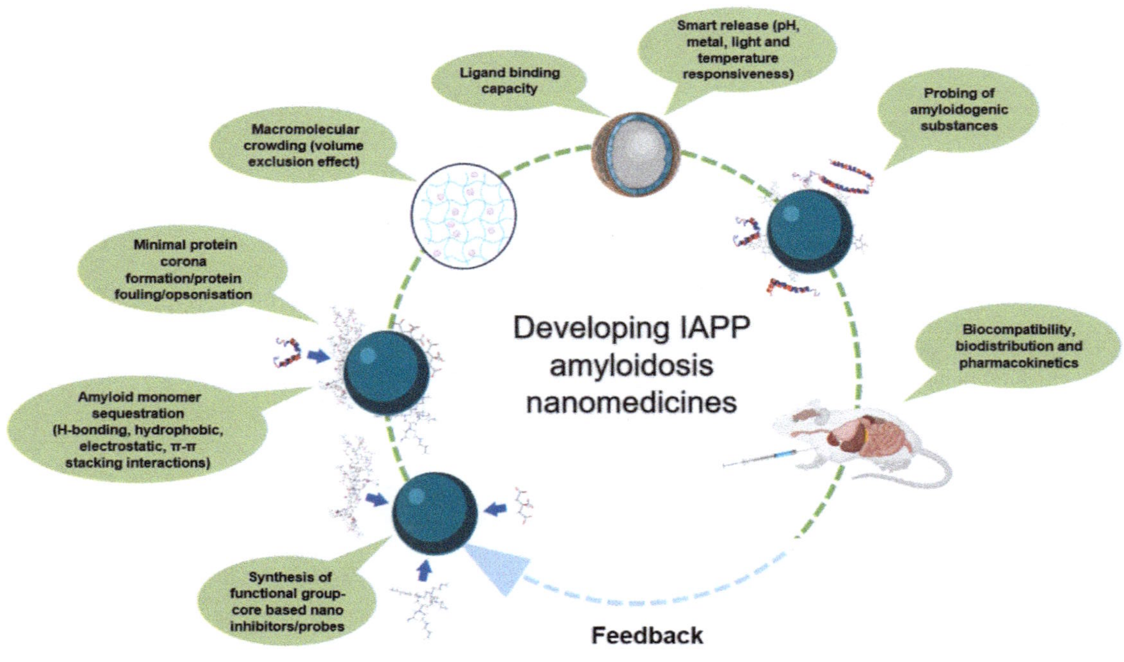

Figure 4.5 Considerations for the development of IAPP amyloidosis nanomedicines.

References

1. L. Marzban, K. Park and C. B. Verchere, *Exp. Gerontol.*, 2003, **38**, 347–351.
2. S. E. Kahn, D. A. D'Alessio, M. W. Schwartz, W. Y. Fujimoto, J. W. Ensinck, G. J. Taborsky, Jr and D. Porte, Jr, *Diabetes*, 1990, **39**, 634–638.
3. S. E. Kahn, S. Andrikopoulos and C. B. Verchere, *Diabetes*, 1999, **48**, 241–253.
4. P. Westermark, *Amyloid*, 1994, **1**, 47–60.
5. A. Clark, C. A. Wells, I. D. Buley, J. K. Cruickshank, R. I. Vanhegan, D. R. Matthews, G. J. Cooper, R. R. Holman and R. C. Turner, *Diabetes Res.*, 1988, **9**, 151–159.
6. J. R. Brender, E. L. Lee, K. Hartman, P. T. Wong, A. Ramamoorthy, D. G. Steel and A. Gafni, *Biophys. J.*, 2011, **100**, 685–692.
7. J. R. Brender, K. Hartman, R. P. R. Nanga, N. Popovych, R. de la Salud Bea, S. Vivekanandan, E. N. G. Marsh and A. Ramamoorthy, *J. Am. Chem. Soc.*, 2010, **132**, 8973–8983.
8. S. Salamekh, J. R. Brender, S. J. Hyung, R. P. Nanga, S. Vivekanandan, B. T. Ruotolo and A. Ramamoorthy, *J. Mol. Biol.*, 2011, **410**, 294–306.
9. X. Ge, A. Kakinen, E. N. Gurzov, W. Yang, L. Pang, E. H. Pilkington, P. Govindan-Nedumpully, P. Chen, F. Separovic, T. P. Davis, P. C. Ke and F. Ding, *Chem. Commun.*, 2017, **53**, 9394–9397.
10. E. T. A. S. Jaikaran, C. E. Higham, L. C. Serpell, J. Zurdo, M. Gross, A. Clark and P. E. Fraser, *J. Mol. Biol.*, 2001, **308**, 515–525.
11. M. R. Nilsson and D. P. Raleigh, *J. Mol. Biol.*, 1999, **294**, 1375–1385.
12. P. Westermark, U. Engström, K. H. Johnson, G. T. Westermark and C. Betsholtz, *Proc. Natl. Acad. Sci. U. S. A.*, 1990, **87**, 5036–5040.
13. R. P. Nanga, J. R. Brender, S. Vivekanandan and A. Ramamoorthy, *Biochem. Biophys. Acta*, 2011, **1808**, 2337–2342.
14. E. A. Mirecka, S. Feuerstein, L. Gremer, G. F. Schröder, M. Stoldt, D. Willbold and W. Hoyer, *Sci. Rep.*, 2016, **6**, 33474.
15. S. B. Padrick and A. D. Miranker, *J. Mol. Biol.*, 2001, **308**, 783–794.
16. P. Marek, A. Abedini, B. Song, M. Kanungo, M. E. Johnson, R. Gupta, W. Zaman, S. S. Wong and D. P. Raleigh, *Biochemistry*, 2007, **46**, 3255–3261.
17. L. Haataja, T. Gurlo, C. J. Huang and P. C. Butler, *Endocr. Rev.*, 2008, **29**, 303–316.
18. E. N. Gurzov, P. C. Ke, U. Ahlgren, R. S. Garcia Ribeiro and M. Gotthardt, *Trends Endocrinol. Metab.*, 2020, **31**, 905–917.
19. J. A. Kulas, K. L. Puig and C. K. Combs, *J. Endocrinol.*, 2017, **235**, 49–67.
20. H. Arai, V. M. Lee, M. L. Messinger, B. D. Greenberg, D. E. Lowery and J. Q. Trojanowski, *Ann. Neurol.*, 1991, **30**, 686–693.
21. J. Wegiel, H. M. Wisniewski, M. Muzylak, M. Tarnawski, E. Badmajew, J. Nowakowski, K. C. Wang, M. Shoji, C. Mondadori and A. Giovanni, *Amyloid*, 2000, **7**, 95–104.
22. M. Vandal, P. J. White, G. Chevrier, C. Tremblay, I. St-Amour, E. Planel, A. Marette and F. Calon, *FASEB J.*, 2015, **29**, 4273–4284.
23. P. Bharadwaj, T. Solomon, B. R. Sahoo, K. Ignasiak, S. Gaskin, J. Rowles, G. Verdile, M. J. Howard, C. S. Bond, A. Ramamoorthy, R. N. Martins and P. Newsholme, *Sci. Rep.*, 2020, **10**, 10356.

24. E. H. Pilkington, M. Lai, X. Ge, W. J. Stanley, B. Wang, M. Wang, A. Kakinen, M.-A. Sani, M. R. Whittaker, E. N. Gurzov, F. Ding, J. F. Quinn, T. P. Davis and P. C. Ke, *Biomacromolecules*, 2017, **18**, 4249–4260.
25. X. Fang, M. Yousaf, Q. Huang, Y. Yang and C. Wang, *Sci. Rep.*, 2018, **8**, 4463.
26. E. N. Gurzov, B. Wang, E. H. Pilkington, P. Chen, A. Kakinen, W. J. Stanley, S. A. Litwak, E. G. Hanssen, T. P. Davis, F. Ding and P. C. Ke, *Small*, 2016, **12**, 1615–1626.
27. B. R. Sahoo, T. Genjo, T. W. Nakayama, A. K. Stoddard, T. Ando, K. Yasuhara, C. A. Fierke and A. Ramamoorthy, *Chem. Sci.*, 2019, **10**, 3976–3986.
28. C. Cabaleiro-Lago, I. Lynch, K. A. Dawson and S. Linse, *Langmuir*, 2010, **26**, 3453–3461.
29. H. Sun, F. Lv, L. Liu and S. Wang, *ACS Appl. Mater. Interfaces*, 2019, **11**, 22973–22978.
30. M. Yousaf, H. Huang, P. Li, C. Wang and Y. Yang, *ACS Chem. Neurosci.*, 2017, **8**, 1368–1377.
31. C. Bai, Z. Lao, Y. Chen, Y. Tang and G. Wei, *Front. Chem.*, 2020, **8**, 51.
32. Y. Mo, S. Brahmachari, J. Lei, S. Gilead, Y. Tang, E. Gazit and G. Wei, *ACS Chem. Neurosci.*, 2018, **9**, 2741–2752.
33. W. H. Brooks, W. C. Guida and K. G. Daniel, *Curr. Top. Med. Chem.*, 2011, **11**, 760–770.
34. A. Faridi, Y. Sun, Y. Okazaki, G. Peng, J. Gao, A. Kakinen, P. Faridi, M. Zhao, I. Javed, A. W. Purcell, T. P. Davis, S. Lin, R. Oda, F. Ding and P. C. Ke, *Small*, 2018, **14**, 1802825.
35. A. R. A. Ladiwala, M. Bhattacharya, J. M. Perchiacca, P. Cao, D. P. Raleigh, A. Abedini, A. M. Schmidt, J. Varkey, R. Langen and P. M. Tessier, *Proc. Natl. Acad. Sci. U. S. A.*, 2012, **109**, 19965–19970.
36. Y. Zhao, J. Cai, Z. Liu, Y. Li, C. Zheng, Y. Zheng, Q. Chen, H. Chen, F. Ma, Y. An, L. Xiao, C. Jiang, L. Shi, C. Kang and Y. Liu, *Nano Lett.*, 2019, **19**, 674–683.
37. F. Huang, J. Wang, A. Qu, L. Shen, J. Liu, J. Liu, Z. Zhang, Y. An and L. Shi, *Angew. Chem., Int. Ed.*, 2014, **53**, 8985–8990.
38. Z. Wang, X. Dong and Y. Sun, *Langmuir*, 2018, **34**, 14419–14427.
39. H. Skaat, R. Chen, I. Grinberg and S. Margel, *Biomacromolecules*, 2012, **13**, 2662–2670.
40. M.-H. Melchor, F.-G. Susana, G.-S. Francisco, B. Hiram I., R.-F. Norma, L.-R. Jorge A., L.-C. Perla Y. and B.-I. Gustavo, *RSC Adv.*, 2018, **8**, 39667–39677.
41. J. Wang, K. Wang, Z. Zhu, Y. He, C. Zhang, Z. Guo and X. Wang, *RSC Adv.*, 2019, **9**, 14126–14131.
42. R. Khurana, C. Coleman, C. Ionescu-Zanetti, S. A. Carter, V. Krishna, R. K. Grover, R. Roy and S. Singh, *J. Struct. Biol.*, 2005, **151**, 229–238.
43. H. B. Rajamohamedsait and E. M. Sigurdsson, in *Amyloid Proteins: Methods and Protocols*, ed. E. M. Sigurdsson, M. Calero and M. Gasset, Humana Press, Totowa, NJ, 2012, pp. 411–424.

44. H. Puchtler, F. Sweat and M. Levine, *J. Histochem. Cytochem.*, 1962, **10**, 355–364.
45. H. Watanabe, K. Kawano, Y. Shimizu, S. Iikuni, Y. Nakamoto, K. Togashi and M. Ono, *Mol. Pharmaceutics*, 2020, **17**, 1293–1299.
46. H. Watanabe, Y. Miki, Y. Shimizu, H. Saji and M. Ono, *Dyes Pigm.*, 2019, **170**, 107615.
47. F. Xin, Y. Li, C. Fu, I. Javed, X. Huang, A. Schaschkow, R. S. Garcia Ribeiro, E. N. Gurzov, T. P. Davis, X. Zhang, P. C. Ke and R. Qiao, *Chem. Mater.*, 2020, **32**, 1080–1088.
48. J. Pansieri, M. Plissonneau, N. Stransky-Heilkron, M. Dumoulin, L. Heinrich-Balard, P. Rivory, J.-F. Morfin, E. Toth, M. J. Saraiva, E. Allémann, O. Tillement, V. Forge, F. Lux and C. Marquette, *Nanomedicine*, 2017, **12**, 1675–1687.
49. W.-L. Chen, S.-T. Ma, Y.-W. Chen, Y.-C. Chao, A.-C. Chan, L.-H. Tu and W.-M. Liu, *Biochemistry*, 2020, **59**, 1482–1492.
50. K. Gudala, D. Bansal, F. Schifano and A. Bhansali, *J. Diabetes Investig.*, 2013, **4**, 640–650.
51. J. N. Fawver, Y. Ghiwot, C. Koola, W. Carrera, J. Rodriguez-Rivera, C. Hernandez, K. T. Dineley, Y. Kong, J. Li, J. Jhamandas, G. Perry and I. V. Murray, *Curr. Alzheimer Res.*, 2014, **11**, 928–940.
52. I. Moreno-Gonzalez, G. Edwards Iii, N. Salvadores, M. Shahnawaz, R. Diaz-Espinoza and C. Soto, *Mol. Psychiatry*, 2017, **22**, 1327–1334.
53. I. Martinez-Valbuena, R. Valenti-Azcarate, I. Amat-Villegas, M. Riverol, I. Marcilla, C. E. de Andrea, J. A. Sánchez-Arias, M. del Mar Carmona-Abellan, G. Marti, M.-E. Erro, E. Martínez-Vila, M.-T. Tuñon and M.-R. Luquin, *Ann. Neurol.*, 2019, **86**, 539–551.
54. T. Cedervall, I. Lynch, S. Lindman, T. Berggård, E. Thulin, H. Nilsson, K. A. Dawson and S. Linse, *Proc. Natl. Acad. Sci. U. S. A.*, 2007, **104**, 2050–2055.
55. R. Cai and C. Chen, *Adv. Mater.*, 2019, **31**, 1805740.
56. P. C. Ke, S. Lin, W. J. Parak, T. P. Davis and F. Caruso, *ACS Nano*, 2017, **11**, 11773–11776.
57. C. Cabaleiro-Lago, F. Quinlan-Pluck, I. Lynch, S. Lindman, A. M. Minogue, E. Thulin, D. M. Walsh, K. A. Dawson and S. Linse, *J. Am. Chem. Soc.*, 2008, **130**, 15437–15443.
58. M. Wang, A. Kakinen, E. H. Pilkington, T. P. Davis and P. C. Ke, *Biomater. Sci.*, 2017, **5**, 485–493.
59. J. Pansieri, M. Plissonneau, N. Stransky-Heilkron, M. Dumoulin, L. Heinrich-Balard, P. Rivory, J. F. Morfin, E. Toth, M. J. Saraiva, E. Allémann, O. Tillement, V. Forge, F. Lux and C. Marquette, *Nanomedicine*, 2017, **12**, 1675–1687.

CHAPTER 5

Phytochemicals in the Management of Diabetes

Q. ABBAS, S. M. AL-THAWADI* AND S. PERNA*

University of Bahrain, Department of Biology, College of Science,
Sakhir Campus P. O. Box 32038, Bahrain
*Email: salthawadi@uob.edu.bh

5.1 Introduction

Diabetes mellitus is defined by chronic hyperglycemia. High blood sugar can cause macroangiopathy and microangiopathy, which can lead to chronic comorbidities such as diabetic foot, diabetic neuropathy, diabetic nephropathy, diabetic retinopathy, and other health problems. As per the International Diabetes Federation, approximately 537 million people worldwide have diabetes.[1]

Type 1 diabetes mellitus (T1DM) and type 2 diabetes mellitus (T2DM) together constitute the most diagnosed prevalent form of diabetes. In T1DM, hyperglycemia develops as the consequence of the drastic insulinopenia in the bloodstream caused by autoimmune β-cell destruction. Insulin replacement by insulin recombinant or islet/pancreas transplantation is the only means of survival for patients with T1DM. T2DM represents 90% of all diabetes forms. Obesity is the first T2DM risk factor. In obesity-associated T2DM, expansion of visceral adiposity is associated with the development of insulin resistance in the skeletal muscles, white adipose tissue, and liver. Systematically, insulin resistance is compensated by an increase of insulin production, thereby achieving normoglycemia. However, for many obese

Nanoscience & Nanotechnology Series No. 59
Nanotechnology for Diabetes Management
Edited by Amar Abderrahmani, Sabine Szunerits, Rabah Boukerroub
and Abdelfattah El Ouaamari

Published by the Royal Society of Chemistry, www.rsc.org

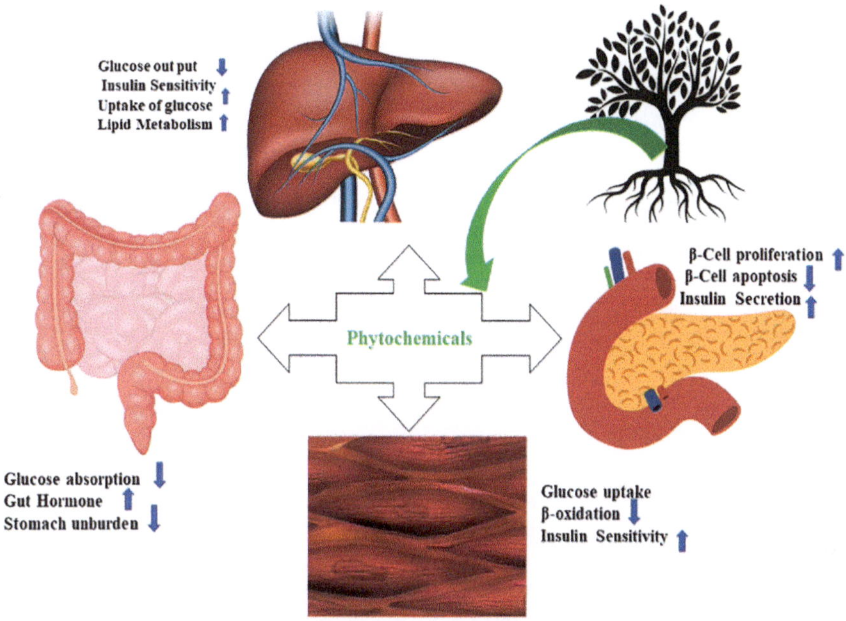

Figure 5.1 General mechanism of phytochemicals for the management of diabetes.

people, blood-circulating insulin is insufficient to counteract insulin resistance. The loss of β-cell function and mass accounts for the impaired insulin production, and therefore for the development of diabetes (Figure 5.1).[2] This chapter deals with plant-based natural substances, also referred to as ethnomedicine, as potential antidiabetic medicines (Table 5.1). Some substances from ethnomedicine have been used for centuries to treat a variety of ailments, of which some would be beneficial for treating insulin resistance, β-cell failure, and diabetes complications.

5.2 Anti-diabetic Effects of Phytochemicals on Humans (Clinical Trials)

Lagerstroemia speciosa is known by the common name "banaba." The primary active ingredients of banaba are corosolic acid (Scheme 5.1) and ellagitannins (Scheme 5.2), both of which have been shown to have significant anti-diabetic properties. In a clinical study, it was discovered that corosolic acid and the aqueous extract of banaba leaves could alleviate the symptoms associated with the metabolic syndrome through a variety of mechanisms, including decreased gluconeogenesis, impaired hydrolysis of starches and sucrose, and increased cellular uptake of glucose.[3]

In nature, catechin is a phenolic molecule found in tea, cocoa, and palms, where it is frequently conjugated with gallic acid to form an antioxidant (Scheme 5.3). It enhances glucose uptake, but catechin glycosides suppress active glucose transit in a different way. By activating transporters,

Table 5.1 The use of herbal active components and their mechanisms in diabetic management.

Herb	Active components	Mechanism	Sample	Dose	References	Level of evidence
Bitter kola (*Garcinia kola*)	Anti-diabetic and anti-hyperlipidemic activities	Against inflammation and reactive oxygen species (ROS) of free radical on the pancreatic beta cells	*In vivo*		72	**6**
Salacia reticulata var. β-diandra Weight. *Salacia reticulate* var. b-diandra Weight (Family, Celestraceae)-SHRUB commonly known as kotala himbatu or himbutu-wel		A reduction of glycated hemoglobin (HbA1C), fasting blood glucose concentration, fructosamine, and an increase in the concentration of serum insulin. A reduction in fasting blood glucose concentration in T2DM patients. Intestinal α-glucosidase inhibitory activity. Inhibition of α-glucosidase enzyme delays the glucose absorption into blood and suppresses postprandial hyperglycemia, resulting in improvement of the blood glucose level.	*In vivo* rat	175 mg kg^{-1}	73 74 75 76	**1, 6**
Trichosanthes cucumerina (L.). *Trichosanthes cucumerina* (L.) (Family, Cucurbitaceae)	Phenolic compounds (flavonoids and tannins) A number of antidiabetic/antioxidant compounds (carotenoids, flavonols, and flavanones)	Induction of β-cells, direct increment of peripheral glucose utilization, and lack of intestinal glucose absorption	*In vivo* rat		77 78	**6**
Gymnema lactiferum, a plant also called kuringnan	Gymnemic acids (gymnemic acids I, II, III, and IV),	Triggering insulin activity and inhibiting intestinal glucose Absorption, as suggested by inhibition of α-amylase and α-glucosidase	*In vivo* rat, a clinical trial	1.25 g kg^{-1}	73 79 80 81	**1, 6**

Curcumin	Polyphenolic compound, syringic acid	A neuroprotective effect and significantly increased the activity of SOD, CAT, GPx, and GSH. Decreased the plasma levels of the oxidative biomarker thiobarbituric acid reactive substances (TBARS) increased the activity of paraoxonase-1 (PON1) in the kidneys and liver increased the levels of advanced glycation end product (AGE) and detoxification system components (AGE-R1 receptor and glyoxalase-1). Bisdemethoxycurcumin significantly inhibits the α-glucosidase	*In vivo* rat	82 83 84 85	6
Leaves of *Ficus deltoidea*	Flavonoids	Increased SOD, CAT, GPx, and GSH levels and reduced MDA levels in the pancreas and livers and cervical cerebral cortex Suppressed the increase in MDA content in the brains of STZ-induced diabetic rats, and improved learning and memory function.	*In vivo* rat		6
The ancient Chinese herb *Coptis chinensis*	Isoquinoline alkaloid	Increased hepatic SOD1 mRNA expression and kidney SOD and CAT activity to normal levels and decreased MDA levels in the liver and brain. Reduce the plasma levels of MDA and TBARS and increased plasma SOD activity. Attenuated memory impairment, axonopathy, and tau hyperphosphorylation.	*In vivo* rat	86 87 88	6
Catharanthus roseus	Vindoline, an indole alkaloid	Improved oxygen radical absorbance capacity (ORAC). Scavenged radicals, and improved SOD activity in the liver	*In vivo*-rat	89	6

Table 5.1 (*Continued*)

Herb	Active components	Mechanism	Sample	Dose	References	Level of evidence
Ginseng	Ginsenosides Korean red ginseng has antioxidant and cardio-protective effects, and therefore, treatment with this herb decreased MDA levels and showed antioxidant activity	Decreased MDA activity and increased SOD activity in the hippocampus and improved learning and memory decline. Increased GPx activity in the plasma	*In vivo* rat		90 91	6
Aphloia theiformis	Saponins (tormentic acid ester glucoside, 23-hydroxytormentic acid ester glucoside, and 6-*b*-hydroxy-tormentic acid ester glucoside)	Inhibitory potential against various carbohydrate hydrolyzing enzymes Decrease in IC50 value againstα-amylase and lipase enzymes	*In vitro*	IC50 = 22.94–939.97 $\mu g\,mL^{-1}$	92	6
Comarum palustre	Agrimoniin-containing eluates ellagitannin	Strong inhibitory activity against α-glucosidase. Increased the plasma glucose and glycosylated hemoglobin level and significantly decreased the levels of insulin and hemoglobin in plasma	*In vivo* rat	IC50 = 52 $\mu g\,mL^{-1}$, IC50 = 204.7 $\mu g\,mL^{-1}$	93	6

S. cumini leaves	Ursolic acid and oleanolic acid	Inhibition of porcine pancreatic α-amylase. However, there is a strong inhibition of α-glucosidase inhibition of α-amylase, and α-glucosidase	*In vitro*	39.9 μg mL^{-1}	94	6
Phoenix dactylifera	Phenolic compounds	Inhibitory activity against α-glucosidase as compared to α-amylase	*In vivo* and *in vitro*	IC50 = 0.6 *vs.* 2.5 mg ml^{-1}	95	6
Bombax ceiba		Decrease in protection of pancreatic β-cells and stimulation of insulin secretion	*In vivo* rat	280 mg kg^{-1}	96	
Moringa oleifera (Moringaceae)	Enhanced serum albumin, globulin and total protein level with a reduction in the concentration of MDA, and alleviation in SOD, CAT, GSH, GPx, TNF-α and IL-6 level	Inhibited alterations in the histoarchitecture of hepatic and pancreatic tissues and normalized the blood glucose levels, decreased hepatic glutathione level, activities of SOD and CAT, and gene expression of GS, PC and caspase Improvement in serum biochemical attributes, but a decrease in lipid peroxidation (MDA) levels in nondiabetic and diabetic mice	*In vivo* rat	250 mg kg^{-1}	97 98	
Ficus carica		Exhibited potent activity on blood glucose, total cholesterol, triglycerides, body weight and hepatic glycogen level. Cytoprotective property of pancreatic β-cells	*In vivo* rat	250 and 500 mg kg^{-1}	99	6

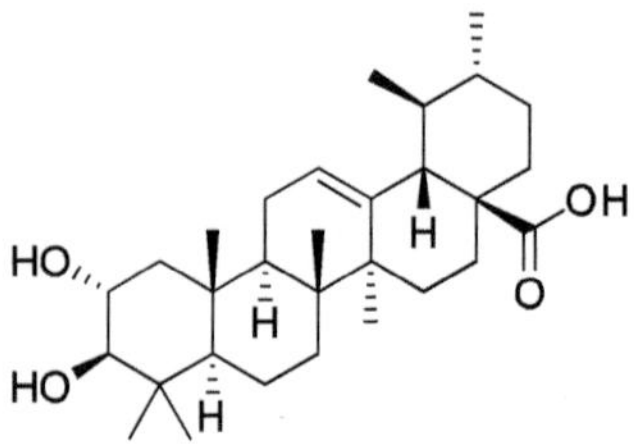

Scheme 5.1 Corosolic acid.

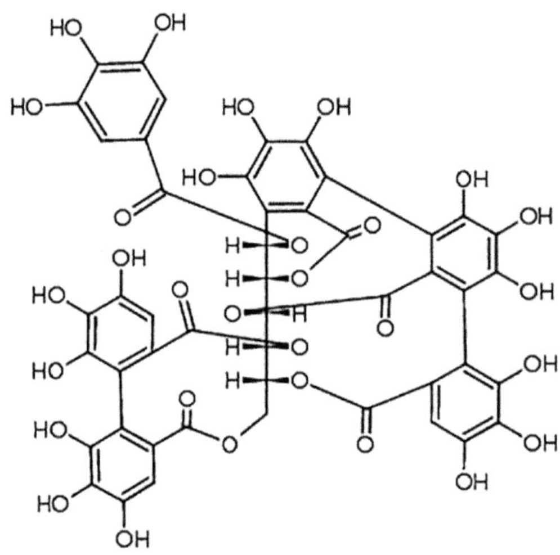

Scheme 5.2 Ellagitannins.

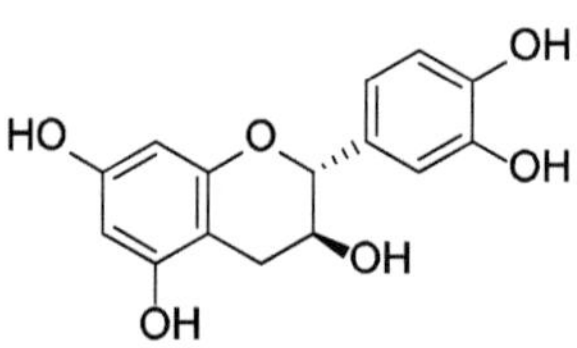

Scheme 5.3 Catechin.

epicatechin gallate (ECG), epigallocatechin gallate (EGCG), and epigallocatechin (EGC), increase the mobility of glucose in the body.[4] Compared to oral medications (Semaglutide or Liraglutide), green tea extract rich in catechins decreased the body weight of overweight people (4.6 percent) and waist-to-hip ratio (4.5 percent), which was equivalent to oral medications.[5]

The levels of triglycerides, total cholesterol, urea, uric acid, low-density lipoprotein (LDL) cholesterol, and creatinine were all lowered by gallic acid, which also dramatically raised insulin levels in the blood and glucose tolerance (Scheme 5.4). Gallic acid is beneficial in both ways in diabetes, acting

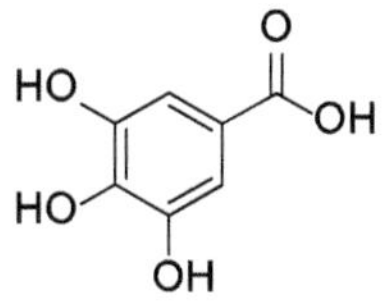

Scheme 5.4 Gallic acid.

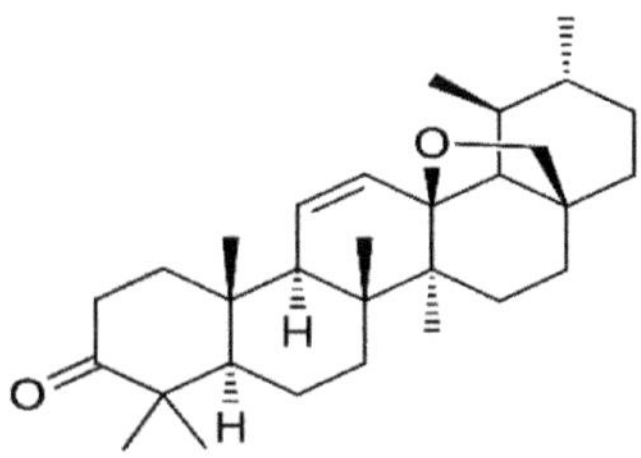

Scheme 5.5 Momordicin and momorcharin.

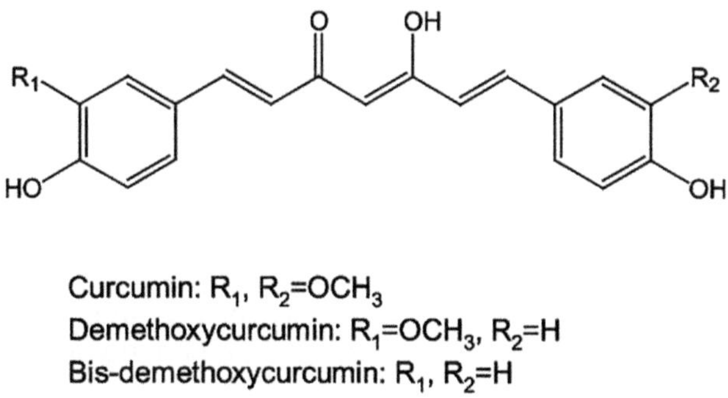

Scheme 5.6 Curcuminoids.

as an antioxidant to minimize oxidative damage in beta-cells while also improving blood glucose and cholesterol levels in those at higher risk of heart disease.[6]

Bitter melon is the common name given to the fruit of *Momordica charantia*, which is a member of the Cucurbitaceae family. It is usually utilized in the treatment of diabetes in traditional Chinese medicine. *M. charantia* contains biologically active compounds, momordicin and momorcharin (Scheme 5.5), which play a critical role as antidiabetic agents in type 2 diabetes.[7]

Curcuminoids are natural polyphenolic compounds derived from the rhizome of the *Curcuma longa* plant, exhibiting potential pharmacological effects, including antioxidant and anti-inflammatory, neuroprotective, cardioprotective, and anti-diabetic properties, among others (Scheme 5.6). Curcumin has been proven in numerous trials to reduce insulin resistance, the risk of obesity, coronary heart disease, and diabetes.[8] As a result of

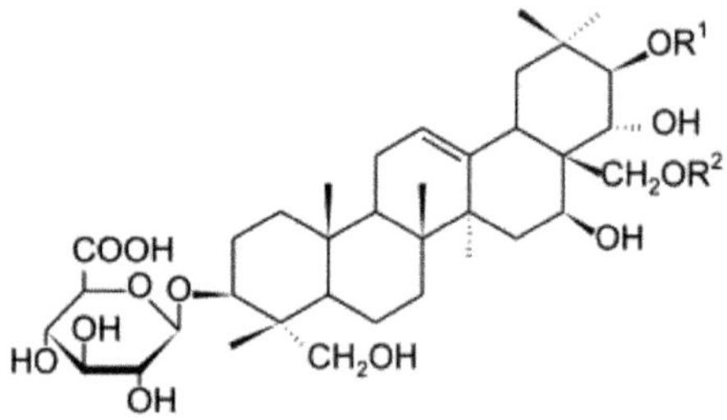

Scheme 5.7 Gymnemic acids (R1 = tigloyl; R2 = acetyl).

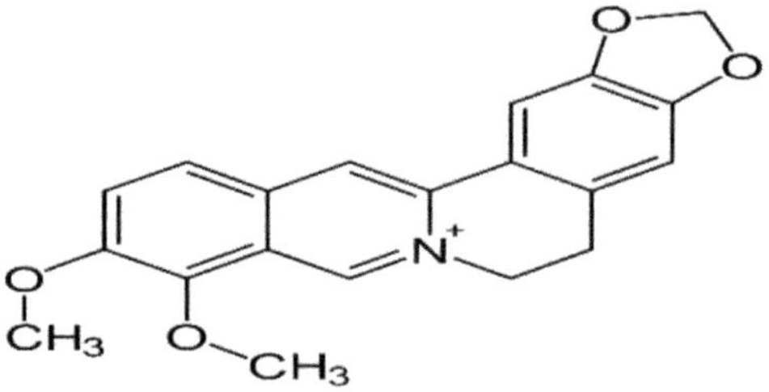

Scheme 5.8 Berberine.

curcuminoids' pleotropic structure, numerous mechanisms have been proposed to explain their anti-diabetic properties, including platelet aggregation, oxidative processes, and a reduction in blood glucose levels.[9]

Gymnemic acid derivatives, obtained from the leaves of *Gymnema inodorum*, have been shown to inhibit glucose absorption from the intestinal tract and, as a result, to reduce the increase in blood glucose levels seen in patients (Scheme 5.7).[10]

5.3 Effects of Phytochemicals *In Vivo*

Berberine treatment of diabetic animal models induced by streptozotocin (STZ) showed significant reductions in circulating levels of the chronic inflammatory biomarker C reactive protein, free fatty acids, cholesterol, and triacylglycerols, primarily through inhibition of the action of dipeptidyl peptidase-4 and protein tyrosine phosphatase-1B (PTP-1B) (Scheme 5.8). As a result, the antioxidant properties of berberine increased insulin sensitivity while simultaneously decreasing insulin resistance, resulting in reduced fasting blood glucose levels.[11]

Resveratol dramatically reduced retinal damage in a STZ-induced diabetic retinopathy model in rats, mostly through its potentiating effect on paraoxonase-1 expression and inhibitory effects on ox-DLD, retinal AGEs, and inflammatory cytokines (IL-1, IL-6, TNF-α, IFN-γ, MCP-1, and VEGF) (Scheme 5.9). When administered to diabetic rats, resveratrol significantly reduced dephosphorylation of retinal AMP-activated protein kinase (AMPK),

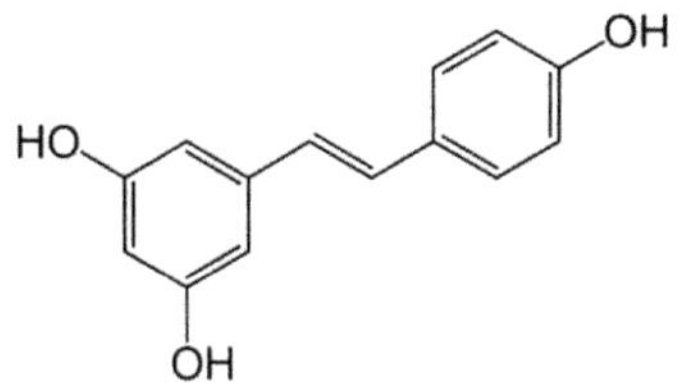

Scheme 5.9 Resveratol.

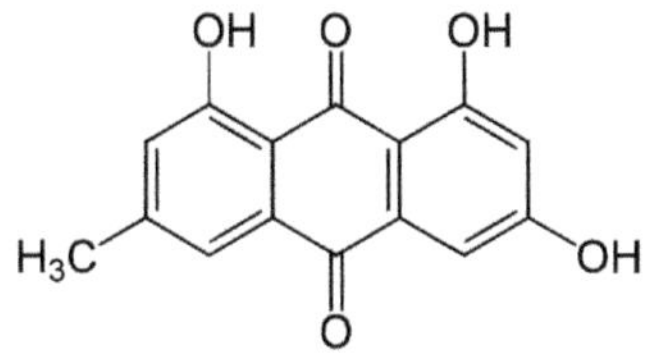

Scheme 5.10 Emodin (1,3,8-trihydroxy-6-methylanthraquinone).

sirtuin 1 (SIRT1) inactivation, and nuclear factor-κB (NF-κB) phosphorylation in the retina.[12]

Emodin (1,3,8-trihydroxy-6-methylanthraquinone), a polyphenol primarily isolated from *Rheum palmatum*, *Polygonum cuspidatum*, *Polygonum multiflorum*, and *Cassia obtusifolia*, among other plants, is well-known as a bioactive component of laxative herbs (Scheme 5.10). Emodin has been shown to have laxative properties in animal studies. To be more specific, a number of studies have revealed the potential of emodin for the treatment of diabetes and its complications.[13] Emodin had a glucose metabolism-regulatory impact, as demonstrated by *in vitro* data showing increased glucose uptake and consumption in C2C12 myotubes and 3T3-L1 adipocytes, which was achieved by modulation of the AMPK signaling system. Furthermore, emodin boosted glycolysis while inhibiting lipolysis in 3T3-L1 adipocytes, and this was in close conjunction with a substantial decrease in the phosphorylation of perilipin.[14]

Ellagic acid (2,3,7,8-tetrahydroxy-chromeno) is a result of hydrolyzed tannins that is mostly found in berries, pomegranates, and dried fruits. Interestingly, current findings suggest that ellagic acid is a significant anti-diabetic component (Scheme 5.11). In a rat model of type 2 diabetes and the resulting kidney damage, ellagic acid therapy markedly reduced hyperglycemia, insulin resistance, and diabetic nephropathy, mostly through decreasing the activation of the nuclear factor-κB (NF-κB) transcription factor.[15]

Epigallocatechin gallate (EGCG) is a naturally occurring compound that can be found in a wide variety of foods, plants, and herbs (Scheme 5.12).[16] Examples of such foods and plants include cranberries, strawberries, blackberries, kiwis, cherries, pears, peaches, apples, avocadoes, pecans, pistachios, and hazelnuts. It is also present in green tea, but this is not the only place it can be found. In recent years, epigallocatechin gallate has also been assessed as a potential anti-diabetic medication treatment option. Treatment with epigallocatechin gallate of

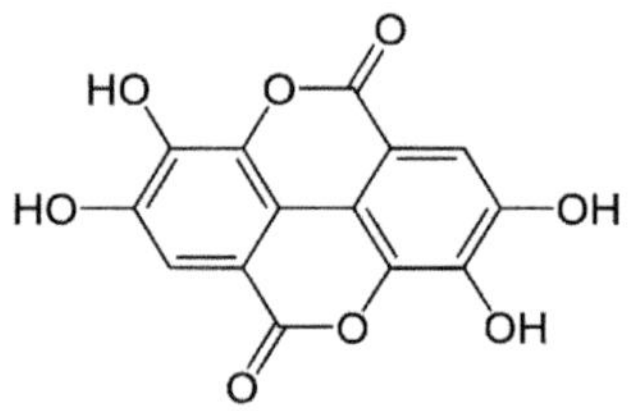

Scheme 5.11 Ellagic acid.

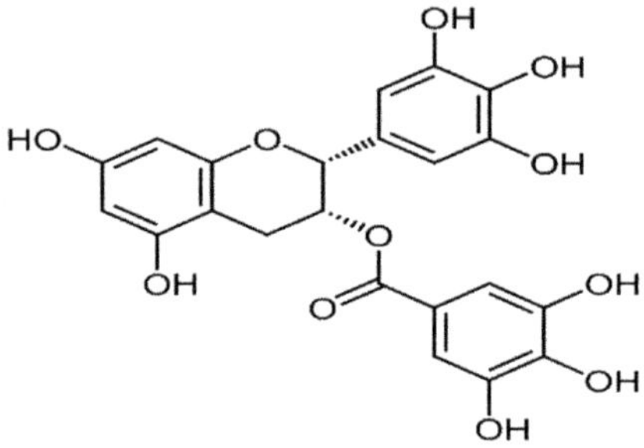

Scheme 5.12 Epigallocatechin gallate.

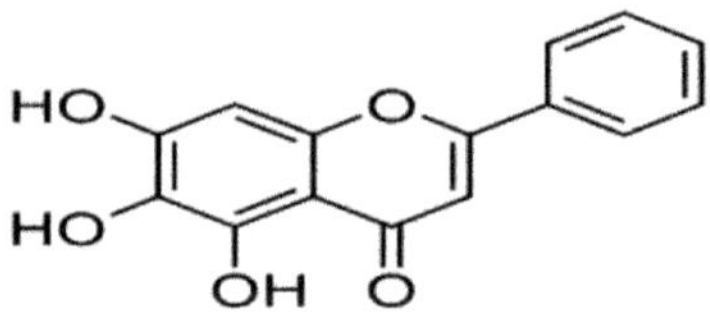

Scheme 5.13 Baicalein.

STZ-induced diabetic mice improved wound healing in close association with decreased levels of pro-inflammatory mediators (including interleukin-1 and interleukin-6) and inhibition of Notch signaling, as well as macrophage accumulation around the wound site.[17]

Baicalein (5,6,7-trihydroxy-2-phenyl-4*H*-1-benzopyran-4-one) is a bioactive flavonoid that was first discovered in the root of *Scutellaria* (Scheme 5.13). It is one of the most important bioactive flavonoids. The biochemical advantages of baicalein, which are some of its multiple biological roles, are currently the subject of intensive research. A 16-week course of therapy with baicalein in rat models of diabetes, caused by a high-fat diet combined with STZ, showed a significant reduction in blood glucose level as well as an increase in insulin sensitivity in the diabetic rats.[18]

Naringenin [5,7,dihydroxy-2-(4,hydroxyphenyl) chroman-4-one] is a flavanone present in high concentrations in grapefruit and oranges (Scheme 5.14). Naringenin appears to be a potential anti-diabetic drug, according to mounting research. When administered to rats with type 2 diabetes, naringenin was effective in reducing hyperglycemia, insulin resistance, and high cholesterol levels.[19]

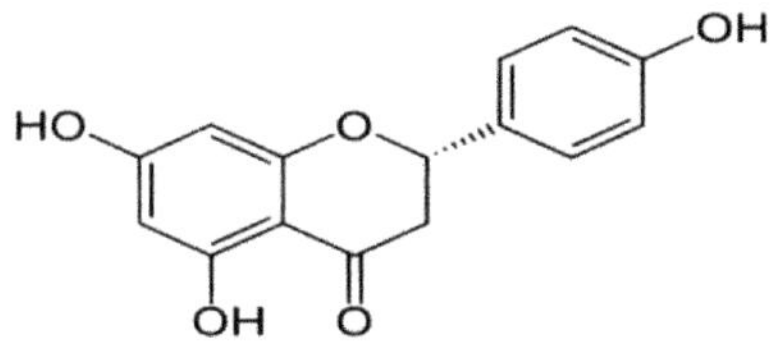

Scheme 5.14 Naringenin.

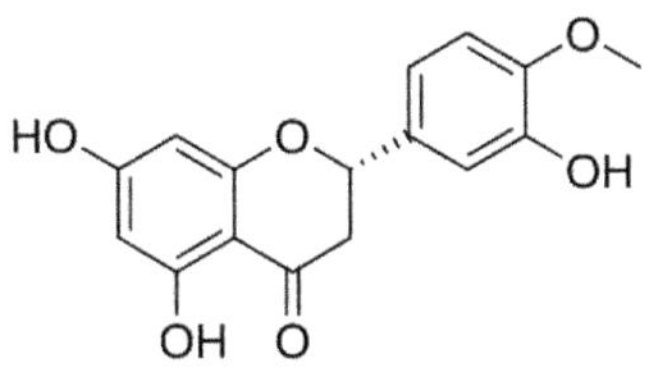

Scheme 5.15 Hesperetin.

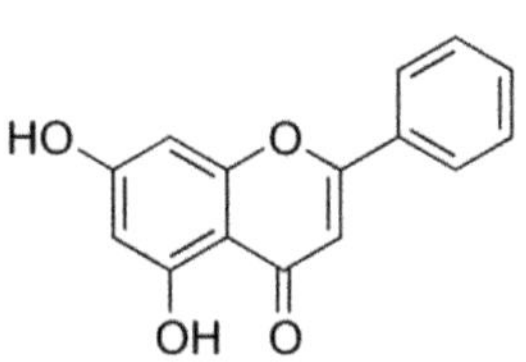

Scheme 5.16 Chysin (5,7-dihydroxyflavone).

Hesperetin (3′,5,7-trihydroxy-4′-methoxyflavanone), the aglycone of hesperidin, is the flavanone that is most prevalent in citrus fruits such as oranges and lemons (Scheme 5.15).[20] Hesperetin exhibits a wide range of pharmacological actions, involving anti-hyperlipidemic qualities as well as anti-inflammatory and anti-oxidative capabilities. It has also been shown to have anti-diabetic, anti-hypertensive, and cardioprotective activities. Hesperetin alleviated hyperglycemia in a mouse model of STZ-induced diabetes by lowering oxidative stress and decreasing the production of pro-inflammatory cytokines TNF-α and IL-6.[21]

The flavonoid chysin (5,7-dihydroxyflavone) is one of the most abundant flavonoids, and can be found in medicinal plants, honey, and propolis (Scheme 5.16). Because of its prospective anti-diabetic and anti-inflammatory capabilities, chysin is gaining increased interest as a drug for the management and cure of diabetes and its consequences. When chrysin therapy was administered to athymic nude diabetic mice, the blood glucose levels were significantly reduced, which was closely correlated with reduced levels of pro-inflammatory cytokines IL-1 and TNF-α in the blood.[22]

Kaempferol [3,5,7-trihydroxy-2-(4-hydroxyphenyl)-4*H*-1-benzopyran-4-one] is a typical dietary flavonoid present in a variety of vegetables and fruits, including leeks, onions, chives, beans, broccoli, mustard, cucumbers,

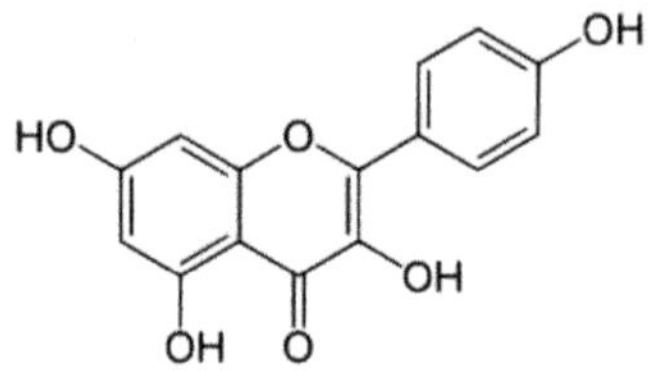

Scheme 5.17 Kaempferol.

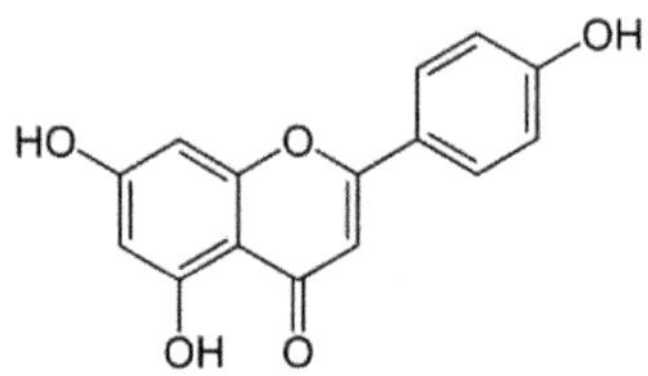

Scheme 5.18 Apigenin.

tomatoes, grapes, raspberries and strawberries (including jam), peach and apple juices, and cowberries (Scheme 5.17). It is also plentiful in several Chinese medicinal herbs, including *Aloe vera*, *Centella asiatica*, *Lycium barbarum*, and *Crocus sativus*,[23] which are used to treat various ailments. Kaempferol has several health benefits, one of which is its ability to treat diabetes, which has been demonstrated in a growing body of research. A study in high-fat diet fed mice revealed that treatment with kaempferol glycoside fractions resulted in decreased body weight gain, blood glucose, and triglyceride levels, as well as reduced insulin resistance.[24]

Apigenin (4′,5,7-trihydroxyflavone) is a flavonoid belonging to the flavone family identified in numerous fruits, vegetables, and herbs, such as grapefruit, oranges, onions, parsley, wheat sprouts, tea, and chamomile (Scheme 5.18).[25] It has many beneficial features such as anti-oxidant, anti-inflammatory, anti-apoptotic, and anti-fibrotic, as well as having anti-diabetic properties. Additionally, apigenin holds significant potential in the prevention of diabetes-related complications, according to the American Diabetes Association. Apigenin therapy alleviated endothelial dysfunction and heart injury in mouse models of diabetic cardiomyopathy by reducing oxidative stress and inflammation in the heart, according to study findings.[26]

Quercetin is a polyphenolic flavonoid found in a wide variety of foods, including apples, black tea, berries, capers, red wine, and onions (Scheme 5.19). Increasing evidence points to the fact that it may have therapeutic promise in the prevention and treatment of diabetes. Several studies have demonstrated that quercetin can reduce oxidative stress and exert protective effects on cytokine-induced damage in pancreatic beta cells, which are the only source of insulin.[27] Because diabetes and non-alcoholic fatty liver disease are frequently associated with one another, the effect of quercetin in T2DM-induced non-alcoholic fatty liver disease in animal models provides more insight into the pleiotropic activities of quercetin in

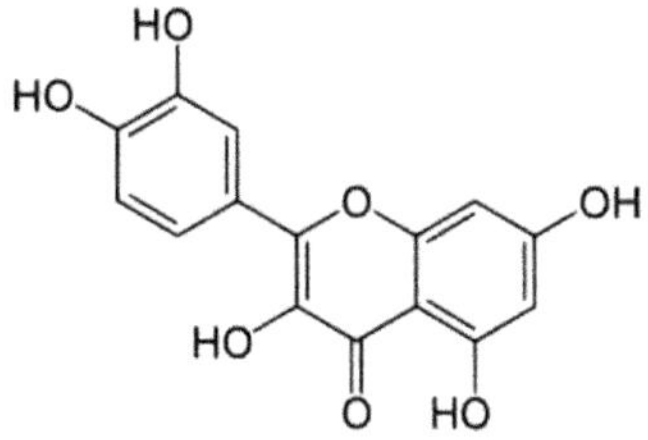

Scheme 5.19 Quercetin.

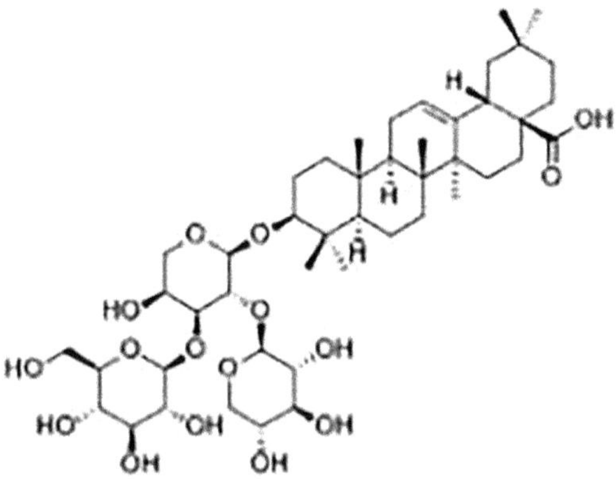

Scheme 5.20 Elatosides E and F.

the treatment of diabetic complications. Furthermore, quercetin therapy had a significant suppressive effect on the release of interleukin-1 and interleukin-18 from the kidney in rats, slowing the advancement of diabetic nephropathy in these animals.[28]

Elatosides E and F are hypoglycemic compounds that were identified in the roots of the *Aralia elata* plant, which belongs to the Araliaceae family (Scheme 5.20). Using an oral sucrose tolerance test, a statistically significant drop in the level of blood glucose in rats was demonstrated.[29]

Cashew is the common name for the tree *A. occidentale*, which belongs to the Anacardiaceae family.[30] Both polar and non-polar extracts of cashew leaves were found to considerably lower the level of blood glucose in diabetic rats that had been induced by streptozotocin. According to some researchers, the occurrence of anacardic acid in cashew nuts may be responsible for the fruit's anti-diabetic properties. Anacardic acid is a physiologically active molecule that increases the consumption and transport of glucose (Scheme 5.21).[31]

Silymarin is a natural substance consisting of a combination of flavonoids including silychristin, silydianin, and silybin (Scheme 5.22). These are the pharmacologically important molecules found in the plant *Silybum marianum*, often known as milk thistle. A study highlighted that silymarin can increase the function of the pancreatic beta cells in rats that had been exposed to alloxan.[32]

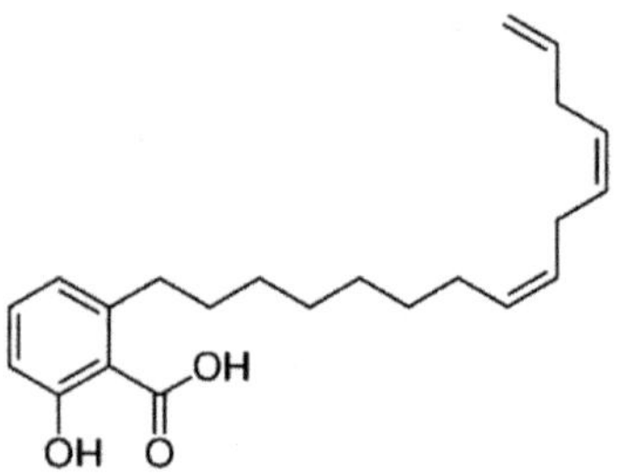

Scheme 5.21 Anacardic acid.

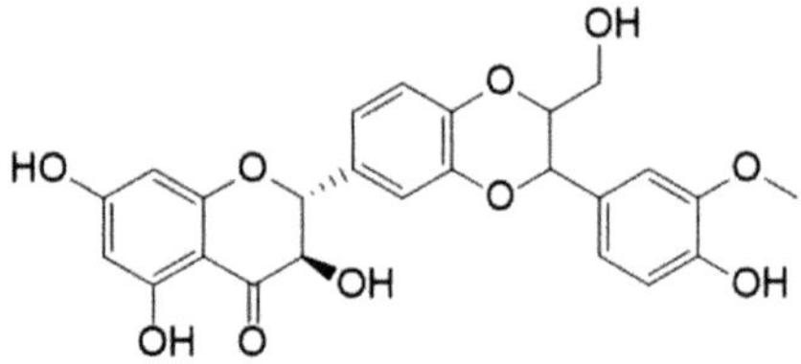

Scheme 5.22 Silymarin.

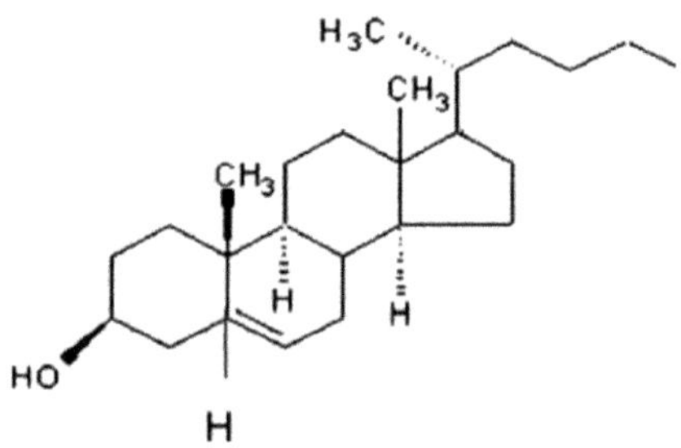

Scheme 5.23 Nymphayol.

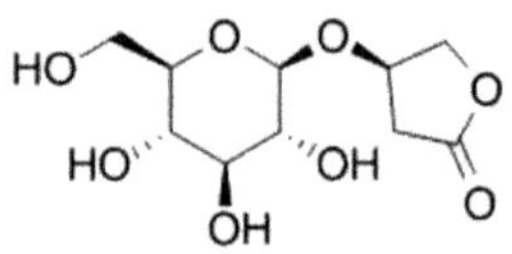

Scheme 5.24 Kinsenoside.

Nymphayol is a plant sterol derived from the *Nymphaea stellata* plant, and is a natural product (Scheme 5.23). This chemical was found to have substantial anti-diabetic properties. The mechanism of its activity as an anti-diabetic medication was assessed in diabetic rats, where it increased the number of β-cells and insulin content, and decreased the level of blood glucose[33].

Kinsenoside is considered to be a major constituent of *Anoectochilus roxburghii* and is isolated from this plant (Scheme 5.24). This drug's ability to operate as an anti-diabetic agent is mediated by its ability to repair and regenerate pancreatic β-cells in mice treated with STZ.[34]

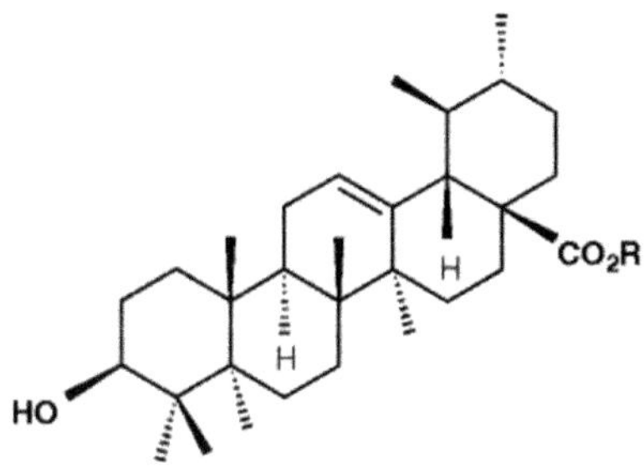

Scheme 5.25 Methyl ursolate.

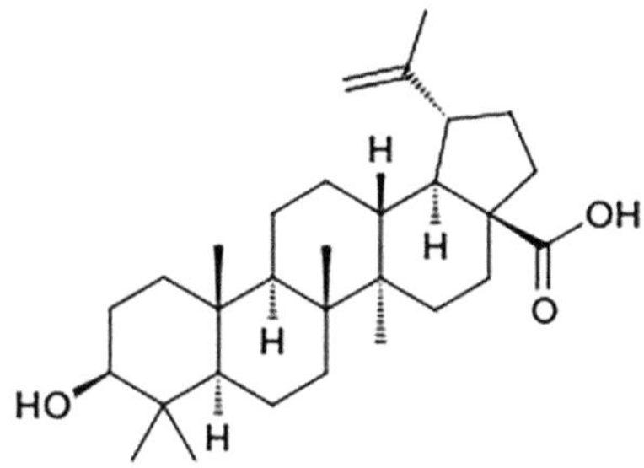

Scheme 5.26 Betulinic acid.

M. alba is commonly called mulberry, and belongs to the family Moraceae. Seventy percent of alcohol extract from its root bark showed a potent hypoglycemic effect in STZ-induced diabetic rats when it was given orally. It exerted its effect *via* reducing lipid peroxidation and protecting the pancreatic β-cells. Its anti-diabetic effect is due to the presence of two triterpenes (methyl ursolate, Scheme 5.25 and betulinic acid, Scheme 5.26) and hydrophobic flavonoids (kuwanon E, morusin, and cyclomorusin).[35]

5.3.1 Oligonol

Oligonol, a highly bioavailable form of low-molecular-weight polyphenols found in lychee fruit, had a powerful antioxidative activity along with hepatoprotective and renal preventative benefits in type 2 diabetic rats.[36]

The extract of *G. elata* demonstrated an anti-diabetic action *via* reducing insulin resistance. It was discovered that the availability of active chemicals such as 4-hydroxybenzaldehyde (Scheme 5.27) and vanillin (Scheme 5.28) was responsible for this action. According to the literature, the mechanism of action of these active compounds as anti-diabetic agents involves the lowering of insulin resistance by enhancing the oxidation of lipids, the decrease of fat storage in adipose tissues, and the potentiation of leptin signaling in obese rats.[37]

Aspalathin is a dihydrochalcone C glucoside found in *Aspalathus linearis*, also known as rooibos, a plant that is popularly used for its tea (Scheme 5.29). When tested in db/db mice, it displayed an effective anti-diabetic effect by boosting insulin production and glucose absorption. In plants and their products, rutin

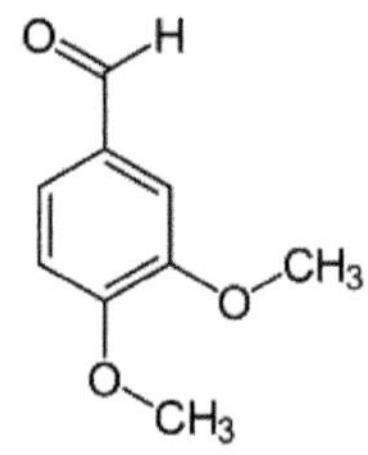

Scheme 5.27 4-Hydroxybenzaldehyde.

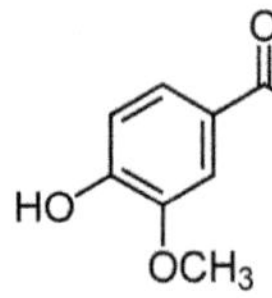

Scheme 5.28 Vanillin.

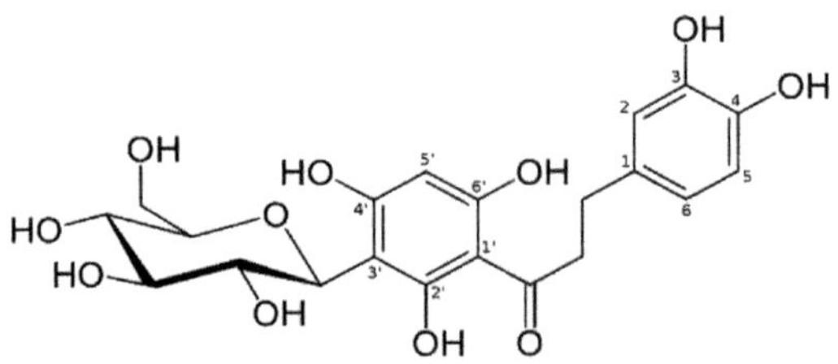

Scheme 5.29 Aspalathin.

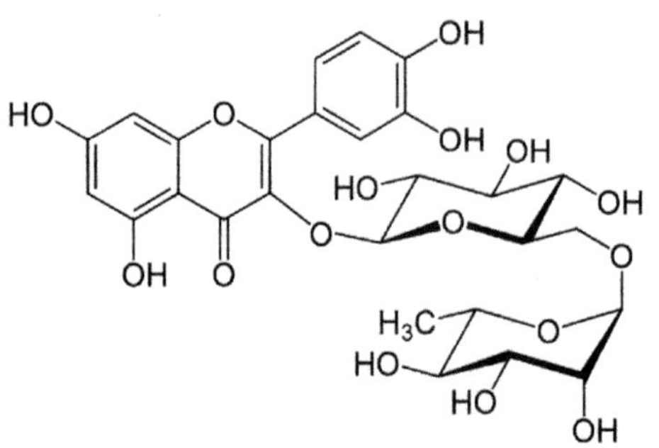

Scheme 5.30 Rutin mixture.

(quercetin-3-*O*-rutinoside) is a flavonoid that has been shown to have a significant anti-diabetic effect through improvement of glucose uptake both experimentally and clinically, increasing insulin release, decreasing the level of blood glucose in STZ-treated rats, and inhibition of α-glucosidase activity (Scheme 5.30).[38] A 1 : 1 mixture of aspalathin and rutin was able to lower the level of blood glucose in diabetic rats caused by STZ.[39]

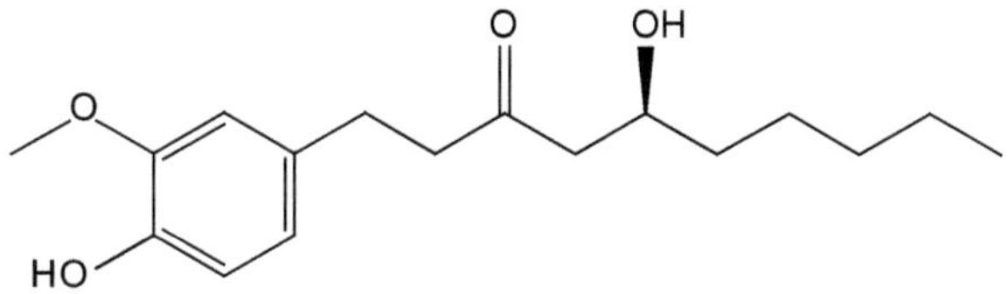

Scheme 5.31 Gingerol.

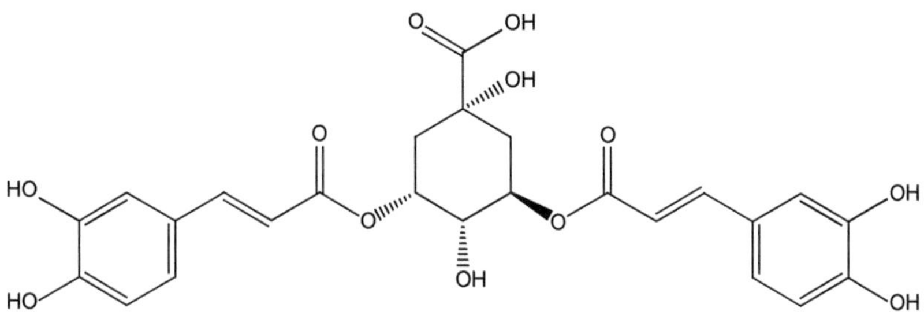

Scheme 5.32 3,5-*O*-dicaffeoylquinic acid.

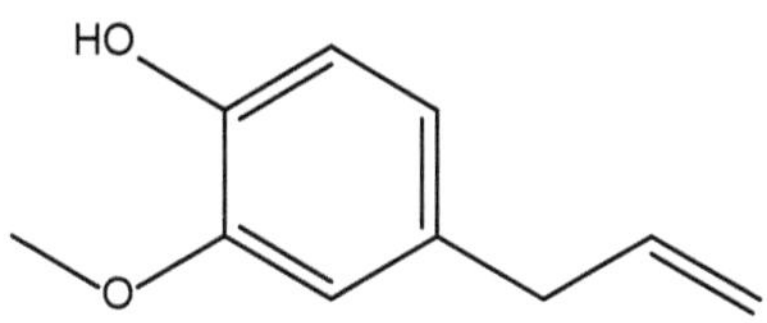

Scheme 5.33 Eugenol.

Gingerol is a physiologically active chemical that has been isolated from the root of the ginger plant (Scheme 5.31). Gingerol had a strong anti-diabetic effect in diabetic rats through increasing insulin sensitivity and glucose uptake.[40]

Maté is the common name for the plant *Ilex paraguariensis*, which belongs to the Aquifoliaceae family. Maté has been shown to have cardioprotective, anti-obesity, anti-inflammation, anti-thrombotic, anti-rheumatoid, hypocholesterolemic, diuretic, hepatoprotective, antioxidant, and neuroprotective properties, according to numerous research reports. 3,5-*O*-dicaffeoylquinic acid is one of the most important ingredients of maté, and it has been shown to have an anti-diabetic effect in ddY mice by increasing the level of serum GLP-1 (Scheme 5.32).[41]

It has been discovered that eugenol, the primary active compound found in *Ocimum sanctum*, is responsible for the anti-diabetic activity of this plant (Scheme 5.33). This is accomplished by decreasing the activity of the enzymes, alkaline phosphatase, AST, ALT, and lactate dehydrogenase, as well as cholesterol, triglyceride, and glucose levels in the blood.[32]

M. koeingii is a member of the Rutaceae family, the leaves of which contain pharmacologically important alkaloids, mahanimbine and carbazole (Schemes 5.34 and 5.35). When administered intraperitoneally, these

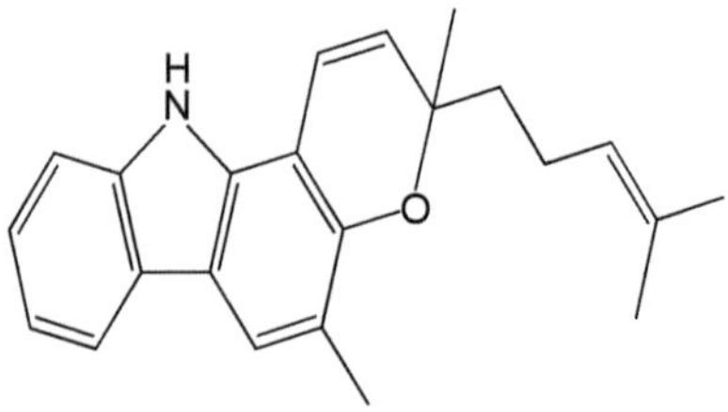

Scheme 5.34 Mahanimbine.

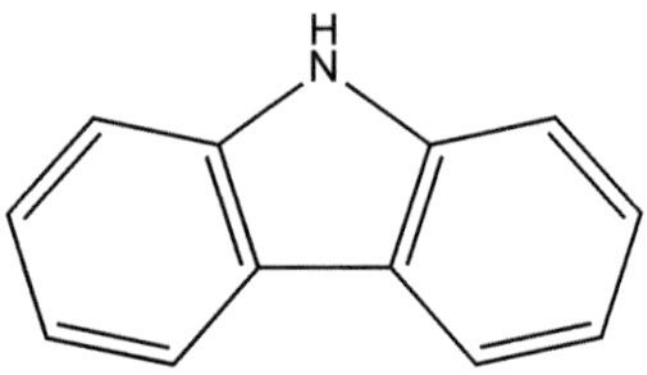

Scheme 5.35 Carbazole.

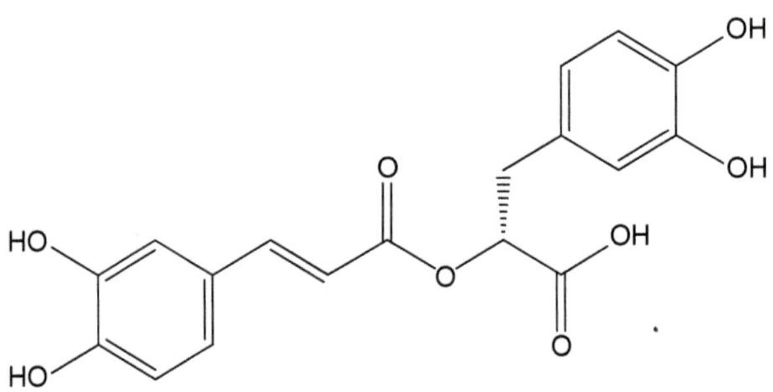

Scheme 5.36 Rosmarinic acid.

alkaloids demonstrated significant hypoglycemic activity by increasing high-density lipoprotein (HDL) levels, while decreasing low-density lipoprotein (LDL), triglycerides, and fasting blood sugar levels.[42]

Rosemary is an aromatic annual herb with a strong scent (family Lamiaceae). It has a significant anti-diabetic impact as a result of the presence of polyphenolic components such as rosmarinic acid, carnosol, and carnosic acid (Schemes 5.36–5.38). A strong anti-diabetic impact of rosemary extract and its polyphenolic components was highlighted in multiple animal models of type 2 diabetes *in vivo*, as well as having a similar effect in insulin target cells *in vitro*.[43]

Conophylline is a chemical substance, derived from the plant *Ervatamia microphylla*, with anti-diabetic properties (Scheme 5.39). The administration of conophylline orally to streptozotocin (STZ)-induced diabetic rats resulted in a reduction in blood glucose levels.[44] Independently, conophylline stimulated the growth of beta cells in the pancreas.[45]

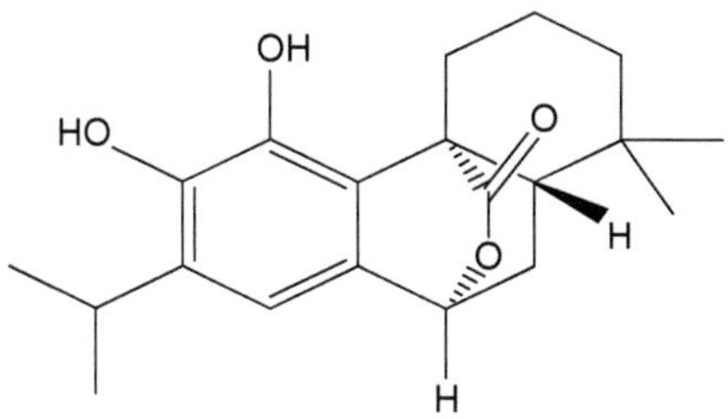

Scheme 5.37 Carnosol.

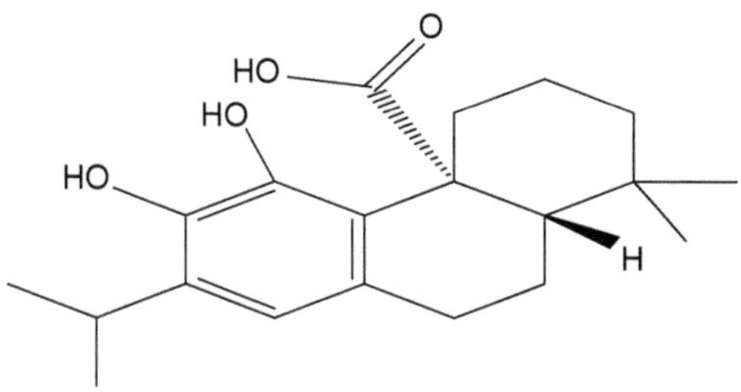

Scheme 5.38 Carnosic acid.

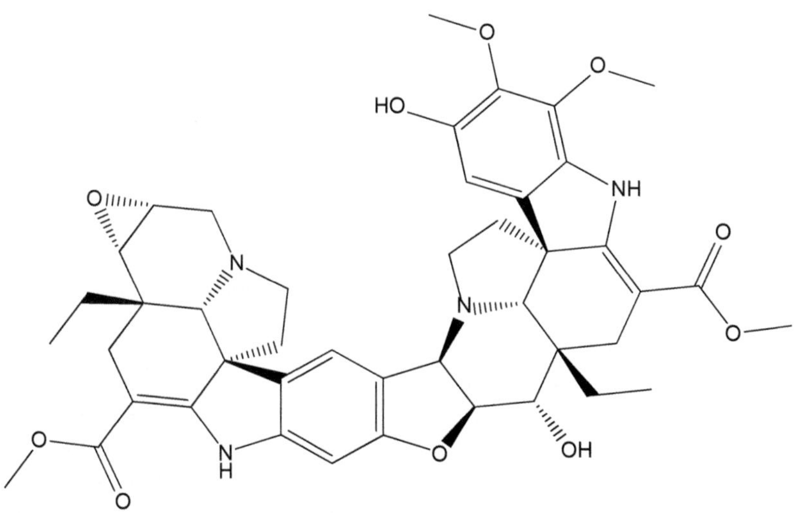

Scheme 5.39 Conophylline.

Kinsenoside is a chemical found in the jewel orchid, *Anoectochilus roxburghii* (Scheme 5.40). The oral treatment of STZ-induced hyperglycemic rats with kinsenoside resulted in a reduction in blood glucose levels. When compared to the control group, the kinsenoside group had higher plasma insulin levels due to enlargement of the pancreatic beta-cells.[46]

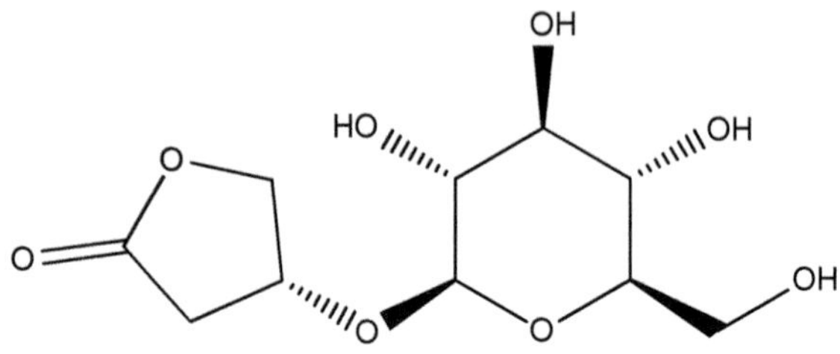

Scheme 5.40 Kinsenoside.

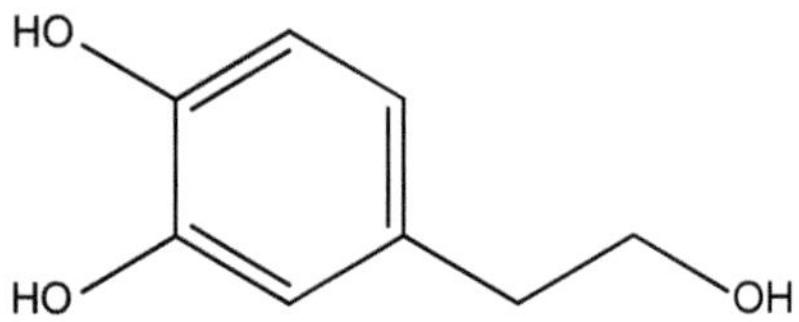

Scheme 5.41 Hydroxytyrosol.

Hydroxytyrosol, a polyphenolic olive-derived compound with a variety of biological activities, particularly anti-oxidant characteristics, has several beneficial effects in humans as well as in animals, where it has been shown to counteract the progression of peripheral diabetic neuropathy in rats (Scheme 5.41).[47]

Stevioside, which is present in the leaves of the plant *Stevia rebaudiana*, is well-known for its intense non-caloric sweetness (Scheme 5.42). In diabetic rats, stevioside displayed hypoglycemic action through modulating the expression of phosphoenolpyruvate carboxykinase (PEPCK), which regulates the generation of glucose from non-carbohydrate molecules in the liver during gluconeogenesis.[48]

Rosmarinic acid is a phenolic constituent found in culinary herbs such as basil, lemon balm, rosemary, sage, thyme, and peppermint (Scheme 5.43). It has been proven to lower the serum glucose co-transporter protein-1 (SGLT1) level in the intestine of diabetic rats after 14 days of treatment.[49]

Curcumin [also known as 1,7-bis(4-hydroxy-3-methoxyphenyl)-6,5-dione], is a polyphenol found largely in the roots of the turmeric plant (*Curcuma longa*, also known as curcuma) (Scheme 5.44). It is also the active element of the spice turmeric, which is used in cooking. Several studies have indicated that curcumin possesses anti-inflammatory, anti-hepatotoxic, and cholesterol-lowering effects. Most intriguingly, there is growing evidence that curcumin may be a successful target for diabetes and its possible implications.[50]

Vijayasar is the common name for the plant *P. marsupium*, which belongs to the Fabaceae family. Various preparations of its bark exhibited a considerable anti-diabetic effect in diabetic rats that had been exposed to alloxan. It has been discovered that the heartwood of *P. marsupium* contains biologically active phenolic chemicals, marsupsin and pterostilbene, which are able to considerably lower the level of blood glucose in a manner similar to that of metformin (Schemes 5.45 and 5.46).[51]

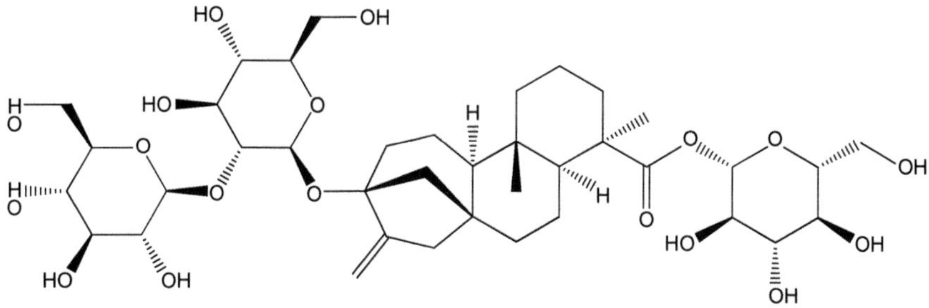

Scheme 5.42 Stevioside.

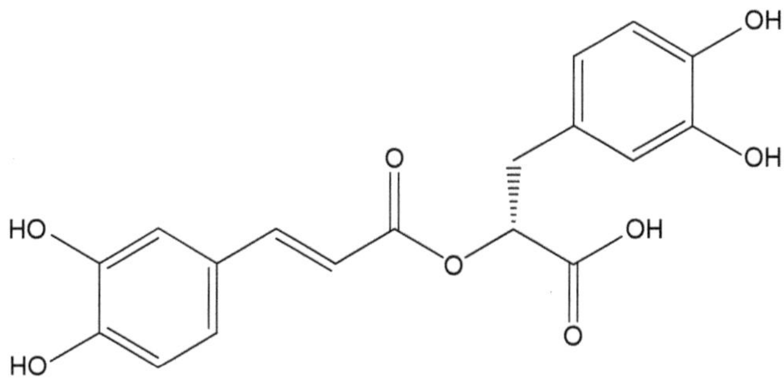

Scheme 5.43 Rosmarinic acid.

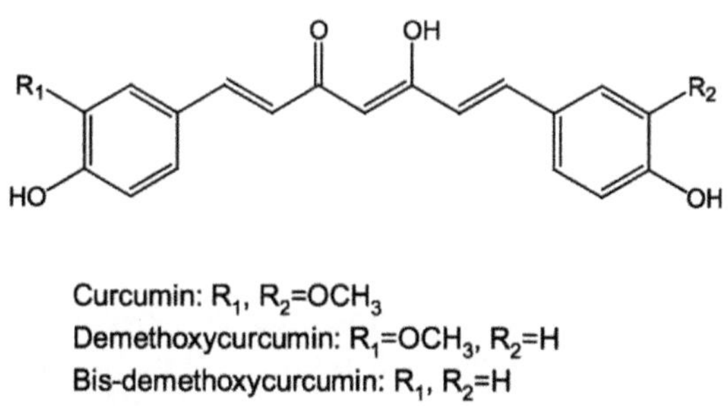

Scheme 5.44 Curcumin.

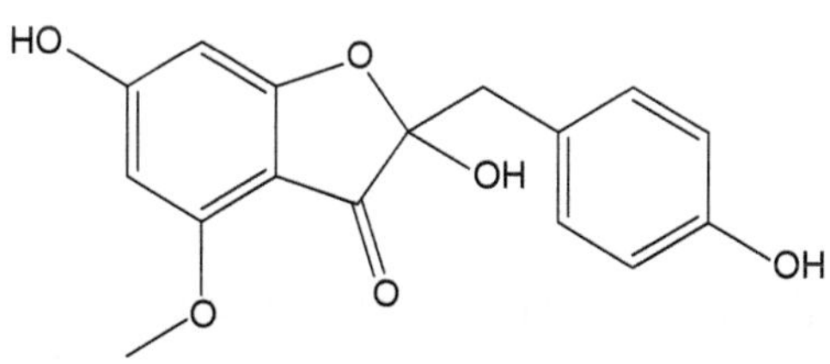

Scheme 5.45 Marsupsin.

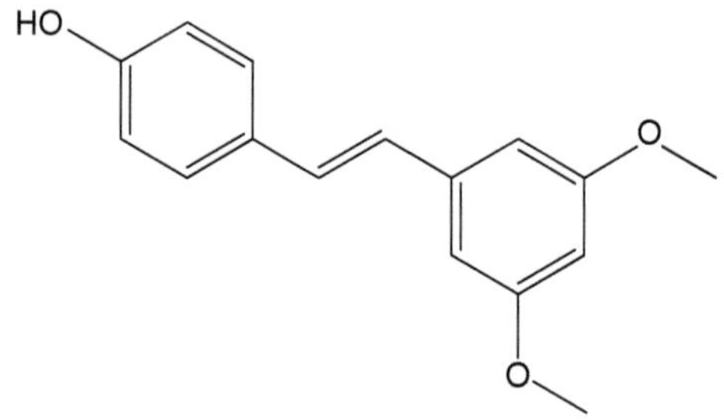

Scheme 5.46 Pterostilbene.

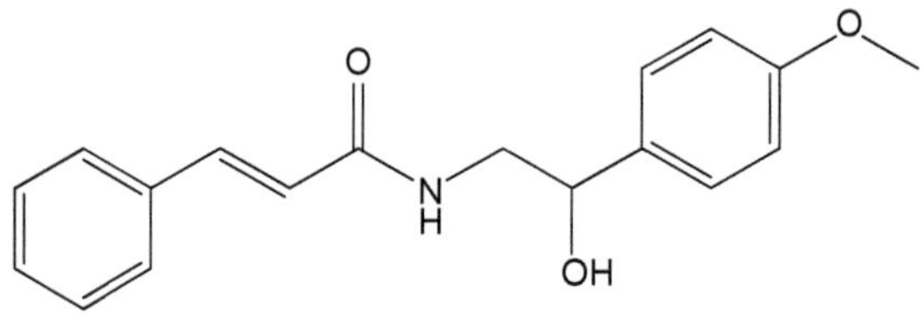

Scheme 5.47 Aegeline.

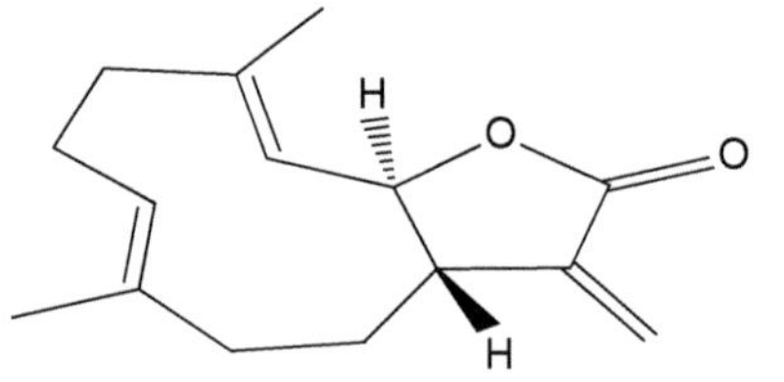

Scheme 5.48 Costunolide.

Aegeline was discovered in the leaves of *Aegeline marmelos*, a plant that has a long history of usage in Ayurvedic medicine (Scheme 5.47). Aegeline was the first compound discovered in nature. Based on animal studies, it has been demonstrated that aegeline has anti-hyperglycemic and anti-dyslipidemic properties. It has been revealed that the use of aegeline reduced the blood glucose levels of sucrose-loaded STZ-induced diabetic rats and improved the lipidemic parameters of hamsters (TG, HDL, and CT).[52]

Costunolide reduced plasma glucose levels in STZ-induced diabetic rats in a dose-dependent manner, and it has also lowered HbA1c, serum total cholesterol, TG, and LDL cholesterol levels in the same animals (Scheme 5.48). In addition, costunolide significantly enhanced the levels of plasma insulin, hepatic and muscle glycogen, and HDL cholesterol in the bloodstream.[53]

The labdane diterpenoid andrographolide was discovered to be the primary bioactive in the anti-diabetic herb *A. paniculata* (Scheme 5.49). Researchers have reported anti-hyperglycemic effects in STZ-induced diabetic rats, with an increase in GLUT4 expression in the soleus muscle, increasing peripheral glucose consumption as a result of the treatment.[54]

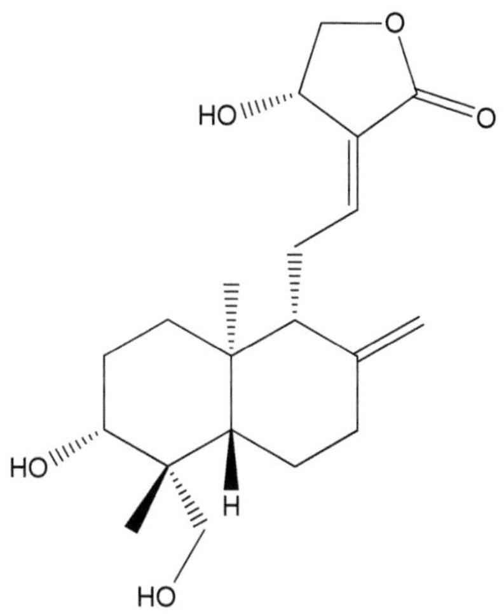

Scheme 5.49 Labdane diterpenoid andrographolide.

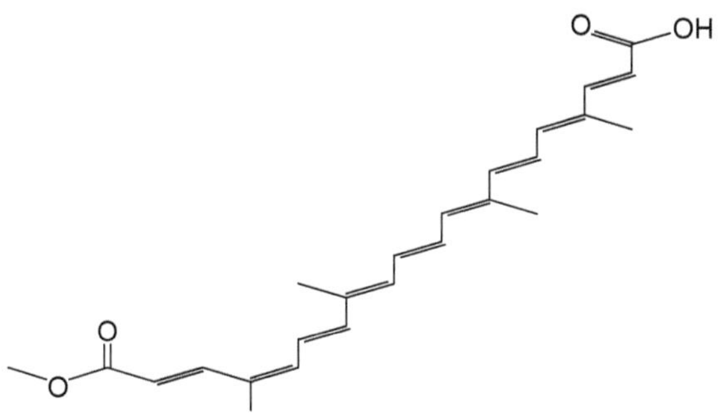

Scheme 5.50 Carotenoid bixin.

Bixin, a carotenoid derived from the seeds of *Bixa orellana*, was discovered by Xu and Kong, in 2017 (Scheme 5.50). Several studies in dogs demonstrated that extracts derived from this plant material have both hypoglycemic and hyperglycemic effects; lipophilic extracts had a hypoglycemic impact, while ethanolic extracts had a hyperglycemic effect.[55]

Corosolic acid, a pentacyclic triterpene acid, discovered in the plant *L. speciose*, displayed anti-diabetic activities both *in vitro* and *in vivo* (Scheme 5.51). It was discovered in 2006 that a single oral administration of GLUT4 in KK-Ay mice (2 mg kg^{-1} body weight) resulted in a reduction in blood glucose levels. The underlying hypoglycemic effect was attributed to GLUT4 translocation in muscle membrane, which decreased insulin resistance.[56]

5.4 Effects of Phytochemicals *In Vitro*

α-Glucosidase (EC#: 3.2.1.106), a membrane-bound enzyme that catalyzes the final step in carbohydrate digestion, has the potential to delay the use of

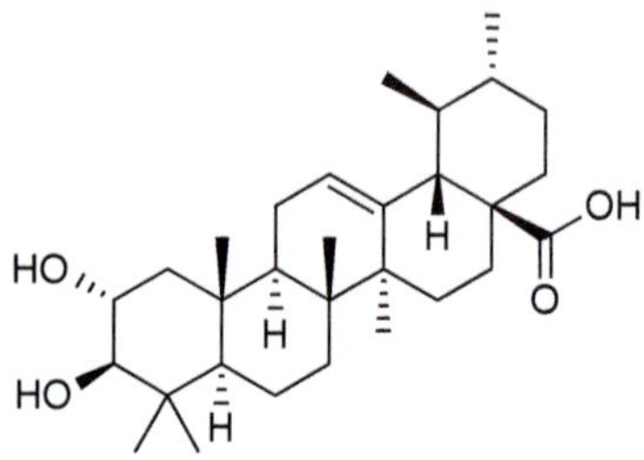

Scheme 5.51 Corosolic acid.

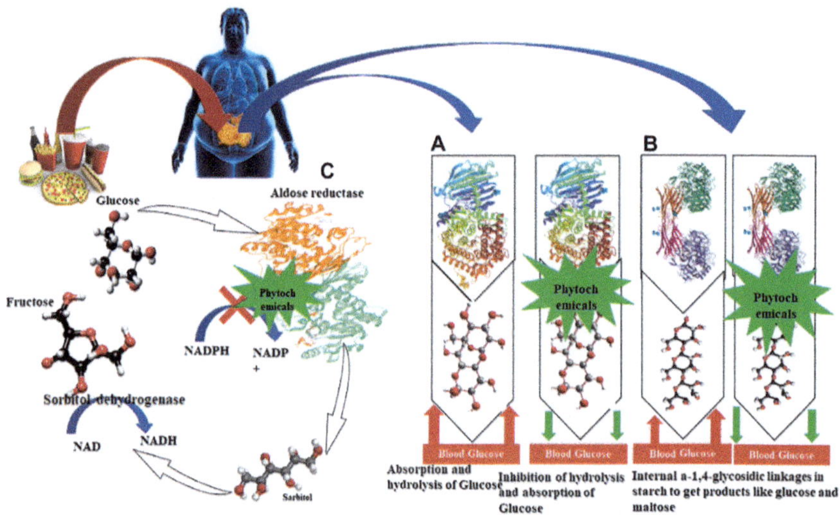

Figure 5.2 *In vitro* inhibition of hyperglycemia (A) α-glucosidase without phytochemicals showed hyperglycemia and with photochemicals showed hypoglycemia; (B) α-amylase, blood glucose decreased when treated with phytochemicals; (C) aldose reductase inhibition by phytochemicals.

dietary carbohydrates and thereby reduce postprandial hyperglycemia.[57] Inhibitors of α-glucosidase have been shown to be effective in reducing postprandial hyperglycemia (Figure 5.2A). In the small intestine, glucosidase catalyzes the breakdown of polysaccharides into monosaccharides, which are then absorbed and induce diabetes mellitus.[58]

It has been proven that inhibiting α-glucosidase activity is one of the most effective and simple approaches to managing hyperglycemia by slowing down the absorption of glucose in the intestine.[59] Voglibose, miglitol, and acarbose are still the most often used α-glucosidase inhibitors, and these inhibitors are also employed as anti-cancer, anti-HIV, and anti-diabetic drugs. To make matters worse, the effectiveness of these inhibitors is less than half that of other anti-diabetic medications like metformin and sulfonylurea. The use of recent medications has been associated with several adverse effects, including flatulence, abdominal discomfort, and diarrhea.[60]

As a result, these adverse effects make it difficult to utilize these medications alone, and they are frequently used in combination with other anti-diabetic treatments to increase their effectiveness.

α-Amylase is an enzyme that catalyzes the hydrolysis of internal α-1,4-glycosidic bonds in starch to produce products such as glucose and maltose.[61]

Carbohydrates are the main elements of a regular diet and play a critical function in the production and distribution of energy. Through the joint action of amylase and alpha glucosidase, the composite elements of dietary carbohydrates should be split down to their monosaccharide forms. It has been demonstrated that the resultant monosaccharides can be absorbed on the intestine's lumen and released into the blood.[62] When α-glucosidase and α-amylase activities are inhibited, carbohydrate digestion is slowed down, leading to a reduction in blood glucose levels. As a result, it may be considered as a viable therapeutic approach for the treatment of diabetes. Indeed, α-amylase inhibitors are important therapeutic agents, and represent one class of medications that has been approved for use in the treatment of diabetes (Figure 5.2B).

Patients with diabetic problems, such as retinopathy, neuropathy, nephropathy, and cataracts, are among the primary triggers of morbidity and death in the general population. Because of this, there is increasing interest in the development of medications that might alleviate the varied symptoms of diabetes problems and their consequences. Several studies have suggested that hyperglycemia may play a significant role in the etiology of diabetes complications, and that this role may be mediated through a variety of pathways. The enhanced aldose reductase (AR)-related polyol pathway (Figure 5.2C) flow is one of the most significant of these.[63] In the polyol pathway, AR is an oxidoreductase that requires NADPH to function. It is one of the most significant enzymes in the conversion of different sugars to sugar alcohols, such as glucose to sorbitol, which is a sugar alcohol. Sorbitol is converted to fructose by sorbitol dehydrogenase, an enzyme that requires NADPH to function properly. While the affinity of cell-based AR for glucose is modest under normal settings, an increased rate of the AR-related polyol pathway leads to a rise in intracellular concentrations of sorbitol and its metabolite fructose in diabetes conditions.[64]

The flavonoid eriodictyol [2-(3,4-dihydroxyphenyl)-5,7-dihydroxy-2,3-dihydrochromen-4-one] is one of the most abundant dietary flavonoids found in high concentration in the plant *Eriodictyon californicum* (Scheme 5.52).

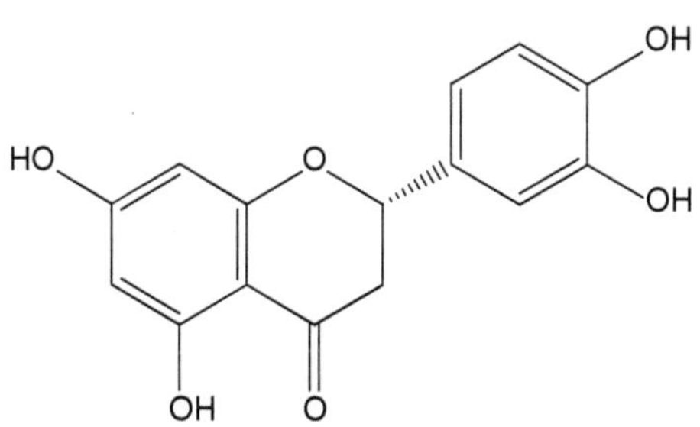

Scheme 5.52 Eriodictyol.

In vitro studies have revealed that eriodictyol has therapeutic potential on the reduced cell viability of rat retinal ganglion cells. In addition, high glucose-induced cell proliferation, as well as the release of ROS and pro-inflammatory cytokines, were inhibited in mesangial cells after incubation with eriodictyol. Moreover, it was found to be a critical element in the pathogenesis of diabetic nephropathy, as it was involved in the creation and maintenance of the glomerular capillary tuft, as well as the preservation of the mesangial matrix.[65]

Conophylline is a plant alkaloid present in *Ervatamia microphylla* and *Tabernaemontana divaricata*, among other plants (Scheme 5.53). Both conophylline and *T. divaricata* are essential in increasing the function of β-cells.[66] It has been shown that conophylline has anti-diabetic properties by lowering the fibrosis of pancreatic islet cells.[67]

Glycyrrhiza foetida Desf. (licorice) and *Amorpha fruticosa* L. contain amorfrutins, which are potent antidiabetic dietary natural compounds (Scheme 5.54). They are benzoic acid derivatives that have been replaced with isoprenoid. It has been demonstrated that amorfrutins have an anti-diabetic action that involves binding to and activating a nuclear receptor

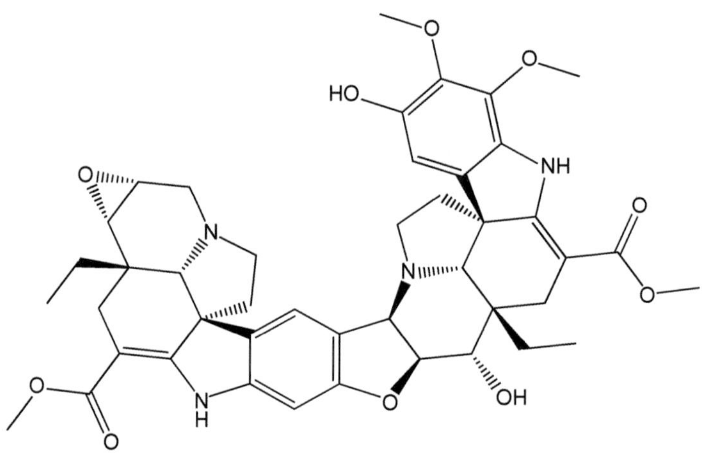

Scheme 5.53 Conophylline.

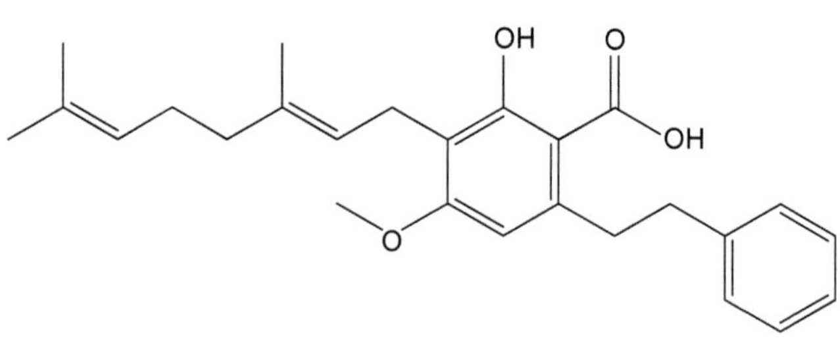

Scheme 5.54 Amorfrutins.

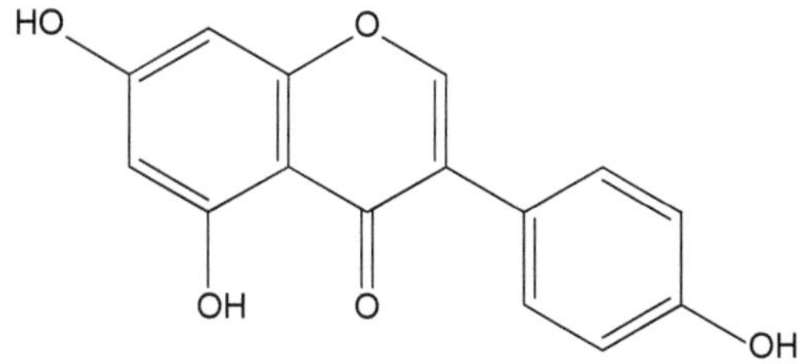

Scheme 5.55 Genistein.

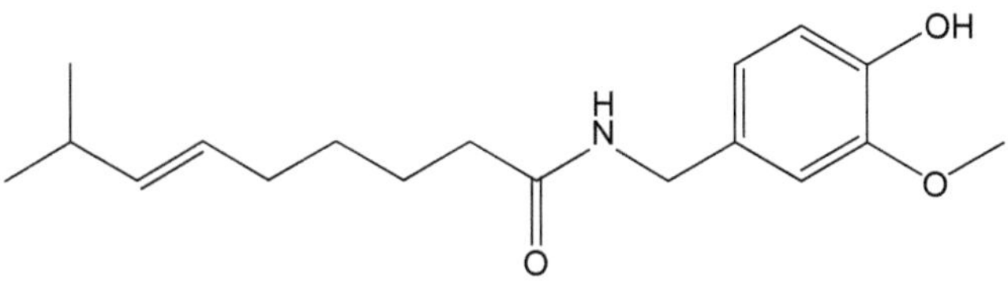

Scheme 5.56 Capsaicin.

(peroxisome proliferator-activated receptor gamma—PPARg), which is important in lipid and glucose metabolism.[68]

Genistein is an isoflavone found in *Glycine max*, which is also known as soybean; it has anti-osteoporosis and osteoporosis-related effects, anti-inflammatory and anticancer effects, and anti-oxidant, anti-diabetic, and obesity-related properties (Scheme 5.55). Because of its therapeutic impact on the pancreatic β-cells, genistein has a significant anti-diabetic effect.[69]

Capsaicin, also known as 8-methyl-*N*-vanillyl-6-nonenamide, is one of the most important natural compounds derived from the chili pepper, which is the popular name for the fruits of the *Capsicum* plant (Scheme 5.56). Chipotle pepper exhibited anti-diabetic, anti-tumor, and anti-obesity properties in both animals and human. Capsaicin is the principal pungent and biologically active component found in peppers, yet it is also the most toxic. The drug's mechanism of action was the modulation of insulin resistance as well as the maintenance of pancreatic β-cells.[70]

Increasing the intake of ellagic acid improves diabetic metabolism as well as structural and functional alterations in cardiac and hepatic regions, which are linked to oxidative stress and inflammatory reactions in the body. Ellagic acid was able to reduce the negative effects of hyperglycemia, glucotoxicity, and inflammation, as well as to enhance renal function (Scheme 5.57). L6 myotubes cells, treated with elagic acid (3.12–50 M), promoted the expression of GLUT-4 and PPAR-.[71]

5.5 Conclusion and Perspectives

It is undeniable that phytochemicals are beneficial in the management of diabetes. These chemicals can inhibit hyperglycemia in a variety of ways, including by inhibiting α-glucosidase, α-amylase, and aldose reductase, as

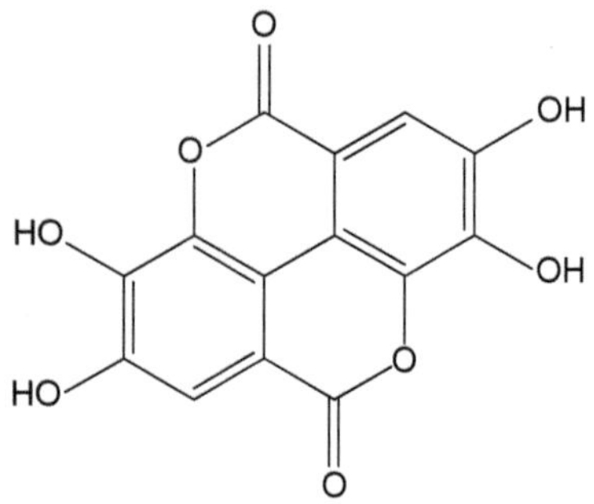

Scheme 5.57 Ellagic acid.

well as by acting on the insulin receptor. The fact that they have multiple modes of action allows them to manage conventional diabetic agents. These plants have the potential to be useful therapeutic agents or as adjuncts to conventional therapies for the control of glucose homeostasis in diabetics. Despite the fact that evidence-based therapeutic applications of many plants are limited, the phytochemicals discussed in this chapter are those that have been traditionally associated with an anti-diabetic effect and that have been studied, either experimentally or clinically.

References

1. International Diabetes Federation, *IDF Diabetes Atlas*, International Diabetes Federation, Brussels, Belgium, 9th edn, 2019.
2. H. Choudhury, M. Pandey, C. K. Hua, C. S. Mun, J. K. Jing, L. Kong, L. Y. Ern, N. A. Ashraf, S. W. Kit, T. S. Yee and M. R. Pichika, An update on natural compounds in the remedy of diabetes mellitus: A systematic review, *J. Tradit., Complementary Med.*, 2018, **8**(3), 361–376.
3. S. J. Stohs, H. Miller and G. R. Kaats, A review of the efficacy and safety of banaba (Lagerstroemia speciosa L.) and corosolic acid, *Phytother. Res.*, 2012, **26**(3), 317–324.
4. K. Johnston, P. Sharp, M. Clifford and L. Morgan, Dietary polyphenols decrease glucose uptake by human intestinal Caco-2 cells, *FEBS Lett.*, 2005, **579**(7), 1653–1657.
5. P. Chantre and D. Lairon, Recent findings of green tea extract AR25 (Exolise) and its activity for the treatment of obesity, *Phytomedicine*, 2002, **9**(1), 3–8.
6. S. S. Patel and R. K. Goyal, Cardioprotective effects of gallic acid in diabetes-induced myocardial dysfunction in rats, *Pharmacogn. Res.*, 2011, **3**(4), 239.
7. J. Singh, E. Cumming, G. Manoharan, H. Kalasz and E. Adeghate, Suppl 2: Medicinal chemistry of the anti-diabetic effects of Momordica charantia: active constituents and modes of actions, *Open Med. Chem. J.*, 2011, **5**, 70.

8. L. T. Marton, L. M. Pescinini-e-Salzedas, M. E. Camargo, S. M. Barbalho, J. F. Haber, R. V. Sinatora, C. R. Detregiachi, R. J. Girio and D. V. Buchaim, Cincotto dos Santos Bueno P. The Effects of Curcumin on Diabetes Mellitus: A Systematic Review, *Front. Endocrinol.*, 2021, **12**, 443.
9. D. W. Zhang, M. Fu, S. H. Gao and J. L. Liu, Curcumin and diabetes: a systematic review, *Evid. -Based Complementary Altern. Med.*, 2013, **2013**, 636053.
10. C. Andrade, N. G. Gomes, S. Duangsrisai, P. B. Andrade, D. M. Pereira and P. Valentao, Medicinal plants utilized in Thai Traditional Medicine for diabetes treatment: ethnobotanical surveys, scientific evidence and phytochemicals, *J. Ethnopharmacol.*, 2020, 113177.
11. Y. Dou, R. Huang, Q. Li, Y. Liu, Y. Li, H. Chen, G. Ai, J. Xie, H. Zeng and J. Chen, Luo C. Oxyberberine, an absorbed metabolite of berberine, possess superior hypoglycemic effect via regulating the PI3K/Akt and Nrf2 signaling pathways, *Biomed. Pharmacother.*, 2021, **137**, 111312.
12. Y. Chen, J. Meng, H. Li, H. Wei, F. Bi, S. Liu, K. Tang, H. Guo and W. Liu, Resveratrol exhibits an effect on attenuating retina inflammatory condition and damage of diabetic retinopathy via PON1, *Exp. Eye Res.*, 2019, **181**, 356–366.
13. J. Xue, W. Ding and Y. Liu, Anti-diabetic effects of emodin involved in the activation of PPARγ on high-fat diet-fed and low dose of streptozotocin-induced diabetic mice, *Fitoterapia*, 2010, **81**(3), 173–177.
14. A. Arvindekar, T. More, P. V. Payghan, K. Laddha, N. Ghoshal and A. Arvindekar, Evaluation of anti-diabetic and alpha glucosidase inhibitory action of anthraquinones from Rheum emodi, *Food Funct.*, 2015, **6**(8), 2693–2700.
15. C. Ceci, P. M. Lacal, L. Tentori, M. G. De Martino, R. Miano and G. Graziani, Experimental evidence of the antitumor, antimetastatic and antiangiogenic activity of ellagic acid, *Nutrients*, 2018, **10**(11), 1756.
16. J. M. Harnly, R. F. Doherty, G. R. Beecher, J. M. Holden, D. B. Haytowitz, S. Bhagwat and S. Gebhardt, Flavonoid content of US fruits, vegetables, and nuts, *J. Agric. Food Chem.*, 2006, **54**(26), 9966–9977.
17. Y. W. Huang, Q. Q. Zhu, X. Y. Yang, H. H. Xu, B. Sun, X. J. Wang and J. Sheng, Wound healing can be improved by (—)-epigallocatechin gallate through targeting Notch in streptozotocin-induced diabetic mice, *FASEB J.*, 2019, **33**(1), 953–964.
18. A. Ahad, M. Mujeeb, H. Ahsan and W. A. Siddiqui, Prophylactic effect of baicalein against renal dysfunction in type 2 diabetic rats, *Biochimie*, 2014, **106**, 101–110.
19. D. J. Den Hartogh and E. Tsiani, Antidiabetic properties of naringenin: A citrus fruit polyphenol, *Biomolecules*, 2019, **9**(3), 99.
20. A. Roohbakhsh, H. Parhiz, F. Soltani, R. Rezaee and M. Iranshahi, Molecular mechanisms behind the biological effects of hesperidin and hesperetin for the prevention of cancer and cardiovascular diseases, *Life Sci.*, 2015, **124**, 64–74.

21. R. Jayaraman, S. Subramani, S. H. Abdullah and M. Udaiyar, Antihyperglycemic effect of hesperetin, a citrus flavonoid, extenuates hyperglycemia and exploring the potential role in antioxidant and antihyperlipidemic in streptozotocin-induced diabetic rats, *Biomed. Pharmacother.*, 2018, **97**, 98–106.
22. J. J. Ramírez-Espinosa, J. Saldaña-Ríos, S. García-Jiménez, R. Villalobos-Molina, G. Ávila-Villarreal, A. N. Rodríguez-Ocampo, G. Bernal-Fernández and S. Estrada-Soto, Chrysin induces antidiabetic, antidyslipidemic and anti-inflammatory effects in athymic nude diabetic mice, *Molecules.*, 2018, **23**(1), 67.
23. M. Calderon-Montano, J. Burgos-Morón, E. Pérez-Guerrero and C. López-Lázaro, M. A review on the dietary flavonoid kaempferol, *Mini Rev. Med. Chem.*, 2011, **11**(4), 298–344.
24. Y. Zang, L. Zhang, K. Igarashi and C. Yu, The anti-obesity and anti-diabetic effects of kaempferol glycosides from unripe soybean leaves in high-fat-diet mice, *Food Funct.*, 2015, **6**(3), 834–841.
25. S. Shukla and S. Gupta, Apigenin: a promising molecule for cancer prevention, *Pharm. Res.*, 2010, **27**(6), 962–978.
26. C. Escande, V. Nin, N. L. Price, V. Capellini, A. P. Gomes, M. T. Barbosa, L. O'Neil, T. A. White, D. A. Sinclair and E. N. Chini, Flavonoid apigenin is an inhibitor of the NAD+ ase CD38: implications for cellular NAD+ metabolism, protein acetylation, and treatment of metabolic syndrome, *Diabetes*, 2013, **62**(4), 1084–1093.
27. X. Dai, Y. Ding, Z. Zhang, X. Cai, L. Bao and Y. Li, Quercetin but not quercitrin ameliorates tumor necrosis factor-alpha-induced insulin resistance in C2C12 skeletal muscle cells, *Biol. Pharm. Bull.*, 2013, **36**(5), 788–795.
28. H. Yang, T. Yang, C. Heng, Y. Zhou, Z. Jiang, X. Qian, L. Du, S. Mao, X. Yin and Q. Lu, Quercetin improves nonalcoholic fatty liver by ameliorating inflammation, oxidative stress, and lipid metabolism in db/db mice, *Phytother. Res.*, 2019, **33**(12), 3140–3152.
29. K. T. Akhilesh, K. B. Pravin, R. B. Jagdish, M. B. Dinesh, S. K. Mayuresh, M. A. Yogesh and B. B. Anand, Herbal antidiabetics: a review, *Int. J. Res. Pharm. Sci.*, 2011, **2**(1), 30–37.
30. A. de Sousa Leite, M. T. Islam, A. L. Júnior, J. M. E Sousa, M. V. de Alencar, M. F. Paz, H. M. Rolim, M. D. de Medeiros, A. A. de Carvalho Melo-Cavalcante and J. A. Lopes, Pharmacological properties of cashew (Anacardium occidentale), *Afr. J. Biotechnol.*, 2016, **15**(35), 1855–1863.
31. L. Tedong, P. Madiraju, L. C. Martineau, D. Vallerand, J. T. Arnason, D. D. Desire, L. Lavoie, P. Kamtchouing and P. S. Haddad, Hydro-ethanolic extract of cashew tree (Anacardium occidentale) nut and its principal compound, anacardic acid, stimulate glucose uptake in C2C12 muscle cells, *Mol. Nutr. Food Res.*, 2010, **54**(12), 1753–1762.

32. C. Soto, R. Mena, J. Luna, M. Cerbon, E. Larrieta, P. Vital, E. Uria, M. Sanchez, R. Recoba, H. Barron and L. Favari, Silymarin induces recovery of pancreatic function after alloxan damage in rats, *Life Sci.*, 2004, **75**(18), 2167–2180.
33. M. K. Apaya, T. F. Kuo, M. T. Yang, G. Yang, C. L. Hsiao, S. B. Chang, Y. Lin and W. C. Yang, Phytochemicals as modulators of β-cells and immunity for the therapy of type 1 diabetes: Recent discoveries in pharmacological mechanisms and clinical potential, *Pharmacol. Res.*, 2020, **156**, 104754.
34. Y. Zhang, J. Cai, H. Ruan, H. Pi and J. Wu, Antihyperglycemic activity of kinsenoside, a high yielding constituent from Anoectochilus roxburghii in streptozotocin diabetic rats, *J. Ethnopharmacol.*, 2007, **114**(2), 141–145.
35. A. N. Singab, H. A. El-Beshbishy, M. Yonekawa, T. Nomura and T. Fukai, Hypoglycemic effect of Egyptian Morus alba root bark extract: effect on diabetes and lipid peroxidation of streptozotocin-induced diabetic rats, *J. Ethnopharmacol.*, 2005, **100**(3), 333–338.
36. J. S. Noh, C. H. Park and T. Yokozawa, Treatment with oligonol, a low-molecular polyphenol derived from lychee fruit, attenuates diabetes-induced hepatic damage through regulation of oxidative stress and lipid metabolism, *Br. J. Nutr.*, 2011, **106**(7), 1013–1022.
37. S. Park, D. S. Kim and S. Kang, Gastrodia elata Blume water extracts improve insulin resistance by decreasing body fat in diet-induced obese rats: vanillin and 4-hydroxybenzaldehyde are the bioactive candidates, *Eur. J. Nutr.*, 2011, **50**(2), 107–118.
38. C. L. Chang, Y. Lin, A. P. Bartolome, Y. C. Chen, S. C. Chiu and W. C. Yang, Herbal therapies for type 2 diabetes mellitus: chemistry, biology, and potential application of selected plants and compounds, *J. Evidence-Based Complementary Altern. Med.*, 2013, **2013**, article number 378657.
39. C. J. Muller, E. Joubert, D. De Beer, M. Sanderson, C. J. Malherbe, S. J. Fey and J. Louw, Acute assessment of an aspalathin-enriched green rooibos (Aspalathus linearis) extract with hypoglycemic potential, *Phytomedicine*, 2012, **20**(1), 32–39.
40. A. S. Rocca, J. LaGreca, J. Kalitsky and P. L. Brubaker, Monounsaturated fatty acid diets improve glycemic tolerance through increased secretion of glucagon-like peptide-1, *Endocrinology*, 2001, **142**(3), 1148–1155.
41. G. M. Hussein, H. Matsuda, S. Nakamura, T. Akiyama, K. Tamura and M. Yoshikawa, Protective and ameliorative effects of maté (Ilex paraguariensis) on metabolic syndrome in TSOD mice, *Phytomedicine*, 2011, **19**(1), 88–97.
42. A. N. Kesari, R. K. Gupta and G. Watal, Hypoglycemic effects of Murraya koenigii on normal and alloxan-diabetic rabbits, *J. Ethnopharmacol.*, 2005, **97**(2), 247–251.

43. M. Naimi, F. Vlavcheski, H. Shamshoum and E. Tsiani, Rosemary extract as a potential anti-hyperglycemic agent: Current evidence and future perspectives, *Nutrients*, 2017, **9**(9), 968.
44. M. Fujii, I. Takei and K. Umezawa, Antidiabetic effect of orally administered conophylline-containing plant extract on streptozotocin-treated and Goto-Kakizaki rats, *Biomed. Pharmacother.*, 2009, **63**(10), 710–716.
45. M. Kawakami, A. Hirayama, K. Tsuchiya, H. Ohgawara, M. Nakamura and K. Umezawa, Promotion of β-cell differentiation by the alkaloid conophylline in porcine pancreatic endocrine cells, *Biomed. Pharmacother.*, 2010, **64**(3), 226–231.
46. Y. Zhang, J. Cai, H. Ruan, H. Pi and J. Wu, Antihyperglycemic activity of kinsenoside, a high yielding constituent from Anoectochilus roxburghii in streptozotocin diabetic rats, *J. Ethnopharmacol.*, 2007, **114**(2), 141–145.
47. G. Ristagno, F. Fumagalli, C. Porretta-Serapiglia, A. Orrù, C. Cassina, M. Pesaresi, S. Masson, L. Villanova, A. Merendino, A. Villanova and L. Cervo, Hydroxytyrosol attenuates peripheral neuropathy in streptozotocin-induced diabetes in rats, *J. Agric. Food Chem.*, 2012, **60**(23), 5859–5865.
48. T. H. Chen, S. C. Chen, P. Chan, Y. L. Chu, H. Y. Yang and J. T. Cheng, Mechanism of the hypoglycemic effect of stevioside, a glycoside of Stevia rebaudiana, *Planta Med.*, 2005, **71**(02), 108–113.
49. M. F. Azevedo, C. F. Lima, M. Fernandes-Ferreira, M. J. Almeida, J. M. Wilson and C. Pereira-Wilson, Rosmarinic acid, major phenolic constituent of Greek sage herbal tea, modulates rat intestinal SGLT1 levels with effects on blood glucose, *Mol. Nutr. Food Res.*, 2011, **55**(S1), S15–25.
50. S. A. Noureddin, R. M. El-Shishtawy and K. O. Al-Footy, Curcumin analogues and their hybrid molecules as multifunctional drugs, *Eur. J. Med. Chem.*, 2019, **182**, 111631.
51. M. Bnouham, A. Ziyyat, H. Mekhfi, A. Tahri and A. Legssyer, Medicinal plants with potential antidiabetic activity-A review of ten years of herbal medicine research (1990-2000), *Int. J. Diabetes Metabol.*, 2006, **14**(1), 1.
52. D. K. Patel, K. Patel, M. Rahman and S. Chaudhary Therapeutic potential of "Aegeline," an important phytochemical of Aegle marmelos: Current health perspectives for the treatment of disease, in *Nanomedicine for Bioactives*, Springer, Singapore, 2020, pp. 383–392.
53. J. Eliza, P. Daisy, S. Ignacimuthu and V. Duraipandiyan, Normo-glycemic and hypolipidemic effect of costunolide isolated from Costus speciosus (Koen ex. Retz.) Sm. in streptozotocin-induced diabetic rats, *Chem. –Biol. Interact.*, 2009, **179**(2–3), 329–334.
54. G. Brahmachari Andrographolide: A molecule of antidiabetic promise, in *Discovery and Development of Antidiabetic Agents from Natural Products*, 2016, ed. G. Brahmachari, pp. 1–27.

55. R. Rivera-Madrid, M. Aguilar-Espinosa, Y. Cárdenas-Conejo and L. E. Garza-Caligaris, Carotenoid derivates in achiote (Bixa orellana) seeds: synthesis and health promoting properties, *Front. Plant Sci.*, 2016, **7**, 1406.
56. T. Miura, N. Ueda, K. Yamada, M. Fukushima, T. Ishida, T. Kaneko, F. Matsuyama and Y. Seino, Antidiabetic effects of corosolic acid in KK-Ay diabetic mice, *Biol. Pharm. Bull.*, 2006, **29**(3), 585–587.
57. Jun Watanabe, *et al.*, "Isolation and identification of α-glucosidase inhibitors from tochu-cha (Eucommia ulmoides).", *Biosci., Biotechnol., Biochem.*, 1997, **61.1**, 177–178.
58. H. Gao, Y. N. Huang, P. Y. Xu and J. Kawabata, Inhibitory effect on α-glucosidase by the fruits of Terminalia chebula Retz, *Food Chem.*, 2007, **105**(2), 628–634.
59. R. Rabasa-Lhoret and J. L. Chiasson α-Glucosidase inhibitors, in *International Textbook of Diabetes Mellitus*, John Wiley & Sons Ltd., UK, 3rd edn, 2004, vol. 1, pp. 901–914.
60. Q. Abbas, M. Hassan, H. Raza, S. J. Kim, K. W. Chung, G. H. Kim and S. Y. Seo, In vitro, in vivo and in silico anti-hyperglycemic inhibition by sinigrin, *Asian Pac. J. Trop. Med.*, 2017, **10**(4), 372–379.
61. A. Sundarram and T. P. Murthy, α-amylase production and applications: a review, *J. Appl. Environ. Microbiol.*, 2014, **2**(4), 166–175.
62. R. T. Dewi, Y. M. Iskandar, M. Hanafi, L. B. Kardono, M. Angelina, I. D. Dewijanti and S. D. Banjarnahor, Inhibitory effect of Koji Aspergillus terreus on a-glucosidase activity and postprandial hyperglycemia, *Pak. J. Biol. Sci.*, 2007, **10**(18), 3131–3135.
63. M. Brownlee, Biochemistry and molecular cell biology of diabetic complications, *Nature*, 2001, **414**(6865), 813–820.
64. C. Veeresham, A. Rama Rao and K. Asres, Aldose reductase inhibitors of plant origin, *Phytother. Res.*, 2014, **28**(3), 317–333.
65. J. Bai, Y. Wang, X. Zhu and J. Shi, Eriodictyol inhibits high glucose-induced extracellular matrix accumulation, oxidative stress, and inflammation in human glomerular mesangial cells, *Phytother. Res.*, 2019, **33**(10), 2775–2782.
66. M. Kawakami, A. Hirayama, K. Tsuchiya, H. Ohgawara, M. Nakamura and K. Umezawa, Promotion of β-cell differentiation by the alkaloid conophylline in porcine pancreatic endocrine cells, *Biomed. Pharmacother.*, 2010, **64**(3), 226–231.
67. R. Saito, S. Yamada, Y. Yamamoto, T. Kodera, A. Hara, Y. Tanaka, F. Kimura, I. Takei, K. Umezawa and I. Kojima, Conophylline suppresses pancreatic stellate cells and improves islet fibrosis in Goto-Kakizaki rats, *Endocrinology*, 2012, **153**(2), 621–630.
68. G. F. Raoof and K. Y. Mohamed Natural products for the management of diabetes, in *Studies in Natural Products Chemistry*, Elsevier, 2018, vol. 59, pp. 323–374.
69. L. Tie, Y. An, J. Han, Y. Xiao, Y. Xiaokaiti, S. Fan, S. Liu, A. F. Chen and X. Li, Genistein accelerates refractory wound healing by suppressing superoxide and FoxO1/iNOS pathway in type 1 diabetes, *J. Nutrit. Biochem.*, 2013, **24**(1), 88–96.

70. M. S. Islam and H. Choi, Dietary red chilli (Capsicum frutescens L.) is insulinotropic rather than hypoglycemic in type 2 diabetes model of rats, *Phytother. Res.*, 2008, **22**(8), 1025–1029.
71. R. P. Nankar and M. Doble, Ellagic acid potentiates insulin sensitising activity of pioglitazone in L6 myotubes, *J. Funct. Foods.*, 2015, **15**, 1–10.
72. N. S. Etukudoh, O. M. Uchejeso and I. I. Etim, Bitter Kola (Garcinia kola) as Herbal Remedy for Diabetes, *Curr. Top. Med. Med. Res.*, 2021, **12**, 61–68.
73. K. G. Wasana, A. P. Attanayake, K. A. Jayatilaka and T. P. Weerarathna, Antidiabetic Activity of Widely Used Medicinal Plants in the Sri Lankan Traditional Healthcare System: New Insight to Medicinal Flora in Sri Lanka, *Evid. -Based Complementary Altern. Med.*, 2021, **2021**, 6644004.
74. N. K. Ruvin Kumara, R. N. Pathirana and C. Pathirana, Hypoglycemic activity of the root and stem of Salacia reticulata. var. β-diandra. in alloxan diabetic rats, *Pharm. Biol.*, 2005, **43**(3), 219–225.
75. H. Matsuda, T. Murakami, K. Yashiro, J. Yamahara and M. Yoshikawa, Antidiabetic principles of natural medicines. IV. Aldose reductase and α-glucosidase inhibitors from the roots of Salacia oblonga Wall. (Celastraceae): structure of a new friedelane-type triterpene, kotalagenin 16-acetate, *Chem. Pharm. Bull.*, 1999, **47**(12), 1725–1729.
76. N. K. Ruvin Kumara, C. Pathirana, S. Lekamwasam and R. N. Pathirana Anti-hyperglycemic activity of Salacia reticulata var. b-diandra in non insulin dependent diabetic (NIDDM) patients, in *Proceedings of the Annual Meeting of the American Society of Pharmacognosy*, 2003, p. 175.
77. P. J. Watkins, *ABC of Diabetes*, BMJ Publishing Group, London, 5th edn, 2003.
78. C. E. Mogensen Pharmacotherapy of diabetes, in *Pharmacotherapy of Diabetes: New Developments*, Springer, Boston, MA, 2007, pp. 277–281.
79. G. Di Fabio, V. Romanucci, M. Zarrelli, M. Giordano and A. Zarrelli, C-4 gem-dimethylated oleanes of Gymnema sylvestre and their pharmacological activities, *Molecules*, 2013, **18**(12), 14892–14919.
80. T. Bandara, B. Rokeya, S. Khan, L. Ali, S. Ekanayake, E. R. Jansz and K. Balasubramanium, Effects of Gymnema lactiferum leaf on glycemic and lipidemic status in type 2 diabetic subjects, *Bangladesh J. Pharmacol.*, 2009, **4**(2), 92–95.
81. J. Poongunran, H. K. Perera, W. I. Fernando, L. Jayasinghe and R. Sivakanesan, α-Glucosidase and α-amylase inhibitory activities of nine Sri Lankan antidiabetic plants, *J. Pharm. Res. Int.*, 2015, 365–374.
82. M. Salazar-García and J. C. Corona, The Use of Natural Compounds as a Strategy to Counteract Oxidative Stress in Animal Models of Diabetes Mellitus, *Int. J. Mol. Sci.*, 2021, **22**(13), 7009.
83. N. M. Faheem and A. El Askary, Neuroprotective role of curcumin on the hippocampus against the structural and serological alterations of streptozotocin-induced diabetes in Sprague Dawely rats, *Iran. J. Basic Med. Sci.*, 2017, **20**(6), 690.

84. F. Pivari, A. Mingione, C. Brasacchio and L. Soldati, Curcumin and type 2 diabetes mellitus: prevention and treatment, *Nutrients*, 2019, **11**(8), 1837.
85. M. Rashedinia, M. Alimohammadi, N. Shalfroushan, M. J. Khoshnoud, M. Mansourian, N. Azarpira and Z. Sabahi, Neuroprotective Effect of Syringic Acid by Modulation of Oxidative Stress and Mitochondrial Mass in Diabetic Rats, *BioMed Res. Int.*, 2020, **2020**, 8297984.
86. W. Chatuphonprasert, T. Lao-Ong and K. Jarukamjorn, Improvement of superoxide dismutase and catalase in streptozotocin–nicotinamide-induced type 2-diabetes in mice by berberine and glibenclamide, *Pharm. Biol.*, 2014, **52**(4), 419–427.
87. C. Liu, Z. Wang, Y. Song, D. Wu, X. Zheng, P. Li, J. Jin, N. Xu and L. Li, Effects of berberine on amelioration of hyperglycemia and oxidative stress in high glucose and high fat diet-induced diabetic hamsters in vivo, *BioMed. Res. Int.*, 2015, **2015**, 313808.
88. A. Kumar, K. Chopra, M. Mukherjee, R. Pottabathini and D. K. Dhull, Current knowledge and pharmacological profile of berberine: an update, *Eur. J. Pharmacol.*, 2015, **761**, 288–297.
89. M. Goboza, Y. G. Aboua, N. Chegou and O. O. Oguntibeju, Vindoline effectively ameliorated diabetes-induced hepatotoxicity by docking oxidative stress, inflammation and hypertriglyceridemia in type 2 diabetes-induced male Wistar rats, *Biomed. Pharmacother.*, 2019, **112**, 108638.
90. Z. Tian, N. Ren, J. Wang, D. Zhang and Y. Zhou, Ginsenoside ameliorates cognitive dysfunction in type 2 diabetic goto-kakizaki rats, *Med. Sci. Monit.: Int. Med. J. Exp. Clin. Res.*, 2018, **24**, 3922.
91. M. A. Hossain, D. Lee, B. Kim, C. W. Kang, N. S. Kim and J. H. Kim, Korean Red Ginseng attenuates type 2 diabetic cardiovascular dysfunction in Otsuka Long-Evans Tokushima Fatty rats, *J. Ginseng Res.*, 2020, **44**(2), 308–311.
92. M. C. Picot and M. F. Mahomoodally, Effects of Aphloia theiformis on key enzymes related to diabetes mellitus, *Pharm. Biol.*, 2017, **55**(1), 864–872.
93. N. I. Kashchenko, N. K. Chirikova and D. N. Olennikov, Agrimoniin, an active ellagitannin from Comarum palustre herb with anti-α-glucosidase and antidiabetic potential in streptozotocin-induced diabetic rats, *Molecules*, 2017, **22**(1), 73.
94. J. Poongunran, H. K. Perera, L. Jayasinghe, I. T. Fernando, R. Sivakanesan, H. Araya and Y. Fujimoto, Bioassay-guided fractionation and identification of α-amylase inhibitors from Syzygium cumini leaves, *Pharm. Biol.*, 2017, **55**(1), 206–211.
95. H. El Abed, B. Khemakhem, I. Fendri, M. Chakroun, M. Triki, N. Drira and H. Mejdoub, Extraction, partial purification and characterization of amylase from parthenocarpic date (Phoenix dactylifera): effect on cake quality, *J. Sci. Food Agric.*, 2017, **97**(10), 3445–3452.
96. X. U. Guang-Kai, Q. I. Xiao-Ying, W. A. Guo-Kai, X. I. Guo-Yong, L. I. Xu-Sen, S. U. Chen-Yu, L. I. Bao-Lin and Q. I. Min-Jian, Antihyperglycemic,

antihyperlipidemic and antioxidant effects of standard ethanol extract of Bombax ceiba leaves in high-fat-diet-and streptozotocin-induced Type 2 diabetic rats, *Chin. J. Nat. Med.*, 2017, **15**(3), 168–177.
97. M. A. Abd Eldaim, A. Shaban Abd Elrasoul and S. A. Abd Elaziz, An aqueous extract from Moringa oleifera leaves ameliorates hepatotoxicity in alloxan-induced diabetic rats, *Biochem. Cell Biol.*, 2017, **95**(4), 524–530.
98. E. I. Omodanisi, Y. G. Aboua and O. O. Oguntibeju, Assessment of the anti-hyperglycaemic, anti-inflammatory and antioxidant activities of the methanol extract of Moringa oleifera in diabetes-induced nephrotoxic male wistar rats, *Molecules*, 2017, **22**(4), 439.
99. S. Stephen Irudayaraj, S. Christudas, S. Antony, V. Duraipandiyan, A. D. Naif Abdullah and S. Ignacimuthu, Protective effects of Ficus carica leaves on glucose and lipids levels, carbohydrate metabolism enzymes and β-cells in type 2 diabetic rats, *Pharm. Biol.*, 2017, **55**(1), 1074–1081.

CHAPTER 6

Inhalation of Insulin for Diabetes Management

F. BAHMAN[a] AND S. TAURIN*[b]

[a] Department of Molecular Medicine, Kuwait Ministry of Health, Kuwait
[b] Department of Molecular Medicine, Princess Al-Jawhara Centre for Molecular Medicine, School of Medicine and Medical Sciences, Arabian Gulf University, Kingdom of Bahrain
*Email: sebastient@agu.edu.bh

6.1 Introduction

The number of people suffering from diabetes mellitus worldwide increased from approximately 200 million in 1990 to 463 million in 2019, and is projected to reach 700 million by 2045;[1] the highest rise in cases is expected to occur on the African continent and in the Middle East.[1,2] The disease's economic burden and its complications are also staggering, with an estimated cost of $2.1 trillion in 2030.[3] The increased incidence is observed for both types 1 and 2 diabetes. The rapid increase in type 1 diabetes, especially in the young population below 15 years of age, is not uniform worldwide and seems to be linked to non-genetic factors, including environmental and/or lifestyle.[4–6] Type 2 diabetes accounts for 90% of cases, and its rise is associated with a sedentary lifestyle, unhealthy diet, rampant obesity, aging, and also better management of the disease, which has improved the survival of people with diabetes.[7,8] Insulin remains the mainstay treatment for type 1 diabetes and is used in type 2 diabetes for individuals whose glucose level is not controlled by lifestyle intervention or oral hypoglycaemic agents.

Nanoscience & Nanotechnology Series No. 59
Nanotechnology for Diabetes Management
Edited by Amar Abderrahmani, Sabine Szunerits, Rabah Boukerroub and Abdelfattah El Ouaamari

Published by the Royal Society of Chemistry, www.rsc.org

The predominant insulin roles regulate and maintain blood glucose levels through cellular uptake and glucose storage in skeletal muscles, liver, and adipocytes. Since the first oral antidiabetic medication, sulfonylurea, was launched, pharmaceutical companies have been investing extensively in the production of antidiabetic drugs.[9] Insulin analogues with various onsets and durations of action are commonly available. The glucose level control requires multiple daily subcutaneous insulin injections (3 to 5 per day), with long-acting insulin to address the hypoglycaemic period between meals and during the night, and short-acting insulin to manage hyperglycaemia after meals.[10] Insulin is commonly taken through subcutaneous injections using syringes, pens, or insulin pumps. However, this injection route is frequently associated with pain at the injection site, lipohypertrophy, iatrogenic hyperinsulinemia, insulin resistance, and a non-negligible needle phobia, leading to suboptimal medication compliance and diabetes-associated complications.[11–15] Also, alternative delivery for the management of type 2 diabetes is required, as the maintenance of a continuous insulin level through injection is not needed.

Furthermore, despite the progress made in pharmacotherapy, the delivery devices, and the self-monitoring systems, a large proportion of diabetic patients fail to reach optimum glucose control.[16,17] For several years, alternative non-invasive insulin delivery routes to mimic endogenous insulin secretion and improve safety and compliance have been tested, including oral, buccal, transdermal, peritoneal, pulmonary, vaginal, rectal, and nasal routes. Inhaled insulin was the first reported alternative to subcutaneous injection and constituted an attractive alternative for insulin absorption and systemic delivery through the large lung alveolar capillary network. The inhalation of drugs could allow their systemic delivery within seconds or minutes.[18] Also, the degradation of proteins in the alveoli is limited, with more than 95% of proteins, including insulin, being absorbed intact. This chapter focuses on inhaled insulin, the inhalation route's physiology, the different inhaled insulin formulations, and their advantages and limitations.

6.2 Inhalation Route

Shortly after the isolation of insulin by Banting and Best at the University of Toronto in 1921,[19] inhaled aerosolized insulin was assessed 3–4 years later in diabetic patients.[20,21] The inhaled insulin decreased the blood glucose levels in a dose-dependent manner.[20–22] This concept of inhalation of insulin for the management of diabetes was revived using a pressurized metered-dose inhaler in the late 1950s and 1960s.[23] A hypoglycaemic effect was elicited, but the delivery of macromolecules by inhalation remained unreliable and raised concerns about the long-term safety of this approach.[23] In 1971, aerosolized bovine or porcine insulin delivered using a nebulizer to diabetic rabbits demonstrated temporary hypoglycaemia.[24] Later, semi-synthetic insulin administered to six diabetic children with a nebulizer allowed them to reach blood glucose control like those following subcutaneous injection.[25] Still, the bioavailability of inhaled insulin remained significantly lower when compared to that subcutaneously injected.[25] Further development of inhaled insulin delivery relied on new insulin formulations and delivery devices. Nektar Therapeutics developed the Pulmosphere™ technology (HMR 4006, Exubera) using a spray-drying process to deliver the aerosolized insulin powder deep into the lungs. The inhaler technology was licensed to Pfizer, which developed the device for the market in collaboration with Sanofi-Aventis. Subsequently, several inhalers were developed and these are discussed in this chapter.

6.2.1 Respiratory System Anatomy and Physiology

In recent years, the progress made in the aerosolization of drugs for systemic drug delivery has revived interest in the management of diabetes using inhaled insulin. The lungs demonstrate several characteristics favourable for insulin absorption, including large surface area, thin epithelium, less proteolytic activity, and extensive vascularization.[26]

The airways are comprised of the conducting and respiratory airways. They differ by their surface area and by the pseudostratified epithelium's nature that constitutes the barrier to absorption into the bloodstream. The conducting airways start at the mouth and nose and include the pharynx, larynx, trachea, bronchi, bronchioles, and terminal bronchioles, as shown

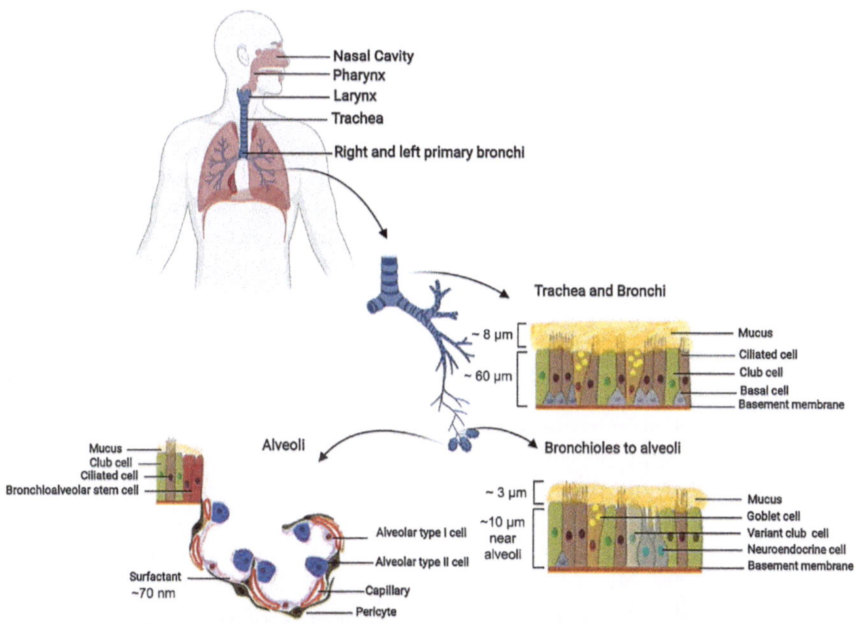

Figure 6.1 The anatomy of respiratory tract and the respiratory epithelium development.

in Figure 6.1. The surface area of the conducting airways is relatively small (2–3 m^2) and characterized by a thick epithelium and fluid layer. The pseudostratified columnar epithelium gradually thins as it gets closer to the respiratory portion of the lungs and is composed of many ciliated and goblet mucus-secreting cells that constitute the mucociliary escalator.[18]

In the nasal cavities, particles larger than 3–10 μm are efficiently trapped by the mucus blanket.[27] Inhaled particles that sediment in the conducting airways are expelled out of the respiratory tract by the mucociliary escalator. Before reaching the alveoli, the airways undergo more than 17 bifurcations.[28]

The respiratory airways facilitate the gas exchange and comprise the respiratory bronchioles, alveolar ducts, alveolar sacs, and alveoli. The alveolar ducts are 1 mm long on average, formed by a collection of connected alveoli with an average diameter of 250 μm. The alveoli of the lungs constitute a large surface area (70–100 m^2), are highly vascularized, and covered by a thin epithelial layer (0.1–0.4 μm) and a liquid lining layer of 70 nm,[29] which allows the rapid absorption of drugs and particles into the systemic circulation and bypassing of the first-pass metabolism.[30] The epithelium of the alveoli is composed of type I and II pneumocytes. The type I cells form a simple squamous epithelium covering 95% of the internal surface of the alveoli cells and facilitate the diffusion and exchange of gas. The type II cells are multifunctional as they participate in the immune response; they act as a precursor of type I and type II if damage occurs in the alveoli and produce the lipid–protein surfactant that reduces the surface tension at the interface of air and liquid.[31,32] Both types of alveolar cells can endocytose particles with sizes up to 200 nm.[33] The insoluble particles reaching the alveoli are eliminated by the resident alveolar macrophages that destroy or sequester the insoluble materials, possibly for years.

6.2.2 Physiological Barriers for Lung Absorption

The inhaled drug absorption rate in the lung depends on several barriers, such as physical clearance mechanisms of the mucus secretory cells, alveolar macrophage clearance, and pulmonary proteases metabolism. These aspects are important to facilitate drug absorption efficiency.

6.2.2.1 Mucociliary Clearance

A set of protective systems have evolved to defend the lungs from potentially dangerous viral and bacterial infections, inhaled particles, and gaseous waste, while keeping the tissue in an almost sterile state.[34] Lung defence relies on several different mechanisms, including cough, anatomical defences, aerodynamic changes, and immune systems, but mucociliary clearance (MCC) is the most effective.[35] MCC is the lungs' primary innate defence mechanism.[35] The essential components of this defence system include the airway surface layer (ASL), composed of the protective mucus layer entrapping the inhaled foreign bodies and particles, and the periciliary layer formed of a low-viscosity liquid layer that lubricates the airway surface and facilitates the movement of the cilia, which are specialized organelles located on the cell surface.[36] The mucus secretory cells or goblet cells are intercalated with the cilia cells and constitute 20% and 60% of the cells lining the trachea, respectively.[35] Goblet cells heavily secrete glycosylated proteins called mucins, forming the mucus layer composed of 97% water and 3% mucins and other molecules.[37] As the lungs bifurcate, the percentage of ciliated and goblet cells decreases and they are replaced by serous and club cells, which produce watery secretions. Club cells can also act as progenitors of goblet cells following bronchiolar injury.[38] The low-viscosity liquid forms the periciliary layer and comprises membrane-associated mucins and other molecules, including glycolipids.[39,40] The periciliary layer lubricates the cilia and protects the cell surface from inhaled particles. The inhaled particles are either rapidly absorbed in the alveoli or cleared by the MCC. The percentage of the inhaled particles reaching the alveoli rarely exceeds 20% when using a pressurized metered-dose inhaler.[41] Other devices, such as spacers and the synchronization of breath and actuation, can elevate the dose reaching the alveoli over 20% of the dose administrated. The inhaled particle deposit is influenced by the characteristics of the aerosolized particle droplets, including shape, size, mass, and density. The velocity and time of residence of the aerosolized particles are also critical and characteristic of lung physiology. The MCC transports most of the non-absorbed inhaled particles to the trachea, where they are then swallowed into the gastrointestinal tract.[41]

6.2.2.2 *Alveolar Macrophages*

As the MCC mainly removes the inhaled particles in the upper airway tract, alveolar macrophages (AMs) are involved in the degradation and/or confinement of inhaled particles in the deep lung. AMs are lung phagocytic mononuclear cells that can polarize into different classes.[42,43] Interstitial macrophages (IMs) and AMs both patrol the lungs. The IMs are mainly located between the vascular endothelium and alveolar epithelium but can migrate into the alveolus.[44] The phagocytic capacity of IMs is less compared to the AMs but they secrete a higher level of interleukin-10, which negatively regulates the recruitment of neutrophils and other mononuclear cells.[45] In humans, one AM per 0.0171 mm^2 is detected in the alveolus, suggesting 14–15 AMs per alveoli.[46] AMs phagocyte particles less than 2 μm with 85–90% efficacy; the clearance of the small particles can take up to 200 days.[46]

6.2.2.3 *Peptidase and Protease Activity*

The absorption of the inhaled particles into the systemic circulation is essentially located in the alveoli. The dissolved particles can be absorbed by diffusion or specific carriers depending on the hydrophobic or hydrophilic nature of the compound. The transports of peptides and proteins are poorly characterized and may involve paracellular transport for small proteins or peptides less than 40 kDa and transcytosis for proteins larger than 40 kDa.[28] The transport of inhaled insulin into the systemic circulation depends on it reaching the deepest part of the lungs and the alveoli.[47] The involvement of dynamin-independent and dynamin-dependent transcytosis, independent of clathrin and caveolin, was demonstrated to transport insulin across both primary rat AT1-like and AT2 cells, respectively.[48] The relative contribution of these protein transport mechanisms for insulin remains to be elucidated as no specific transporter for insulin has been identified to date.[18] The proteolytic activity in the alveolar cells is insufficient to degrade the aerosolized proteins rapidly, including insulin.[49,50] The AT2 cells are likely to degrade the inhaled insulin slowly.[50] Although the AMs express higher quantities of proteolytic enzymes than the epithelial cells, their contribution to the degradation of proteins, including insulin, remains negligible.[50] The activity of these proteolytic enzymes is also profoundly affected by external factors such as smoking.[51]

6.3 Deep Lung Deposit of Inhaled Particles

Understanding the architecture and complex lung bifurcations, the progress achieved in aerosol, formulations, and delivery devices made the deep deposition of insulin possible. The lung deposition of inhaled aerosolized particles is essentially controlled by their aerodynamic diameter.[52] Inhaled aerosolized particles less than 5 μm can migrate into the lungs. Particles with a diameter of 1–2 μm, if slowly and deeply inhaled, can deposit in the lungs with an efficacy of 90%, primarily in the alveoli.[18,52] Particles less than 1 μm are exhaled without deposition, and larger ones (5–10 μm) are trapped by the mucus before reaching the lungs, as mentioned previously. Several additional factors contribute to the deposition of the inhaled particles, including density, shape, airflow velocity and volume, respiratory rate, and interpatient variations.[29] Multiple mechanisms have been involved in the deposition of inhaled particles, such as diffusion, impaction, and sedimentation.[53] The impaction is dependent on the density, size greater than 2 μm, and velocity of the inhaled particles, and is expected to take place in the upper airways due to the inertial forces.[54] Particles less than 0.5 μm are likely to deposit by diffusion due to Brownian motion, intensifying as the particles get smaller.[55] Sedimentation is the main mechanism for the deposition of particles between 0.5–2 μm; the mechanism is dependent on the settling velocity and residence time greater than the distance required for the contact to the surface airway.[29] The particle deposition is also largely dependent on the airflow pattern instigated by the inhalation.[56] Nebulizers use a passive inhalation method, where slow inhalation may increase lung deposition.[57] The pressurized metered-dose inhaler produces a high-velocity aerosol that requires the timing of the breathing pattern, which can increase the lung deposition from 7% to 20%.[58] Dry-powder inhalers necessitate sufficient airflow for deep lung deposition.[29] Each delivery system requires a specific breathing pattern to improve the deep lung deposition. The variation in the deposition is also affected by person-to-person lung anatomical differences that, combined with the breathing pattern, can create a tidal volume from 460 to 900 mL in adults and a variation in particle deposition by twofold.[29,59]

6.4 Inhaled Insulin Delivery Devices

Several devices have been developed for the deep lung delivery of insulin (see Table 6.1). These inhalers were designed to deliver a consistent and optimum insulin dose to maintain a normal blood glucose level. Also, these devices were designed to be easily used and portable, to facilitate patient compliance and overcome the psychological barrier of the subcutaneous insulin injection. The insulin formulations developed for inhalation were either based on liquid or dry powder. These devices differed by the droplet sizes generated, the insulin release mechanisms, and the mechanical or electronic regulations of the insulin released. Based on these differences, the bioavailability of the inhaled insulin was specific to each device and varied between 10% and 46% for the dry powder and liquid insulin, respectively.[60] Most of the inhaled insulin is lost in the oropharynx or upper airways.[60]

6.4.1 Exubera

Exubera was the first rapid inhaled insulin on the market, approved by the Food and Drugs Administration (FDA) and European Medicines Agency's Committee (EMA) in 2006 to treat diabetes types 1 and 2.[61] Exubera was initially developed by Nektar Therapeutics, and the technology was later licensed to Pfizer, which partnered with Sanofi-Aventis to propel the device onto the market after 15 years of development. The device delivers a dry insulin powder containing sodium citrate, mannitol, glycine, and sodium hydroxide as excipients.[62] The powder is held in pre-packaged blister packets containing 1 or 3 mg of recombinant human insulin produced by a non-pathogenic laboratory strain of *Escherichia coli* (K12) and equivalent to 3 and 8 IU of subcutaneous short-acting insulin, respectively.[63] Up to 45% of the 1 mg blister content, and up to 25% of the 3 mg blister content, may be retained in the blister.[64] The pharmacokinetics of inhaled insulin delivered by Exubera in healthy volunteers is similar to that of postprandial insulin secretion.[65] The inhaled insulin has a faster onset of action than the regular insulin or fast-acting insulin (Lispro).[65] Inhaled insulin also has a longer duration of action than fast-acting insulin, and is comparable to regular insulin.[65] The safety and efficacy of Exubera were further demonstrated in 53 adults with type 2 diabetes for 11 years. The randomized open-labelled clinical trial had two arms: subcutaneous insulin or preprandial inhaled insulin combined with bedtime subcutaneous long-acting ultralente insulin injection for 12 weeks.[66,67]

Table 6.1 Insulin inhalation systems: past and present.[a]

Inhalation system	Company	Insulin formulation	Insulin dose	Particle size (μm)	Inhaler device/ method of inhalation	Device size	Current status	References
Exubera	Nektar Therapeutics and Pfizer	Dry powder Blisters	1 mg = 3 IU 3 mg = 8 IU	3	Mechanical Breath actuated	20 cm × 4 cm	FDA approved in 2006 then withdrawn in 2007	62, 63
AERx iDMS	Aradigm and Novo Nordisk A/S	Liquid insulin Blisters	1 AERx unit = 1 IU	2–3	Electromechanical Guided system	8 cm × 4 cm	Cancelled	77, 78
AIR	Alkermes Inc. and Eli Lilly	Dry powder Capsules	0.9 mg = 2 IU 2.6 mg = 6 IU	5–30	Mechanical Breath actuated	7 cm × 2 cm	Cancelled	84, 86
Technosphere (Afrezza)	Mannkind Corporation	Dry powder Microspheres Cartridges	4 TU = 10 IU 8 TU = 20 IU 12 TU = 30 IU	2.5	Mechanical Breath actuated	10 cm × 5 cm	FDA approved in 2014	90,91

[a]IU = International Unit, TU = Technosphere Unit, mg = milligram.

The inhaled insulin treatment improved haemoglobin A1c (HbA1c) compared with the baseline and achieved a glycaemic control comparable to that attained using conventional subcutaneous injected insulin.[66,67] A similar study involving type 1 diabetes demonstrated a similar improvement of HbA1c and a glycaemic control indissociable to the conventional subcutaneous injected insulin.[68] In phase III clinical trials enrolling patients with types 1 and 2 diabetes, the inhaled insulin decreased HbA_{1c} to the same extent as the subcutaneous insulin.[69–71] In type 1 diabetes patients, inhaled insulin was also more efficient in reducing the fasting and postprandial glucose levels than subcutaneous insulin.[69–71] Hypoglycaemia and cough were the most frequent inhaled insulin adverse effects, hypoglycaemia being observed as frequently as with subcutaneous insulin.[69,70] Less than 1% of the patients suffering from cough had to discontinue the treatment.[69] Anti-insulin antibodies mean titres were also higher in the inhaled insulin group as compared to the subcutaneous insulin injection and higher when comparing type 1 to type 2 diabetes patients.[69,70] The pulmonary function was slightly (less than 2%) but consistently reduced in the inhaled insulin group, being early, non-progressive, and reversible if treatment was ceased.[72] The patient satisfaction rate was also very high, as most patients (85%) opted to continue, and 75% decided to use inhaled insulin.[73] The FDA and EMA approved Exubera to control types 1 and 2 diabetes in adults in 2006. The use of inhaled insulin was not approved for smokers and those who had smoked in the previous 6 months as Pfizer and Nektar Pharmaceutics reported an increased incidence of lung cancers in the group treated with Exubera.[10,74] The side effects commonly observed from multiple clinical trials included cough, pharyngitis, respiratory infection, and rhinitis.[75,76] Less frequent side effects included sinusitis, shortness of breath, otitis media, ear pain, dry mouth, pneumonia, asthma, and other seasonal illnesses.[75] One year after its approval, Pfizer decided to remove Exubera from the market due to poor sales, totalling only 12 million for the first 9 months of 2007.[77] The inhaled insulin delivered by Exubera required a higher dose and showed similar activity to the subcutaneous insulin, but the excess cost outweighed the needle phobia for most patients.[10] Exubera was more expensive than the injectable insulin, $5 *versus* $2–3 per day, respectively. The inhaler was also large and inconvenient, and the dosing was confusing. The discontinuation of Exubera cost Pfizer a pre-tax loss of $2.6 billion.[78]

6.4.2 AERx iDMS

The AERx Insulin Diabetes Management System (iDMS) was developed by a collaboration between Aradigm Corporation and Novo Nordisk AS. Subsequently, Novo Nordisk AS obtained all the development and manufacturing rights.[79] AERx iDMS is an inhalation-activated electromechanical system delivering aerosolized liquid human insulin deep into the lungs. The microprocessor-controlled actuator allows for a single-unit dose increment.[80] The extrusion of the liquid insulin from single-use strips under pressure through laser-generated perforations generates droplets of 2–3 μm. The system then guides the patient to breathe at the optimal speed and depth to maximize the deep lung deposition of insulin.[81] Mechanical pressure is used to generate the aerosol without the need for a propellant. The emitted dose accounts for 57% of the loaded dose into the strip. Each AERx strip contains 2.6 mg or 75 IU of insulin, approximately equivalent to 10 IU of subcutaneously injected insulin.[81] The system can deliver between 1–10 AERx units of insulin with each inhaled AERx unit mimicking the effect of 1 IU of subcutaneously delivered insulin.[80] In a randomized, open-label, five-period crossover trial, type 1 diabetes patients were treated with various doses of inhaled insulin (0.3, 0.6, 1.2, and 1.8 IU kg^{-1} body weight) using the AERx IDMS device compared to subcutaneous insulin injection (0.12 IU kg^{-1} body weight).[82] The inhaled insulin had a more rapid onset, like short-acting insulin, when compared to subcutaneous insulin administration; the system efficiency of the inhaled insulin averaged 13% of the loaded dose during the first 6 h.[82] The clinical efficacy of AERx iDMS was assessed in a randomized, 12-week, open-label, parallel, multicentre, multinational trial in 107 non-smoking patients with type 2 diabetes for an average of 11.9 years.[83] Patients received either fast-acting inhaled insulin using AERx iDMS immediately before meals or fast-acting subcutaneous insulin, both in combination with Neutral Protamine Hagedorn (NPH) insulin given subcutaneously in the evening. The AERx iDMS-delivered insulin showed similar efficacy to subcutaneous insulin for the glycaemic control over 3 months, the HbA1c levels were similar, and fasting serum glucose was lower in the AERx iDMS group.[83] The adverse effects were also similar in both groups; pulmonary function was not altered by the inhaled insulin delivered by AERx iDMS.[83] A similar study conducted with type 1 diabetes compared AERx iDMS inhaled insulin to aspart insulin, combined with NPH as basal insulin.[84] After a year, the fasting plasma glucose and HbA1c levels were similar in both groups.[84] The risk of hypoglycaemia and pulmonary disfunction were similar in both groups, but increased incidence in nocturnal hypoglycaemic events and higher insulin antibody mean titres were observed in the AERx iDMS group.[84] The cough associated with inhaled insulin delivery was reported in 10% of the patients in the AERx iDMS group, with 5% related to the inhaled insulin device itself.[84] The incidence of cough with liquid inhaled insulin was lower than that reported with the usage of Exubera (27%) and the inhalation of dry insulin powder.[69,84] The compliance of the patients using the AERx iDMS device was also above 90%.[85] The development of the fast-acting inhaled insulin product AERx iDMS device was discontinued by Novo Nordisk AS in January 2008 while in a phase III clinical trial based on economic considerations, just 3 months after the announcement of the discontinuation of Exubera by Pfizer.[86]

6.4.3 AIR Insulin

The AIR insulin system was developed through a collaboration between Alkermes Inc. and Eli Lilly and Co. The system used dry insulin powder and delivered regular insulin in large porous particles of 5–30 μm to the deep lung.[87] Porous particles of small mass density but with large size are delivered efficiently to the deep lung.[88] The large size of the porous particles decreases the cohesion and promotes the dispersion of the powder, and the low density of the particles allows them to behave aerodynamically like small particles.[88] The inhaler was breath-activated and did not require large input energy to form an aerosol powder.[87] The insulin is filled into interchangeable unit-dose cellulose capsules supplied in blister strips, with each capsule containing either 2 or 6 IU of injected rapid-acting insulin.[87] The evaluation of the AIR insulin system in healthy volunteers demonstrated insulin absorption within 45 minutes, comparable to lispro.[89] The 6 IU capsule contained 2.6 mg of insulin and had a bioavailability of around 8% compared to Lispro.[89] In a phase II clinical trial, type 1 diabetes patients experienced a higher frequency of nocturnal hypoglycaemia following inhalation of insulin using the AIR system.[90] The delivery of insulin by the AIR system did not affect pulmonary function.[87] In a randomized, open-label, active-comparator, parallel group phase III clinical trial, the AIR insulin system was evaluated in type 1 diabetes patients and patients receiving subcutaneous regular insulin or lispro at mealtimes; all groups received Glargine once daily.[91] The results demonstrated that only 20% of patients reached the targeted HbA1c in both treatment groups; the HbA1c was significantly lower by 0.44% in the group treated with subcutaneous insulin.[91] There was no difference between the groups in the insulin doses or hypoglycaemic events at the endpoint.[91] Both groups experienced a decrease in lung function; the inhaled insulin group was associated with decreased carbon monoxide (CO) diffusion and increased coughing incidence.[91] The development and licence agreement of the AIR system was terminated while in a phase III clinical trial in 2008 by Eli Lilly for reasons independent of the safety or efficacy of the device.[91]

6.4.4 Technosphere (Afrezza)

At a time when several companies, some mentioned above, withdrew from the development of inhaled insulin, Technosphere, later renamed Afrezza, continued its development. Technosphere was developed by Mannkind Corp., providing the dry powder of recombinant human insulin combined with the MedTone breath-actuated inhaler developed by Pharmaceutical Discovery Corporation. The MedTone inhaler did not require propulsion synchronization by the user. Since 2010, Mannkind Corp. has subsequently developed the Gen2 inhaler to decrease the size of the device and the capability to deliver equivalent doses using a lower cartridge load.[92] Cartridges contain either 1.05 mg (30 IU) insulin (9.9 mg inhalation powder), 0.7 mg (20 IU) insulin

(6.7 mg inhalation powder), or 0.35 mg (10 IU) insulin (3.3 mg inhalation powder), corresponding to 12, 8, and 4 units, respectively.[93,94]

Technosphere insulin is very rapid-acting inhaled insulin similar to subcutaneous prandial insulin. Technosphere insulin is composed of recombinant human insulin mixed with the excipient bis-3,6(4-fumarylaminobutyl)-2,5-diketopiperazine (FDKP) and polysorbate 80.[95] The FDKP forms microcrystalline plates in slightly acidic conditions, then self-assembles into microparticles onto which proteins like insulin are absorbed.[95] The insulin is encapsulated into the FDKP microspheres and then freeze-dried to form a fine powder. The insulin microspheres generated have an average diameter of 2.5 μm (range 0.5–5.8 μm), suitable for deep lung insulin delivery as the particles rapidly dissolved at neutral pH upon contact with the alveolar cells.[95,96] The Technosphere insulin developed by Mannkind differed from other delivery systems as the insulin used was monomeric rather than the hexameric one used by the other delivery systems, and improved its rapid action.[74] Technosphere insulin was initially evaluated in healthy volunteers and type 2 diabetes patients and revealed rapid systemic uptake of the inhaled insulin (approximately 12–14 min), fast rapid action (maximum activity 20–30 min), and short duration of action (2–3 h).[97] The relative bioavailability of the MedTone-delivered inhaled insulin (100 IU) was 50% for the first 3 h and 30% for 6 h compared to subcutaneous human regular insulin injection.[97] The repeated inhalation of Technosphere insulin (100 IU) in type 2 diabetes subjects demonstrated less variability than that observed with the subcutaneous injection of regular insulin in healthy volunteers.[97] This study did not report any adverse effects associated with using the Technosphere inhaled insulin.[97] The Technosphere insulin/MedTone inhaler, assessed in multiple clinical trials, overall was demonstrated to be safe and well-tolerated in healthy volunteers and type 2 diabetes patients. The Technosphere insulin/MedTone inhaler was efficacious without hypoglycaemic events or weight gain reported in the subcutaneous injection of regular or fast-acting insulin.[98] The Technosphere insulin/MedTone inhaler was also evaluated in type 2 diabetic non-smokers ($n = 12$) and smokers ($n = 12$) in a phase I, open-label, single-dose clinical trial.[99] Each group received 30 units of Technosphere insulin while fasting and 4–6 h after beginning an intravenous insulin infusion.[99] The absorption of the Technosphere insulin did not seem to affect type 2 diabetic smokers.[99] Since 2010, a second-generation inhaler, Gen2 inhaler, has been used and evaluated in two phase III clinical trials (Affinity 1 and Affinity 2).[100,101] The Affinity 1 trial was a 24-week, open-label, noninferiority study that compared the efficacy of Technosphere insulin using the Gen2 inhaler to a subcutaneous rapid-acting insulin aspart in type 1 diabetes patients. The endpoint of the study was the HbA1c level. Both groups were treated with basal insulin at a dose used before the enrolment. At the end of the study, the HbA1c levels were higher in the aspart group (−0.4%) than in those using Technosphere insulin (−0.21%). The percentage of patients reaching an HbA1c level of less than 7% was also higher in the aspart-treated group. The Affinity 2 trial was a 24-week, randomized, double-blind, placebo-controlled study involving insulin-naïve type 2 diabetes patients whose disease was uncontrolled. After 6 weeks on the previous regimen, patients were assigned to Technosphere insulin or Technosphere placebo. At week 24, the Technosphere insulin reduced the HbA1c level by 0.4% compared to the placebo control, and more patients reached the target HbA1c level of 7% (37.7% *versus* 19%) in the Technosphere insulin group. A third phase III clinical trial compared prandial Technosphere inhaled insulin plus insulin glargine to twice daily biaspart in type 2 diabetes patients.[102] The clinical trial was a multinational, randomized, open-label, parallel-group study, and enrolled adults with type 2 diabetes and poor glycaemic control despite insulin therapy, with or without oral anti-diabetes drugs.[102] Patients were randomly allocated in two groups in a 1 : 1 ratio and received 52 weeks' treatment with prandial Technosphere inhaled insulin powder plus bedtime insulin glargine subcutaneous injection, or twice-daily premixed biaspart insulin (70% insulin aspart protamine suspension and 30% insulin aspart of recombinant DNA origin) by subcutaneous injection. The primary endpoint was the HbA1c level. The change in HbA1c level in the Technosphere inhaled insulin-treated group was similar and non-inferior to the biaspart insulin-treated group.[102] Patients had fewer mild-to-moderate and severe hypoglycaemic events and significantly less weight gain on inhaled insulin plus insulin glargine than on biaspart insulin. Both treatment regimens had similar safety and tolerability profile, but intermittent or single-episode cough and change in asymptomatic pulmonary function were observed in the group receiving inhaled insulin plus insulin glargine.[102] In a phase I clinical trial using a Gen2 inhaler, Afrezza had a relative bioavailability of 33% compared to the subcutaneously injected insulin Lispro, but double the relative bioavailability achieved with Exubera.[103,104] Overall, Afrezza has a quicker onset but a shorter duration of action, which supports its classification as an ultra-rapid-acting inhaled insulin.[104,105]

Afrezza side effects commonly reported in clinical trials were cough, hypoglycaemia, and throat pain or irritation.[106] The rate of coughing is less with Afrezza as compared to Exubera, but was observed in 27% of patients.[107,108] The forced expiratory volume in 1 second (FEV1) declined by 40 mL more in patients treated with inhaled insulin than those with injected insulin.[108] FEV1 started to decline within 3 months of treatment initiation and persisted over 2 years.[108] Acute bronchospasm had also been reported in patients with asthma or COPD, and follow-up recommendations contraindicated the use of Afrezza in patients with chronic obstructive pulmonary disease or FEV1 decline by more than 20%.[106,108,109] Afrezza is contraindicated for patients with hypersensitivity to regular insulin.[106,109] Also, Afrezza did not exhibit carcinogenic potential after a 2-year follow-up study in rats in a preclinical study.[110] In the registration trial, two heavy smoker patients in the Afrezza group developed lung cancer (0.8 cases per 1000 patient-years).[108,111] After completing the study, two additional cases of squamous lung cancer in non-smokers were reported, but the overall incidence was within the normal range of the general population.[108,111]

An original new drug application by Mannkind to the FDA for Afrezza (Technosphere) was initially submitted in 2009 using the MedTone inhaler; a modified version of the new drug application using Gen2 was later submitted in 2013.[100,101] In June 2014, the FDA approved Afrezza to treat types 1 and 2 diabetes with a Risk Evaluation and Mitigation Strategy (REMS) for a communication plan to inform healthcare providers about the risk of acute bronchospasm associated with Afrezza. The FDA lifted the REMS after revision in 2015. Afrezza is the only insulin inhaler on the US market. Afrezza was not recommended for smokers or those with a recent history of smoking (less than 6 months), people experiencing chronic lung problems such as asthma, cancer, or chronic obstructive pulmonary disease, and the treatment of diabetic ketoacidosis.[100,101] The cost of Afrezza inhaled insulin ranges from US$371.69 to US$1141.42 for 90 cartridges of 4 and 12 units, respectively, as seen on the GoodRx website (https://www.goodrx.com/Afrezza), with some insurance companies covering part of the cost. The costs of fast-acting injectable insulin Lispro (Humalog) (https://www.goodrx.com/humalog) and Aspart (Novolog) (https://www.goodrx.com/novolog) are US$163 and US$314, respectively, for 1000 units. The relative cost per unit of inhaled insulin is nearly 3–5 times the cost of fast-acting injectable insulin without insurance. By 2017, Afrezza was included by the majority of US insurance companies; the inhaler's small size and safety profile, and the perspective of a history-tracking device bluhale, were all factors in attracting consumers. A favourable payer coverage is essential to create a dynamic market. Afrezza was approved in Brazil in 2019,[112] and Mannkind established several distribution agreements for distribution outside the US.[113] Afrezza is still not approved in Canada or by the EMA.

6.4.5 Other Inhaled Insulin Delivery Devices

Several other devices have been developed and discontinued, such as Spiros manufactured by Dura Pharmaceuticals and Eli Lilly that reached a phase I clinical trial, ProMaxx manufactured by Epic Therapeutic-Baxter Healthcare Corporation that reached a phase II clinical trial, Aerodose Insulin developed by Aerogen Inc. that reached a phase II clinical trial. Aerodose Insulin entered a phase II clinical trial and showed promising results[114,115] but was discontinued due to a lack of funding.

Another system, named Dance-501, manufactured by Aerami Therapeutics (Dance Biopharm), is in development and has entered phases I and II clinical trials. Dance 501 is a battery-operated handheld breath-synchronized nebulizer using a vibrating mesh micropump developed by Aerogen Ltd to aerosolize a liquid insulin formulation.[116] The device generates aerosol only when the inspiratory flow is within 7–14 $L\,min^{-1}$ to provide optimal lung deposition.[116] The aerosol particle sizes generated were between 3.5 and 6 μm.[116] In a clinical trial, Dance 501 demonstrated favourable glucodynamic properties, with an earlier onset of action and higher early effect than Lispro administered subcutaneously.[117]

6.5 Conclusion

The delivery of insulin by inhalation is an exciting concept and has generated many expectations. The data obtained from various clinical trials using Exubera, Afrezza, and others have demonstrated that inhaled insulin is adequate for managing types 1 and 2 diabetes and are better tolerated than subcutaneously injected insulin. Inhaled insulin treatment was also well-accepted among users. The cost of treatment with inhaled insulin is also expected to be higher, mainly associated with the cost of development and, to a lesser extent, the larger insulin doses used. The economic failure of Exubera and concerns about the long-term safety of inhaled insulin for smokers have profoundly impacted the development of this route of insulin delivery.[118] Effective inhalation depends on several factors, including the device, the drug, and the patient. Drug formulations and devices are continuously being improved to promote deep lung deposition and rapid absorption. Patients' and healthcare providers' understanding and training are critical for inhaled insulin delivery to succeed. Only short-acting formulations of inhaled insulin are currently offered; most patients still are required to inject regular or long-acting insulin. It is doubtful that inhaled insulin will overcome the need for subcutaneous injection. Nearly 100 years since the first attempt at inhaled insulin, much progress is still needed to understand the factors for efficient deep lung deposition and absorption in the systemic circulation. With the advances in glucose monitoring, the design of better dosing algorithms, and improved delivery devices, it is hoped that inhaled insulin can be part of the safer and more effective management of diabetes.

References

1. International Diabetes Federation, *IDF Diabetes Atlas-9th Edition Brussels*, International Diabetes Federation, Belgium, 2019.
2. X. Lin, Y. Xu, X. Pan, J. Xu, Y. Ding and X. Sun, *et al.*, Global, regional, and national burden and trend of diabetes in 195 countries and territories: an analysis from 1990 to 2025, *Sci. Rep.*, 2020, **10**(1), 1–11.
3. C. Bommer, V. Sagalova, E. Heesemann, J. Manne-Goehler, R. Atun and T. Bärnighausen, *et al.*, Global economic burden of diabetes in adults: projections from 2015 to 2030, *Diabetes Care*, 2018, **41**(5), 963–970.
4. K. Vehik and D. Dabelea, The changing epidemiology of type 1 diabetes: why is it going through the roof?, *Diabetes/Metabol. Res. Rev.*, 2011, **27**(1), 3–13.
5. L. A. DiMeglio, C. Evans-Molina and R. A. Oram, Type 1 diabetes, *Lancet*, 2018, **391**(10138), 2449–2462.
6. N. Cho, J. Shaw, S. Karuranga, Y. Huang, J. da Rocha Fernandes and A. Ohlrogge, *et al.*, IDF Diabetes Atlas: Global estimates of diabetes prevalence for 2017 and projections for 2045, *Diabetes Res. Clin. Pract.*, 2018, **138**, 271–281.

7. S. Basu, P. Yoffe, N. Hills and R. H. Lustig, The relationship of sugar to population-level diabetes prevalence: an econometric analysis of repeated cross-sectional data, *PLoS One*, 2013, **8**(2), e57873.
8. V. Bellou, L. Belbasis, I. Tzoulaki and E. Evangelou, Risk factors for type 2 diabetes mellitus: an exposure-wide umbrella review of meta-analyses, *PLoS One*, 2018, **13**(3), e0194127.
9. S. Raptis and G. Dimitriadis, Oral hypoglycemic agents: insulin secretagogues, α-glucosidase inhibitors and insulin sensitizers, *Exp. Clin. Endocrinol. Diabetes*, 2001, **109**(Suppl 2), S265–S287.
10. L. D. Mastrandrea, Inhaled insulin: overview of a novel route of insulin administration, *Vasc. Health Risk Manage.*, 2010, **6**, 47–58.
11. R. P. Radermecker, G. E. Piérard and A. J. Scheen, Lipodystrophy reactions to insulin: effects of continuous insulin infusion and new insulin analogs, *Am. J. Clin. Dermatol.*, 2007, **8**(1), 21–28.
12. J. M. Gregory, A. D. Cherrington and D. J. Moore, The peripheral peril: Injected insulin induces insulin insensitivity in type 1 diabetes, *Diabetes*, 2020, **69**(5), 837–847.
13. S. Wright, M. Yelland, K. Heathcote, S.-K. Ng and G. Wright, Fear of needles-nature and prevalence in general practice, *Aus. Fam. Physician*, 2009, **38**(3), 172–176.
14. P. M. Ho, J. S. Rumsfeld, F. A. Masoudi, D. L. McClure, M. E. Plomondon and J. F. Steiner, *et al.*, Effect of medication nonadherence on hospitalization and mortality among patients with diabetes mellitus, *Arch. Intern. Med.*, 2006, **166**(17), 1836–1841.
15. L. E. Egede, M. Gebregziabher, C. Echols and C. P. Lynch, Longitudinal effects of medication nonadherence on glycemic control, *Ann. Pharm.*, 2014, **48**(5), 562–570.
16. M. K. Ali, K. M. Bullard, J. B. Saaddine, C. C. Cowie, G. Imperatore and E. W. Gregg, Achievement of goals in US diabetes care, 1999–2010, *N. Eng. J. Med.*, 2013, **368**(17), 1613–1624.
17. G. Carls, J. Huynh, E. Tuttle, J. Yee and S. V. Edelman, Achievement of glycated hemoglobin goals in the US remains unchanged through 2014, *Diabetes Ther.*, 2017, **8**(4), 863–873.
18. J. S. Patton and P. R. Byron, Inhaling medicines: delivering drugs to the body through the lungs, *Nat. Rev. Drug Discovery*, 2007, **6**(1), 67–74.
19. M. Fralick and B. Zinman, The discovery of insulin in Toronto: beginning a 100 year journey of research and clinical achievement, *Diabetologia*, 2021, **64**(5), 947–953.
20. W. Heubner, Uber inhalation von insulin, *Klin. Wochenschr.*, 1924, **51**, 2342–2343.
21. M. Gänsslen, Uber inhalation von insulin, *Klin. Wochenschr.*, 1925, **4**, 71.
22. G. Scheuch and R. Siekmeier, Novel approaches to enhance pulmonary delivery of proteins and peptides, *J. Physiol. Pharmacol.*, 2007, **58**(5), 615–625.

23. S. W. Stein and C. G. Thiel, The History of Therapeutic Aerosols: A Chronological Review, *J. Aerosol Med. Pulm. Drug Delivery*, 2017, **30**(1), 20–41.
24. F. M. Wigley, J. H. Londono, S. H. Wood, J. C. Shipp and R. H. Waldman, Insulin across respiratory mucosae by aerosol delivery, *Diabetes*, 1971, **20**(8), 552–556.
25. R. B. Elliott, B. W. Edgar, C. C. Pilcher, C. Quested and J. McMaster, Parenteral absorption of insulin from the lung in diabetic children, *J. Paediatric. Child Health*, 1987, **23**(5), 293–297.
26. M. Sakagami, In vivo, in vitro and ex vivo models to assess pulmonary absorption and disposition of inhaled therapeutics for systemic delivery, *Adv. Drug Delivery Rev.*, 2006, **58**(9), 1030–1060.
27. R. Dahl and N. Mygind, Anatomy, physiology and function of the nasal cavities in health and disease, *Adv. Drug Delivery Rev.*, 1998, **29**(1–2), 3–12.
28. J. S. Patton, Mechanisms of macromolecule absorption by the lungs, *Adv. Drug Delivery Rev.*, 1996, **19**(1), 3–36.
29. K. P. O'Donnell and H. D. Smyth. *Macro-and Microstructure of the Airways for Drug Delivery*, Controlled Pulmonary Drug Delivery, Springer, 2011. p. 1-19.
30. E. Fröhlich, A. Mercuri, S. Wu and S. Salar-Behzadi, Measurements of Deposition, Lung Surface Area and Lung Fluid for Simulation of Inhaled Compounds, *Front. Pharmacol.*, 2016, **7**, 181.
31. H. Ward and T. Nicholas, Alveolar type I and type II cells, *Aust. N. Z. J. Med.*, 1984, **14**, 731–734.
32. A. V. Andreeva, M. A. Kutuzov and T. A. Voyno-Yasenetskaya, Regulation of surfactant secretion in alveolar type II cells, *Am. J. Physiol.: Lung Cell. Mol. Physiol.*, 2007, **293**(2), L259–L271.
33. A. Lankoff, W. J. Sandberg, A. Wegierek-Ciuk, H. Lisowska, M. Refsnes and B. Sartowska, *et al.*, The effect of agglomeration state of silver and titanium dioxide nanoparticles on cellular response of HepG2, A549 and THP-1 cells, *Toxicol. Lett.*, 2012, **208**(3), 197–213.
34. R. P. Dickson and G. B. Huffnagle, The Lung Microbiome: New Principles for Respiratory Bacteriology in Health and Disease, *PLoS Pathog.*, 2015, **11**(7), e1004923.
35. X. M. Bustamante-Marin and L. E. Ostrowski, Cilia and mucociliary clearance, *Cold Spring Harbor Perspect. Biol.*, 2017, **9**(4), a028241.
36. M. R. Knowles and R. C. Boucher, Mucus clearance as a primary innate defense mechanism for mammalian airways, *J. Clin. Investig.*, 2002, **109**(5), 571–577.
37. Z. Chen, M. Zhong, Y. Luo, L. Deng, Z. Hu and Y. Song, Determination of rheology and surface tension of airway surface liquid: a review of clinical relevance and measurement techniques, *Respir. Res.*, 2019, **20**(1), 274.
38. Y. Zhu, C. Ehre, L. H. Abdullah, J. K. Sheehan, M. Roy and C. M. Evans, *et al.*, Munc13-2 −/− baseline secretion defect reveals

source of oligomeric mucins in mouse airways, *J. Physiol.*, 2008, **586**(7), 1977–1992.
39. S. H. Randell and R. C. Boucher, Effective mucus clearance is essential for respiratory health, *Am. J. Respir. Cell Mol. Biol.*, 2006, **35**(1), 20–28.
40. B. Button, L. H. Cai, C. Ehre, M. Kesimer, D. B. Hill and J. K. Sheehan, *et al.*, A periciliary brush promotes the lung health by separating the mucus layer from airway epithelia, *Science*, 2012, **337**(6097), 937–941.
41. T. C. Carvalho, J. I. Peters and R. O. Williams, Influence of particle size on regional lung deposition – What evidence is there?, *Int. J. Pharm.*, 2011, **406**(1), 1–10.
42. P. Allavena and A. Mantovani, Immunology in the clinic review series; focus on cancer: tumour-associated macrophages: undisputed stars of the inflammatory tumour microenvironment, *Clin. Exp. Immunol.*, 2012, **167**(2), 195–205.
43. E. Weagel, C. Smith, P. Liu, R. Robison and K. O'Neill, Macrophage polarization and its role in cancer, *J. Clin. Cell. Immunol.*, 2015, **6**(4), 338.
44. C. E. Boorsma, C. Draijer and B. N. Melgert, Macrophage heterogeneity in respiratory diseases, *Mediat. Inflammat.*, 2013, **2013**, 769214.
45. H. Kawano, H. Kayama, T. Nakama, T. Hashimoto, E. Umemoto and K. Takeda, IL-10-producing lung interstitial macrophages prevent neutrophilic asthma, *Int. Immunol.*, 2016, **28**(10), 489–501.
46. M. Geiser, Update on macrophage clearance of inhaled micro-and nanoparticles, *J. Aerosol Med. Pulm. Drug Delivery*, 2010, **23**(4), 207–217.
47. V. Codrons, F. Vanderbist, B. Ucakar, V. Préat and R. Vanbever, Impact of formulation and methods of pulmonary delivery on absorption of parathyroid hormone (1–34) from rat lungs, *J. Pharm. Sci.*, 2004, **93**(5), 1241–1252.
48. M. Ikehata, R. Yumoto, Y. Kato, J. Nagai and M. Takano, Mechanism of insulin uptake in rat alveolar type II and type I-like epithelial cells, *Biol. Pharm. Bull.*, 2009, **32**(10), 1765–1769.
49. K. Okumura, S. Iwakawa, T. Yoshida, T. Seki and F. Komada, Intratracheal delivery of insulin absorption from solution and aerosol by rat lung, *Int. J. Pharm.*, 1992, **88**(1–3), 63–73.
50. G. L. Finch, Nonclinical pharmacology and safety studies of insulin administered to the respiratory tract, *Inhalation Toxicol.*, 2005, 609.
51. E. Fröhlich, Toxicity of orally inhaled drug formulations at the alveolar barrier: parameters for initial biological screening, *Drug Delivery*, 2017, **24**(1), 891–905.
52. P. R. Byron and J. S. Patton, Drug delivery via the respiratory tract, *J. Aerosol Med.*, 1994, **7**(1), 49–75.
53. D. C. Muir, Aerosol Deposition in the Lungs of Space Travellers, *Nature*, 1966, **209**(5026), 921.

54. M. Lippmann, D. Yeates and R. Albert, Deposition, retention, and clearance of inhaled particles, *Occup. Environ. Med.*, 1980, **37**(4), 337–362.
55. P. Hofemeier and J. Sznitman, Revisiting pulmonary acinar particle transport: convection, sedimentation, diffusion, and their interplay, *J. Appl. Physiol.*, 2015, **118**(11), 1375–1385.
56. L. Borgström, B. Olsson and L. Thorsson, Degree of throat deposition can explain the variability in lung deposition of inhaled drugs, *J. Aerosol Med.*, 2006, **19**(4), 473–483.
57. M. Svartengren, K. Svartengren, F. Aghaie, K. Philipson and P. Camner, Lung deposition and extremely slow inhalations of particles. Limited effect of induced airway obstruction, *Exp. Lung Res.*, 1999, **25**(4), 353–366.
58. S. P. Newman, A. Weisz, N. Talaee and S. Clarke, Improvement of drug delivery with a breath actuated pressurised aerosol for patients with poor inhaler technique, *Thorax*, 1991, **46**(10), 712–716.
59. W. Hofmann, L. Morawska and R. Bergmann, Environmental tobacco smoke deposition in the human respiratory tract: differences between experimental and theoretical approaches, *J. Aerosol. Med.*, 2001, **14**(3), 317–326.
60. J. S. Patton, J. G. Bukar and M. A. Eldon, Clinical pharmacokinetics and pharmacodynamics of inhaled insulin, *Clin. Pharmacokinet.*, 2004, **43**(12), 781–801.
61. J. Lenzer, Inhaled insulin is approved in Europe and United States, *Br. Med. J.*, 2006, 69–77.
62. N. Sadrzadeh, D. P. Miller, D. Lechuga-Ballesteros, N. J. Harper, C. L. Stevenson and D. B. Bennett, Solid-state stability of spray-dried insulin powder for inhalation: chemical kinetics and structural relaxation modeling of Exubera above and below the glass transition temperature, *J. Pharm. Sci.*, 2010, **99**(9), 3698–3710.
63. S. Bellary and A. H. Barnett, Inhaled insulin (Exubera®): combining efficacy and convenience, *Diabetes Vascular Dis. Res.*, 2006, **3**(3), 179–185.
64. R. Jani, C. Triplitt, C. Reasner and R. A. DeFronzo, First approved inhaled insulin therapy for diabetes mellitus, *Expert Opin. Drug Delivery*, 2007, **4**(1), 63–76.
65. K. Rave, S. Bott, L. Heinemann, S. Sha, R. H. Becker and S. A. Willavize, *et al.*, Time-action profile of inhaled insulin in comparison with subcutaneously injected insulin lispro and regular human insulin, *Diabetes Care*, 2005, **28**(5), 1077–1082.
66. W. T. Cefalu, J. S. Skyler, I. A. Kourides, W. H. Landschulz, C. C. Balagtas and S.-L. Cheng, *et al.*, Inhaled human insulin treatment in patients with type 2 diabetes mellitus, *Ann. Intern. Med.*, 2001, **134**(3), 203–207.
67. R. A. Gerber, J. C. Cappelleri, I. A. Kourides and R. A. Gelfand, Treatment satisfaction with inhaled insulin in patients with type 1 diabetes: a randomized controlled trial, *Diabetes Care*, 2001, **24**(9), 1556–1559.

68. J. S. Skyler, W. T. Cefalu, I. A. Kourides, W. H. Landschulz, C. C. Balagtas and S.-L. Cheng, *et al.*, Efficacy of inhaled human insulin in type 1 diabetes mellitus: a randomised proof-of-concept study, *Lancet*, 2001, **357**(9253), 331–335.
69. T. Quattrin, A. Bélanger, N. J. Bohannon and S. L. Schwartz, Efficacy and safety of inhaled insulin (Exubera) compared with subcutaneous insulin therapy in patients with type 1 diabetes: results of a 6-month, randomized, comparative trial, *Diabetes Care*, 2004, **27**(11), 2622–2627.
70. P. A. Hollander, L. Blonde, R. Rowe, A. E. Mehta, J. L. Milburn and K. S. Hershon, *et al.*, Efficacy and safety of inhaled insulin (exubera) compared with subcutaneous insulin therapy in patients with type 2 diabetes: results of a 6-month, randomized, comparative trial, *Diabetes Care*, 2004, **27**(10), 2356–2362.
71. R. A. DeFronzo, R. M. Bergenstal, W. T. Cefalu, J. Pullman, S. Lerman and B. W. Bode, *et al.*, Efficacy of inhaled insulin in patients with type 2 diabetes not controlled with diet and exercise: a 12-week, randomized, comparative trial, *Diabetes Care*, 2005, **28**(8), 1922–1928.
72. A. H. Barnett and S. Bellary, Inhaled human insulin (Exubera®): clinical profile and patient considerations, *Vasc. Health Risk Manag.*, 2007, **3**(1), 83.
73. J. Rosenstock, J. C. Cappelleri, B. Bolinder and R. A. Gerber, Patient satisfaction and glycemic control after 1 year with inhaled insulin (Exubera) in patients with type 1 or type 2 diabetes, *Diabetes Care*, 2004, **27**(6), 1318–1323.
74. J. Kling, Inhaled insulin's last gasp?, *Nat. Biotechnol.*, 2008, **26**(5), 479–80.
75. P. F. Hite, A. M. Barnes and P. E. Johnston, Exhuberance Over Exubera. Clinical, *Diabetes*, 2006, **24**(3), 110–114.
76. J. Oleck, S. Kassam and J. D. Goldman, Commentary: Why Was Inhaled Insulin a Failure in the Market?, *Diabetes Spectrum.*, 2016, **29**(3), 180–184.
77. G. S. Mack, Pfizer dumps Exubera, *Nat. Biotechnol.*, 2007, **25**(12), 1331–1332.
78. Wonder. Inhaled Insulin Products 2018 [Available from: https://askwonder.com/research/ inhaled-insulin-products-63gkne97z].
79. Insulin Inhalation:NN 1998, *Drugs R&D*, 2004, **5**(1), 46–49.
80. S. Mudaliar, Inhaled insulin using AERx® insulin Diabetes Management System (AERx® iDMS), *Expert Opin. Invest. Drugs*, 2007, **16**(10), 1673–1681.
81. J. Thipphawong, B. Otulana, P. Clauson, J. Okikawa and S. J. Farr, Pulmonary insulin administration using the AERx® insulin diabetes system, *Diabetes Technol. Ther.*, 2002, **4**(4), 499–504.
82. G. Brunner, B. Balent, M. Ellmerer, L. Schaupp, A. Siebenhofer and J. Jendle, *et al.*, Dose-response relation of liquid aerosol inhaled insulin in type I diabetic patients, *Diabetologia*, 2001, **44**(3), 305–308.
83. K. Hermansen, T. Rönnemaa, A. H. Petersen, S. Bellaire and U. Adamson, Intensive therapy with inhaled insulin via the AERx

insulin diabetes management system: a 12-week proof-of-concept trial in patients with type 2 diabetes, *Diabetes Care*, 2004, **27**(1), 162–167.
84. R. G. Moses, P. Bartley, H. Lunt, R. O'Brien, T. Donnelly and M. A. Gall, *et al.*, Safety and efficacy of inhaled insulin (AERx® iDMS 1) compared with subcutaneous insulin therapy in patients with Type 1 diabetes: 1-yeardata from a randomized, parallel group trial, *Diabetic Med.*, 2009, **26**(3), 260–267.
85. J. Cramer, J. Okikawa, S. Bellaire and P. Clauson, Compliance with inhaled insulin treatment using the AERx® iDMS insulin diabetes management system, *Diabetes Technol. Ther.*, 2004, **6**(6), 800–807.
86. A. Opar, Another blow for inhaled protein therapeutics, *Nat. Rev. Drug Discovery*, 2008, **7**(3), 189–190.
87. D. B. Muchmore, B. Silverman, A. D. L. Peña and J. Tobian, The AIR® Inhaled Insulin System: System Components and Pharmacokinetic/Glucodynamic Data, *Diabetes Technol. Ther.*, 2007, **9**(s1), S-41–S-47.
88. D. A. Edwards, J. Hanes, G. Caponetti, J. Hrkach, A. Ben-Jebria and M. L. Eskew, *et al.*, Large porous particles for pulmonary drug delivery, *Science*, 1997, **276**(5320), 1868–1871.
89. K. M. Rave, L. Nosek, A. De La Peña, M. Seger, C. S. Ernest and L. Heinemann, *et al.*, Dose response of inhaled dry-powder insulin and dose equivalence to subcutaneous insulin lispro, *Diabetes Care*, 2005, **28**(10), 2400–2405.
90. S. L. Ellis, K. A. Gemperline and S. K. Garg, Review of phase 2 studies utilizing the AIR® particle technology in the delivery of human insulin inhalation powder versus subcutaneous regular or lispro insulin in subjects with type 1 or type 2 diabetes, *Diabetes Technol. Ther.*, 2007, **9**(S1), S-48–S-56.
91. S. K. Garg, C. Mathieu, N. Rais, H. Gao, J. A. Tobian and J. R. Gates, *et al.*, Two-Year Efficacy and Safety of AIR Inhaled Insulin in Patients with Type 1 Diabetes: An Open-Label Randomized Controlled Trial, *Diabetes Technol. Ther.*, 2009, **11**(s2), S-5–S-16.
92. J. Kling, Dreamboat sinks prospects for fast approval of inhaled insulin, *Nat. Biotechnol.*, 2011, **29**(3), 175–176.
93. M. M. Al-Tabakha, Future prospect of insulin inhalation for diabetic patients: The case of Afrezza versus Exubera, *J. Controlled Release*, 2015, **215**, 25–38.
94. J. Chan and A. Cheng-Lai, Inhaled Insulin, *Cardiol. Rev.*, 2017, **25**(3), 140–146.
95. E. Potocka, J. P. Cassidy, P. Haworth, D. Heuman, S. van Marle and R. A. Baughman, Jr., Pharmacokinetic characterization of the novel pulmonary delivery excipient fumaryl diketopiperazine, *J. Diabetes Sci. Technol.*, 2010, **4**(5), 1164–1173.
96. A. Pfützner and T. Forst, Pulmonary insulin delivery by means of the Technosphere™ drug carrier mechanism, *Expert Opin. Drug Delivery*, 2005, **2**(6), 1097–1106.

97. A. Pfützner, A. E. Mann and S. S. Steiner, Technosphere™/Insulin- A New Approach for Effective Delivery of Human Insulin Via the Pulmonary Route, *Diabetes Technol. Ther.*, 2002, **4**(5), 589–594.
98. J. J. Neumiller and R. K. Campbell, Technosphere® insulin, *BioDrugs*, 2010, **24**(3), 165–172.
99. R. Baughman, J. Cassidy, B. Levy, M. Hompesch, M. Diaz and A. Boss, *et al.*, Technosphere (R) insulin (TI) pharmacokinetics unchanged in subjects who smoke, *Diabetes*, 2008, **57**, A128.
100. T. Goldberg and E. Wong, Afrezza (Insulin Human) Inhalation Powder: A New Inhaled Insulin for the Management Of Type-1 or Type-2 Diabetes Mellitus, *Phamacy Ther.*, 2015, **40**(11), 735–741.
101. E. S. Kim and G. L. Plosker, AFREZZA®(insulin human) inhalation powder: a review in diabetes mellitus, *Drugs*, 2015, **75**(14), 1679–1686.
102. J. Rosenstock, D. L. Lorber, L. Gnudi, C. P. Howard, D. W. Bilheimer and P. C. Chang, *et al.*, Prandial inhaled insulin plus basal insulin glargine versus twice daily biaspart insulin for type 2 diabetes: a multicentre randomised trial, *Lancet*, 2010, **375**(9733), 2244–2253.
103. W. de Kruijf and C. Ehrhardt, Inhalation delivery of complex drugs-the next steps, *Curr. Opin. Pharmacol.*, 2017, **36**, 52–57.
104. L. Heinemann, R. Baughman, A. Boss and M. Hompesch, Pharmacokinetic and Pharmacodynamic Properties of a Novel Inhaled Insulin, *J. Diabetes Sci. Technol.*, 2017, **11**(1), 148–156.
105. A. H. Boss, R. Petrucci and D. Lorber, Coverage of prandial insulin requirements by means of an ultra-rapid-acting inhaled insulin, *J. Diabetes Sci. Technol.*, 2012, **6**(4), 773–779.
106. L. Fala, Afrezza (Insulin Human) Inhalation Powder Approved for the Treatment of Patients with Type 1 or Type 2 Diabetes, *Am. Health Drug Benefits*, 2015, **8**(Spec Feature), 40–43.
107. R. Y. K. Chang, P. C. L. Kwok, S. Ghassabian, J. D. Brannan, H. O. Koskela and H. K. Chan, Cough as an adverse effect on inhalation pharmaceutical products, *Br. J. Pharmacol.*, 2020, **177**(18), 4096–4112.
108. M. Abramowicz, G. Zuccotti and J. M. Pflomm, An Inhaled Insulin (Afrezza), *JAMA*, 2015, **313**(21), 2176–2177.
109. I. B. Hirsch, R. Juneja, J. M. Beals, C. J. Antalis and E. E. Wright Jr., The evolution of insulin and how it informs therapy and treatment choices, *Endocr. Rev.*, 2020, **41**(5), 733–55.
110. S. F. Greene, K. J. Nikula, D. Poulin, K. McInally and J. A. Reynolds, Long-Term Nonclinical Pulmonary Safety Assessment of Afrezza, a Novel Insulin Inhalation Powder, *Toxicol. Pathol.*, 2021, **49**(2), 334–348.
111. D. C. Klonoff, Afrezza inhaled insulin: the fastest-acting FDA-approved insulin on the market has favorable properties, *J. Diabetes Sci. Technol.*, 2014, **8**(6), 1071–1073.
112. A. C. Silva, C. P. Costa, H. Almeida, J. N. Moreira and J. M. S. Lobo, Hormones, Blood Products, and Therapeutic Enzymes, *Current*

Applications of Pharmaceutical Biotechnology, Springer, 2019, pp. 115–153.
113. MannKind. MannKind Announces Afrezza® Distribution Agreement in Australia 2019 [Available from: https://www.globenewswire.com/en/news-release/2019/05/16/1826526/29517/en/MannKind-Announces-Afrezza-Distribution-Agreement-in-Australia.html.
114. A. D. Perera, C. Kapitza, L. Nosek, R. S. Fishman, D. A. Shapiro and T. Heise, *et al.*, Absorption and metabolic effect of inhaled insulin: intrapatient variability after inhalation via the Aerodose insulin inhaler in patients with type 2 diabetes, *Diabetes Care*, 2002, **25**(12), 2276–2281.
115. D. Kim, S. Mudaliar, S. Chinnapongse, N. Chu, S. M. Boies and T. Davis, *et al.*, Dose-response relationships of inhaled insulin delivered via the Aerodose insulin inhaler and subcutaneously injected insulin in patients with type 2 diabetes, *Diabetes Care*, 2003, **26**(10), 2842–2847.
116. J. B. Fink, L. Molloy, J. S. Patton, V. C. Galindo-Filho, J. de Melo Barcelar and L. Alcoforado, *et al.*, Good things in small packages: an innovative delivery approach for inhaled insulin, *Pharmacal Res.*, 2017, **34**(12), 2568–2578.
117. E. Zijlstra, L. Plum-Moerschel, M. Ermer, O. Klein, L. Porter and B. Bueche, *et al.*, Dance 501 Inhaled Human Insulin (INH): Linear Dose Response, Earlier Onset of Action, and Higher Early Effect than sc Insulin Lispro (LIS), *Diabetes*, 2019, **68**, 1085P.
118. R. R. Mohanty and S. Das, Inhaled Insulin – Current Direction of Insulin Research, *J. Clin. Diagn. Res.*, 2017, **11**(4), OE01–OE02.

CHAPTER 7

Microfluidic and Organ-on-a-chip-based Technologies for Diabetes Therapy and Research

LISA MORISSEAU,[a,b,†] TAHA MESSELMANI,[a,†] AMAL ESSAOUIBA,[a,b,†] YASUYUKI SAKAI,[b,c] ANNE LE GOFF,[a] CÉCILE LEGALLAIS,[a] ERIC LECLERC*[a,b] AND RACHID JELLALI*[a]

[a] Université de technologie de Compiègne, CNRS, Biomechanics and Bioengineering, Centre de recherche Royallieu CS 60319, 60203 Compiègne Cedex, France; [b] CNRS IRL 2820, Laboratory for Integrated Micro Mechatronic Systems, Institute of Industrial Science, University of Tokyo, 4-6-1 Komaba, Meguro-ku, Tokyo 153-8505, Japan; [c] Department of Chemical Engineering, Faculty of Engineering, University of Tokyo, 7-3-1 Hongo, Bunkyo-ku, Tokyo, Japan
*Emails: rachid.jellali@utc.fr; eric.leclerc@utc.fr

7.1 Introduction to Diabetes Mellitus (DM)

The islets of Langerhans represent approximately 2% of the total pancreas and play a key role in maintaining glucose homeostasis within a narrow physiological range.[1] Blood glucose levels are controlled by two antagonistic hormones secreted by pancreatic α- and β-cells (Figure 7.1). In fasting

†Authors with equal contribution.

Nanoscience & Nanotechnology Series No. 59
Nanotechnology for Diabetes Management
Edited by Amar Abderrahmani, Sabine Szunerits, Rabah Boukerroub and Abdelfattah El Ouaamari

Published by the Royal Society of Chemistry, www.rsc.org

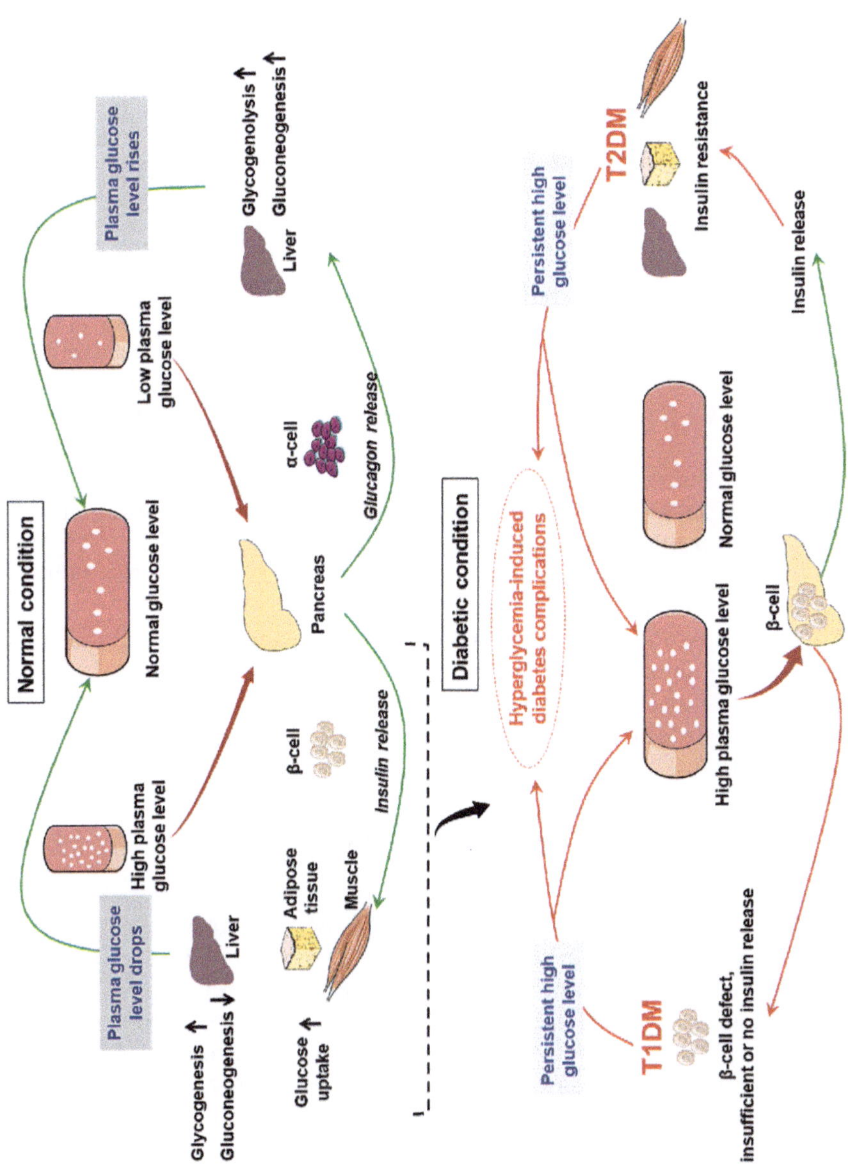

Figure 7.1 Mechanisms of glycaemia regulation by the pancreas and other tissues under normal and diabetic conditions.

periods with low plasma glucose levels, α-cells secrete glucagon to stimulate hepatic glycogenolysis and gluconeogenesis, thereby increasing blood glucose levels. Insulin secretion from β-cells is stimulated when glucose levels are elevated. Insulin release activates the uptake and storage of glucose in the muscle, fatty tissue, and liver through glycogenesis, thus lowering blood glucose levels.[2,3] Disrupted insulin secretion and/or insulin–target organ interaction leads to permanent hyperglycaemia (high levels of blood glucose) and diabetes (Figure 7.1).[2]

Diabetes mellitus (DM) is the most significant dysfunction of the pancreas. According to the World Health Organization (WHO) definition, DM is a chronic disease occurring when the pancreas does not produce enough insulin, or none at all, or when the body cannot use the insulin secreted.[4,5] In the most recent report (Diabetes Atlas 2019), published by the International Diabetes Foundation (IDF), it was estimated that approximately 463 million people (1 in 11 adults) have diabetes worldwide and, worryingly, that 578 and 700 million will be affected by 2030 and 2045, respectively.[6,7] Diabetes is now one of the largest global health concerns. In 2019, the IDF estimated that 4.2 million deaths among adults (20–79 years) can be attributed to diabetes.[6,8] The WHO ranked diabetes among the top ten causes of death in 2019.[5] Diabetes is also associated with multiple complications, such as blindness, kidney failure, cardiovascular disease, sexual dysfunction, neuropathy, lower limb amputations, and peripheral vascular disease.[9–11] The annual healthcare cost of diabetes was estimated at approximately 760 billion USD in 2019 and is predicted to reach 845 billion USD by 2045.[6,11]

DM can be classified into four categories: type 1 DM (T1DM), type 2 DM (T2DM), gestational diabetes mellitus (GDM), and a fourth category which includes specific types of diabetes due to other causes such as genetic mutations, diseases of the exocrine pancreas and drug exposure.[12] T1DM and T2DM alone account for 90% to 95% of all diagnosed cases worldwide.[13]

7.1.1 Type 1 Diabetes Mellitus (T1DM)

T1DM is an autoimmune disorder in which insulin-secreting pancreatic β-cells are destroyed, usually by autoimmune inflammatory mechanisms, resulting in a lack of insulin production.[3,4] This autoimmunity is due to a combination of genetic susceptibility and a host of environmental triggers such as diet, viruses, stress, and exogenous toxins.[11] The incidence of T1DM is increasing worldwide and is estimated to account for up to 10% of diabetic patients, mostly children and young adults.[13] In addition to high blood sugar, T1DM patients present the classic symptomatology including polyurea, polydipsia, fatigue, loss of weight and blurred vision.[6]

T1DM is an incurable disease. Currently, the most effective treatment is the daily administration of exogenous insulin.[11] Conventionally, insulin is administered in subcutaneous injections (with syringes) several times per day.[3,13] The blood glucose level is constantly monitored using a glucometer to help adjust the doses of insulin. Insulin can be also administered *via* an insulin pump which provides constant control of the insulinemia, mimicking the glucose concentration patterns provided by a healthy pancreas.[13,14] Among other strategies for T1DM treatments, whole pancreas and islet of Langerhans transplants are an effective solution for restoring normoglycaemia.[3,14] However, due to the shortage of donors, the reduced number of viable extracted islets and the need for continuous immunosuppression, the use of these strategies remains insignificant compared to the total diabetic population.[3,11,13,14]

7.1.2 Type 2 Diabetes Mellitus (T2DM)

T2DM, the most common form of diabetes, comprises approximately 90% of total diabetic patients (415 million people worldwide), particularly adults aged 20–79 years.[6] It is named the disease of the century, because its prevalence has increased exponentially in recent decades. T2DM is a complex pathology caused by insulin resistance and impaired insulin secretion.[15] Insulin resistance is characterized by the incapacity of insulin-sensitive organs (mainly the liver, muscles and adipose tissue) to uptake and properly metabolize glucose caused by impaired insulin action.[11,15] In the history of T2DM, the plasma insulin level usually augments in response to insulin resistance as the result of a rise in insulin production and β-cell mass. An increase in insulin production is crucial for compensating the hormone demand caused by insulin resistance. T2DM ensues when the pancreatic β-cell fails to produce insulin at sufficient quantity for lowering blood glucose levels (loss of β-cell mass, severe β-cell dysfunction).[13,15,16] Several factors play a role in increasing the risk of developing T2DM: genetic factors, age, obesity, unhealthy lifestyles and lack of physical activity.[11,15] The treatment of T2DM involves lifestyle modifications and antidiabetic drug therapy including orally delivered medicines such as metformin, sulfonylureas, glinides, gliptins, glitazones and SGLT2 inhibitors, and glucagon-like peptide-1 (GLP-1) receptor agonists. Insulin injection is used when drug therapy fails to control glycaemia.[15]

7.1.3 Gestational Diabetes Mellitus (GDM)

GDM is defined by impaired glucose tolerance that is discovered for the first time during pregnancy. It is generally diagnosed in the second or third trimester of pregnancy.[17,18] According to the most recent IDF Diabetes Atlas (2019), 14.4% of pregnant women develop GDM (approximately 17 million births affected worldwide annually).[6] In most cases of GDM, the glucose intolerance is related to insulin resistance and β-cell dysfunction.[18,19] The gestational diabetes risk factors include obesity, advanced age, excessive gestational weight gain, ethnicity, history of previous GDM and a family history of TD2M.[19] The treatment of GDM consists of lifestyle adjustments and the pathology usually resolves once the pregnancy ends.[4] However, GDM is associated with several adverse outcomes, such as pregnancy complications, and an increased risk of developing T2DM and cardiovascular disease for both mother and child.[18]

7.1.4 Other Specific Types of Diabetes

In addition to common T1DM/T2DM and GDM, there are several other rare forms of diabetes with other causes. These forms of diabetes are caused by genetic defects in β-cell function or insulin action (monogenic diabetes), diseases of the exocrine pancreas (cystic fibrosis-related diabetes), endocrinopathies, drugs and chemicals, infection (congenital rubella, cytomegalovirus) and uncommon autoimmune forms (different from those implicated in T1DM, such as latent autoimmune diabetes in adults, LADA).[6,17,20] Of the rare forms of diabetes, monogenic diabetes (neonatal diabetes and maturity-onset diabetes of the young, MODY) is the most common, representing approximately 2% of total diabetes cases.[6,20]

7.2 Microfluidic Technology

Microfluidics refers to the manipulation and study of small-scale volumes (between microlitres and femtolitres), using devices composed of microstructures with dimensions in the tens to hundreds of micrometres range.[21] Microfluidic technology makes it possible to miniaturize conventional laboratories in a micro-scale version with the same functionality.[22] Thanks to miniaturization, the increase in the surface-to-volume ratio, and the continuous mode of operation, microfluidic devices have many advantages: reduced time and cost of the experiments, rapid analysis, precise control, optimization of interactions, and enhanced functionality and reliability.[21,23–25] Furthermore, microdevices can be easily tuned by incorporating sensors, detectors, valves, pumps and electronics.[24,25]

Microfluidic technology has grown exponentially and has made considerable progress in the last 20 years. Nowadays, it covers a wide range of fields in the chemical, pharmaceutical, healthcare and food industries. Depending on the application and/or functionality, microfluidic devices can be called microreactors (chemistry), lab-on-chip (analysis, diagnosis) or organ-on-chip (cell, tissue and organoid culture).[26]

7.2.1 Materials for Microfluidic Devices

In microfluidics, the choice of material is fundamental and depends on the application. Several parameters regarding material properties must be considered: ease of fabrication, physical and chemical properties, chemical inertness, transparency, biocompatibility and gas permeability.[27–29] Since the emergence of microfluidic technology, the materials used to construct microdevices have been varied and can be classified into three groups: inorganic, organic and hybrid/composite materials.[24,30] Inorganic materials, especially silicon and glass, were among the earliest materials used in microfluidics as they are widely used in micro-electro-mechanical systems (MEMS).[31] Silicon and glass have several suitable characteristics, such as stability at high temperatures, excellent resistance to solvents and pressure, and good surface stability.[25,30] Both materials have been used to manufacture microdevices in several fields, particularly for applications requiring the use of high temperatures and pressure, and aggressive solvents.[24,26,32,33] Nevertheless, silicon and glass present some limitations including high costs, manufacturing and bonding difficulties, and no permeability to gas (unsuitable for biological/medical applications).[26,33]

Polymers (organic materials) have played an important role in microfabrication and prototyping in the last 15 years.[32] Polymers have many advantages for manufacturing microfluidic devices as they are inexpensive, easy to access, and suitable for mass production processes. They also offer a large choice of physical, chemical and surface properties through adaptable formulations and easy chemical modifications.[24,26,32] The most commonly used polymers in microfluidic device fabrication are polydimethylsiloxane (PDMS), poly(methyl methacrylate) (PMMA), cyclic olefin copolymer (COC), polycarbonate (PC), polystyrene (PS), thermoset polyester (TPE), poly(vinyl chloride) (PVC), polyurethane (PU) and polytetrafluoroethylene (PTFE).[24,26,30,32–34] Paper, a natural, cellulose-based and biodegradable material, is also a promising low-cost substrate for microfluidics, especially in point-of-care (POC) diagnosis device manufacturing.[35] Among organic materials, PDMS is the most frequently used in microfluidic devices for biological and biomedical applications due to its key properties: high gas permeability (suitable for long-term cell cultures), biocompatibility, and good transparency.[30] However, several disadvantages are associated with using PDMS: non-specific adsorption of molecules, absorption of less hydrophobic molecules and incompatibility with many solvents.[32] The elastomer perfluoropolyether (PFPE, fluorinated polymer) appears to be an excellent alternative to PDMS in biological applications, because of its gas permeability and chemical inertness.[36–38]

In recent years, microfluidic devices from hybrid materials have been developed to improve device functionality and overcome the limitations of certain materials.[24,26,30] This approach consists in combining several materials to develop more advanced microdevices such as PDMS-glass, PDMS-PC and PDMS-NOA81 (Norland Optical Adhesive 81) devices.[39–42]

7.2.2 Fabrication Techniques for Microfluidics Devices

Depending on the material used, a variety of microfabrication methods are available to create the patterns in microfluidic devices such as etching, micromilling, blasting, laser ablation, optical and soft lithography, stereolithography, hot embossing and injection molding.[26,43,44] To manufacture microstructures in silicon and glass substrates, etching is the most commonly used technique.[26,45] In this process, the undesirable material is removed using corrosive liquid chemicals (*e.g.*, hydrofluoric acid, wet etching) or gases (*e.g.*, octafluorocyclobutane C_4F_8 or carbon tetrafluoride CF_4, dry etching).[46] Several other techniques can be used to pattern silicon and glass: substrate erosion by powder blasting, micromilling and laser drilling.[45,46] There are several methods for sealing glass and silicon microfluidic devices, including thermal compression, anodic and adhesive bonding.[45]

For polymer-based microdevices, the microfabrication process depends on the polymer type: elastomer or thermoplastic. PDMS elastomer, the most common microfluidic substrate, is mainly micro-structured by means of simple soft lithography, a low-cost, easy-to-use process that produces high-resolution replicas (Figure 7.2).[26,44] In soft lithography, the desired patterns are first fabricated on moulds (master) which serve as the pattern transfer agent for the PDMS. Then, the PDMS pre-polymer is poured on to the master and heat-cured to replicate the shape. The patterned PDMS layers can easily be irreversibly sealed to each other or to glass with plasma treatment.[26,47] Polymer photoresists (*e.g.*, epoxy SU-8) are usually used to produce the mould using the photolithography process (Figure 7.2).[31,48] Photolithography can also be used for the direct manufacture of microfluidic devices from photosensitive materials such as SU8 and thiolene polymers.[24,29]

Various processes have been used to manufacture microfluidic devices with thermoplastic polymers (PMMA, PC, PS, COC, PFA), including hot embossing, injection moulding, micromilling and laser ablation. The most commonly used methods are injection moulding and hot embossing, which are adapted for mass production.[43,44] Hot embossing involves pressing the mould into a thermoplastic sheet (using a hydraulic press) at a temperature higher than the softening temperature for the plastic (Figure 7.2). The resulting thermoplastic microdevice is the exact mirror of the mould. In the injection moulding process, thermoplastic pellets are introduced into the mould's heated cavity under pressure. The microfluidic device with the desired patterns is obtained after cooling the mould (Figure 7.2).[26,48]

7.3 Microfluidic Lab-on-chip for Analysis and Diagnosis in Diabetes

Labs-on-chips (LoC) are micro-engineered devices with integrated biosensors minted to replace conventional laboratory tests with their integrated, automated and parallelized biochemical assays. The aims of those miniaturized platforms are to provide high sensitivity, simplicity of use and instant results with a cost-effective scheme.[49–51] Since the publication of the first biosensor concept in 1962, consisting of an oxygen electrode for glucose measurement, glucose sensors have become the point-of-care devices (PoC) that are the most studied and used worldwide.[52] Biosensors can be defined as analytical devices with a biocatalyst capable of detecting a biological signal coupled to a transducer that converts the signal produced during the interaction of the biocatalyst and the analyte into a detectable parameter (Figure 7.3).[53]

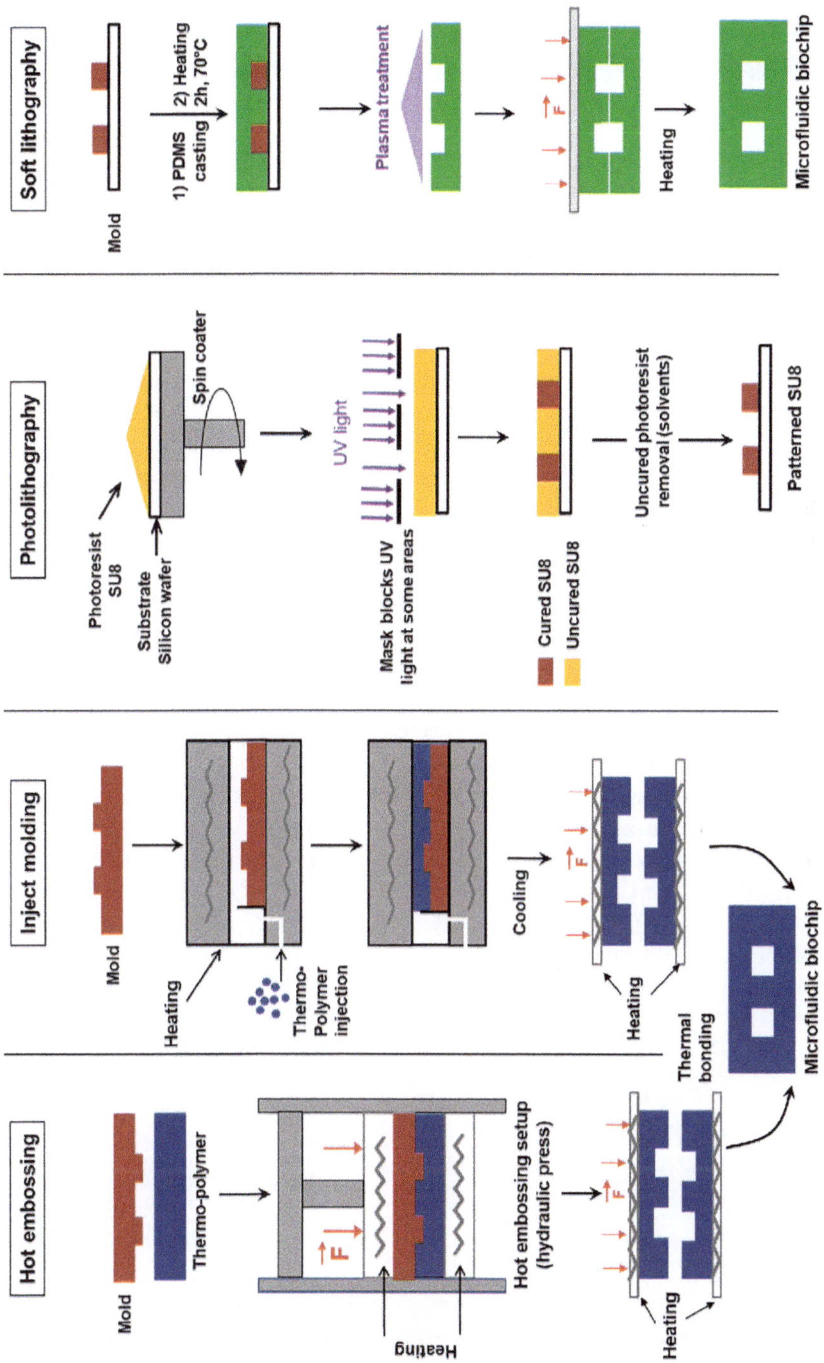

Figure 7.2 Schematic illustration of the main fabrication methods used in microfluidic devices manufacturing.

The bioanalyte can come from a wide range of molecules, such as RNA, DNA, enzymes, metabolites and oligonucleotides. These analytes will interact with specifically designed biorecognition elements consisting of enzymes, antibodies, aptamers, nucleic acids or cells. It is noteworthy that aptamers belong to the new generation of bioreceptors composed of artificial single-stranded DNA or RNA ligands that can bind to amino acids, proteins or large molecules. These synthetic macromolecules have high affinity and specificity thanks to the stable secondary structure that can be functionalized to attain a stronger binding force.[54–56]

On the other hand, the transducer part of the biosensor can be categorized as optical, piezoelectric, calorimetric, electrochemical, acoustic, *etc.*, and the bioreceptors immobilized on various substrates such as electrodes, nanowire arrays, transistors or nanoparticles (Figure 7.3). Then, the signal is converted to digital, and then amplified and processed before being displayed.[55] Thanks to integration of analytical tools, LoC platforms have become the most important field of application for microfluidics, providing a sophisticated approach for sample preparation, separation and detection.[50]

LoCs are trending PoC diagnosis devices due to their versatility, reliability, simplicity, cost-effectiveness and portability. In recent years, microfluidics have played a crucial role in the development of PoC devices for chronic diseases, with numerous approaches covering both diagnosis and management applications.[57] Currently, the gold standard in diabetes detection and aetiopathogenic group classification is quantification of glucose and insulin levels in blood samples.[58] Unfortunately, these two parameters are not considered specific biomarkers for early detection of DM.[59–61] This is particularly true of measuring blood insulin levels, as the test is often carried out when the patient already presents symptoms typical of the disease.[62,63] However, studies have shown that markers such as C-peptide (a by-product of insulin production), glycosylated haemoglobin (HbA1c), glycated human serum albumin (HSA), C-reactive protein (CRP) and adiponectin are key indicators for early DM detection.[57,64,65]

7.3.1 Microfluidic-based Methods for DM Diagnosis

7.3.1.1 Paper-based Analytical Device (μPAD) for Glucose Measurement

To tackle one of the major drawbacks of current glucose measurement methods, paper-based analytical devices (μPAD) have shown great potential for rapid and low-cost detection. Their compatibility with commercial printing technology and good limit of detection (LOD) ranges make them a fast diagnostic tool for glucose detection and monitoring.[58,66] The main advantage of μPADs compared to glucose meter test strips is their ability to confine a small amount of sample thanks to the hydrophobic barrier in a hydrophilic cellulose-based device. This drastically improves the efficiency

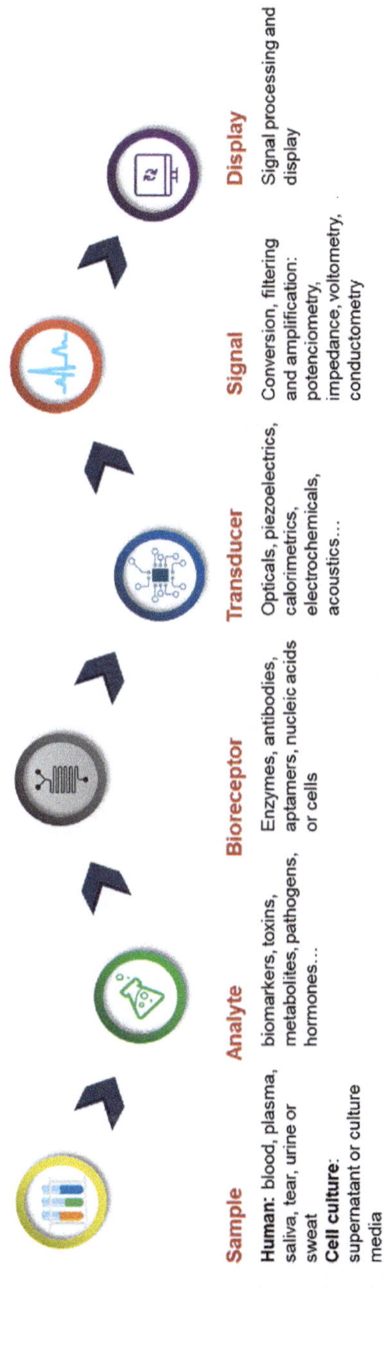

Figure 7.3 Schematic overview of the components of a biosensor integrated in a LoC platform.

and control of reagents, making it possible to parallelize and measure multiple analytes at the same time.[57]

Many original approaches have been reported in the last decade to increase accuracy or reduce the assay time and amount of sample (Table 7.1).[67–70] For example, Evans *et al.* developed a filter paper-based μPAD functionalized with silica nanoparticles with the purpose of enhancing the colour homogeneity of colorimetric glucose detection. Their device uses a small size sample (10 μL) and the assay time is around 30 min.71 Meanwhile, Shechi *et al.* reported a shorter assay time (20 min) and an even smaller size sample (3 μL) but they did not mention the LOD of their 3D μPAD device.[72] Palazzo *et al.* used gold nanoparticles to functionalize the μPAD with the aim of avoiding the colour bleaching that usually happens with these enzyme-based devices.[73] Another approach, proposed by Kugumiya *et al.*, combines spectrometric measurement of hydrogen peroxide resulting from the enzymatic reaction in the microfluidic chip.[74] A recent breakthrough microfluidic chip was reported by Zhang *et al.*, where the detection method in the colorimetric assay was performed with a smartphone application instead of the conventional scanner. They also used an innovative inkjet printing method for the fabrication, to enhance the performance of the μPAD, obtaining an LOD of 0.01 mg/mL and reducing the assay time to only 5 min .[75]

μPADs are definitely recognized as a breakthrough analytical platform for multiple biosensing applications as pointed out by Nadar *et al.*, in a recent review, where they provide an extensive insight into the design, fabrication and enzyme immobilization strategies for developing enzyme-μPADs.[76] However, despite their excellent qualities, such as the high-speed and low-cost production processes, μPADs still have some practical issues to overcome. These are mainly the fragility of the material and the difficulty in separating the hydrophilic regions from the hydrophobic ones.[77,78]

7.3.1.2 Lab-on-chip-based Devices for Glucose and Immune Marker Detection

With the aim of revealing a reliable indicator for diabetes, an accurate and precise indicator is glycated haemoglobin (HbA1c). HbA1c levels are the ratio of glycated haemoglobin and total haemoglobin in the bloodstream, which reflects the average blood glucose level for the preceding 2–3 months.[5,64,79] Usually, HbA1c measurement is carried out with immunoassays, high-performance liquid chromatography (HPLC) or mass spectrometry (MS). However, these methods are time-consuming and not cost-effective. The limitations related to the high cost and handling complexity of MS have been overcome thanks to the development of small, low-cost, and simple operating LoCs with integrated mass spectrometers which have the potential to make MS more accessible to clinical laboratories (Table 7.1).[57] For instance, Mao *et al.*, reported a silicon nanoLC-MS microfluidic device to determine

Table 7.1 Summary of LoC-based technologies developed for DM diagnosis and management.

Target analyte	Technology (group)	Technique	Assay time	LOD[a]	Sample volume	Detector	References
Glucose	Paper µPAD	Colorimetric immunoassay	20 min	—	3 µL	Hp Scanjet 6300c	72
Glucose	Paper µPAD (silica nanoparticles)	Colorimetric immunoassay	30 min	0.5 mM	10 µL	Flatbed scanner	71
Glucose	Paper µPAD	Colorimetric immunoassay	2–3.5 h	0.25 mmol L^{-1}	40 µL	Epson Perfection V600	67
Glucose	Paper µPAD	Colorimetric immunoassay	5 min	0.01 mg mL^{-1}	15 µL	Epson Perfection V19	75
Glucose	Paper µPAD	Colorimetric immunoassay	10 min	13 mg dL^{-1}	5 µL	—	70
Glucose	Paper µPAD (silane/hexane ink)	Colorimetric immunoassay	94 min	5.5 mM	—	Desktop scanner	68
Glucose	Paper µPAD	Colorimetric immunoassay	10 min	0.7 mmol L^{-1}	40 µL	Hewlett–Packard, model F4280	69
Glucose	Optical gold nanoparticles assay	Colorimetric immunoassay		5 µM		Absorbance reader	73
Glucose	Integrated electrode	Electrochemical sensor	Online	1.44 mg dL^{-1}	—	Ammeter	88
Glucose	Enzyme-doped thread	Electrochemical sensor	—	0.1 mM	2 µL	Ammeter	78
Glucose	Long-period grating sensor	Optical fibre sensor	Online	10 µM	—	Wavelength detector	89
HbA1c Glycated HSA apoA-I	Multinozzle Emitter array chip	Top-down proteomics	2.5 h	—	10 µL	Mass spectrometer	64

Table 7.1 (*Continued*)

Target analyte	Technology (group)	Technique	Assay time	LOD[a]	Sample volume	Detector	References
HbA1c Glycated HSA	—	Capillary electrophoresis and mass spectrometry	—	—	10 μL	Mass spectrometer	80
HbA1c	Chemiluminescent aptamer–antibody sandwich assay	Nucleic-acid-based aptamer	25 min	—	50 μL	Luminometer	82
HbA1c	Chemiluminescent antibody sandwich assay	Antibody-based automated chip	15 min	—	1 μL	Luminometer	79
HbA1c Total Hb	Aptamer–antibody sandwich assay	Antibody-based automated chip	30 min	—	100 μL	Photomultiplier tube	81
Serine	—	Enzymatic reaction	30 min	2 mM	250 mM	Thermal lens detector	74
RBC fragility	—	Osmotic lysis kinetics	—	5 mM	—	CDD camera + Microscope	84
RBC deformability	—	Viscoelastic particle focusing	—	—	—	Optical microscope	85
RBC deformability	—	Equilibrium velocity of a cell in a microchannel	—	—	—	Optical microscope	86
Insulin	PDMS device	Chemiluminescent assay	10 min	4×10^{-10} mol L^{-1}	—	Photometer	65
Insulin, GAD65 IA2	Plasmonic gold substrate for NIR-FE detection	Plasmonic chip	2 h		2 μL	Licor Odyssey scanner	83

[a]Limit of detection.

healthy patients from T2DM patients using only 5 μL of blood sample. The platform makes it possible to quantify glucose, HbA1c, glycated HSA and glycated apolipoprotein A1 (apoA-1) within the same microfluidic device.[64] In another study, Redman *et al.* developed a capillary electrophoresis–mass spectrometry (CE–MS) device to indirectly measure HbA1c through glycated β-Hb and HAS, also in a blood sample. Capillary electrophoresis simplifies the sample mixture and improves the resolution of the signal intensity.[80] Independently, Li *et al.* developed a unique aptamer-based microfluidic system for automatic HbA1c and Hb simultaneous measurement *via* magnetic beads. The high performance of their technology relies on reducing reagent consumption by 75% and time of analysis from 3h 30 min to 30 min when compared with traditional HPLC.81 Similarly, Chang *et al.* used an aptamer-based microfluidic system to detect HbA1c and Hb with high sensitivity and specificity, even for a smaller sample (50 μL) in a relatively shorter time (25 min).[82]

In addition to HbA1c measurement for DM diagnosis, it was imperative to develop PoC systems for the aetiopathogenic group classification. Therefore, Zhang *et al.* proposed a plasmonic gold chip for the detection of autoantibodies against several pancreatic islet antigens [insulin, glutamic acid decarboxylase (GAD65) and tyrosine phosphatase islet antigen 2 (IA2)]. The plasmonic gold platform enabled simultaneous detection and quantification of autoantibodies to islet antigens and their isotypes with exceptionally high specificity and sensitivity that cannot be reached by classic ELISA assay using only a 2-μL sample.[83]

7.3.1.3 Red Blood Cell (RBC) Deformability Test

Researchers have discovered that, in addition to the molecular biomarkers, DM detection can be also carried out by analysing the stability and deformability of erythrocytes or red blood cells (RBCs) (Table 7.1). For example, Zhan *et al.* proposed a microfluidic platform to study RBC fragility based on the characterization of osmotic lysis kinetics from the point of view of fluid mechanics.[84] The lysis kinetics were traced by measuring the release of intracellular contents using a CDD camera coupled with a microscope. The change in erythrocyte fragility after exposure to glucose was detected using this LoC at high sensitivity by recording the light intensity of the RBC at several locations in the microchannel.[84] Cha *et al.* proposed a tool for measuring cell stretching based on the focusing of three-dimensional viscoelastic particles, with high accuracy. Moreover, the authors used the platform to monitor the decrease in deformability due to nutrient starvation in human mesenchymal stem cells.[85] In a different approach, Tsai *et al.* developed a new method based only on stiffness of the "equilibrium velocity" of erythrocytes in a microchannel. The value of this approach lies in the elimination of viscosity, which is a time-dependent feature making it possible to evaluate cell stiffness alone.[86] An extensive review by Bento *et al.*

compared the different approaches proposed in the literature to characterize RBC deformability in a microfluidic device.[87]

7.3.2 LoC-based Methods for DM Management

In DM, the continuous monitoring of blood glucose, insulin, c-peptide and other markers at home is crucial for disease management. For this reason, telemedicine and the progress made in miniaturized cyber technology are revolutionising PoC devices for diabetes management (Table 7.1). Accordingly, Yao *et al.* developed a PDMS microfluidic-based telemedicine system for real-time insulin detection. The method for monitoring insulin is based on sandwich immunoassay combined with luminal-hydrogen peroxide (H_2O_2) chemiluminescence. With a photometer, the microfluidic device detects the peak value of the luminous intensity, which indicates the insulin concentration in the patient's blood sample.[65] Nowadays, enzyme electrode sensors are not widely used for continuous glucose monitoring in clinics because of their short lifetime, interference with the body's bioelectricity, and, most importantly, their invasiveness.[57] For instance, Yang *et al.* developed a multiple enzyme-doped thread-based microfluidic system with a polyvinylchloride (PVC)-coated membrane for rapid and low-cost electrophoresis separation and electrochemical detection of glucose in whole human blood.[78] Meanwhile, Pu *et al.* reported a different microfluidic chip with an integrated sensor based on an electrochemical electrode functionalized with graphene and gold nanoparticles (AuNPs).[88] The purpose of the coating combination on the working electrode was to enhance the resolution of glucose sensitivity. This sensor could measure glucose precisely in the linear range from 0 to 162 $mg\,dL^{-1}$ with an LOD of 1.44 $mg\,dL^{-1}$; this novel approach has the potential for tackling the current clinical challenge of continuous glucose monitoring for hypoglycaemia diagnosis.[88] In return, Yin *et al.* reported an optical fibre sensor integrated microfluidic chip for ultrasensitive glucose detection using long-period grating inscribed in a small-diameter single-mode fibre as an optical refractive-index sensor. Their experimental results have shown that such a powerful device has not only an ultralow detection limit (1 nM), but also a remarkably fast response time (70 s).[89]

7.4 Microfluidic Biochips for Islet or β-cell Encapsulation

Transplanting islets of Langerhans or β-cells is an attractive alternative to exogenous insulin in the treatment of T1DM as it maintains coherent and sustained control of glucose homeostasis. However, the need for permanent immunosuppression limits its application.[3,90] The encapsulation of islets/β-cells within a selectively permeable membrane is an increasingly prevalent method for protecting cells from immune rejection. This concept, called the bioartificial pancreas (BAP), provides efficient immune-isolation, while allowing the exchange of oxygen, insulin, hormones and nutrients through the membrane.[91,92] Depending on the dimensions of the BAP device, the encapsulation process can be classified in two categories: micro- (device size 250–1000 μm) and macroencapsulation (device larger than 1 mm).[91]

The microencapsulation process is particularly interesting as it generates microcapsules with a high surface-to-volume ratio, making possible a high diffusion capacity and easy implantation procedures. However, there are several limitations due to the methods (extrusion, dripping and emulsification) used to generate microcapsules, including high polydispersity, large microcapsules (generally close to 1 mm) and poor repeatability.[93,94] The size of the BAP plays a key role in the functionality and survival of the transplanted cells. Large microcapsules lead to limited exchange of oxygen, nutrients and insulin between the host and BAP (the distance between encapsulated cells and vasculature must not exceed 150–200 μm).[95,96] The limited oxygen and nutrient supply causes hypoxia and apoptosis, and reduces insulin secretion, resulting in failure of the transplant.

Microfluidic technology offers several advantages for cell encapsulation, with a variety of biocompatible polymers such as polysaccharides, proteins and polyethylene glycol (PEG). Microfluidic devices can produce uniform microcapsules in sub-100 micron diameters, with thin membranes and very low variation coefficients (<3%).[93,94] The capsules' diameter and membrane thickness can easily be tuned by varying the dimensions and geometry of the channels, plus the flow rate and viscosity of the polymer solution.[97] Furthermore, microfluidic-based encapsulation offers the capacity for high-throughput production with high reproducibility. In recent years, several studies reporting cell (including insulin-secreting cell) encapsulation using microfluidic technology have been published[94,95,98–103] (Figure 7.4A,B). Tomei *et al.* reported conformal coating of islets with PEG and PEG-alginate using a microfluidic flow-focusing method. This method minimizes capsule size and thickness, which improves oxygen, and insulin exchange and graft volume. Transplanting microcapsules into the renal subcapsular space in mice makes it possible to maintain euglycaemia for 102 days.[95] Microfluidic devices were also used for hiPSC-derived pancreatic cell high-throughput encapsulation within alginate/chitosan. The process makes it possible to generate uniform microcapsules (70 μm in diameter) containing islet organoids with high expression of α- and β-cell markers and responsive to glucose challenges.[103]

Although small microcapsules enhance diffusion, viability and cell functionalities, they are difficult to trace, replace and retain within the implantation site. Another approach for developing a BAP consists of encapsulating the islets or β-cells in microfibers. The use of microfibers with a low diameter offers the advantage of working with devices that are easy to handle, while good diffusion properties are preserved.[104,105] Of the methods used in microfiber engineering, microfluidic technology offers several advantages compared to conventional processes (such as electrospinning and extrusion): high reproducibility, easy customization of the dimensions and

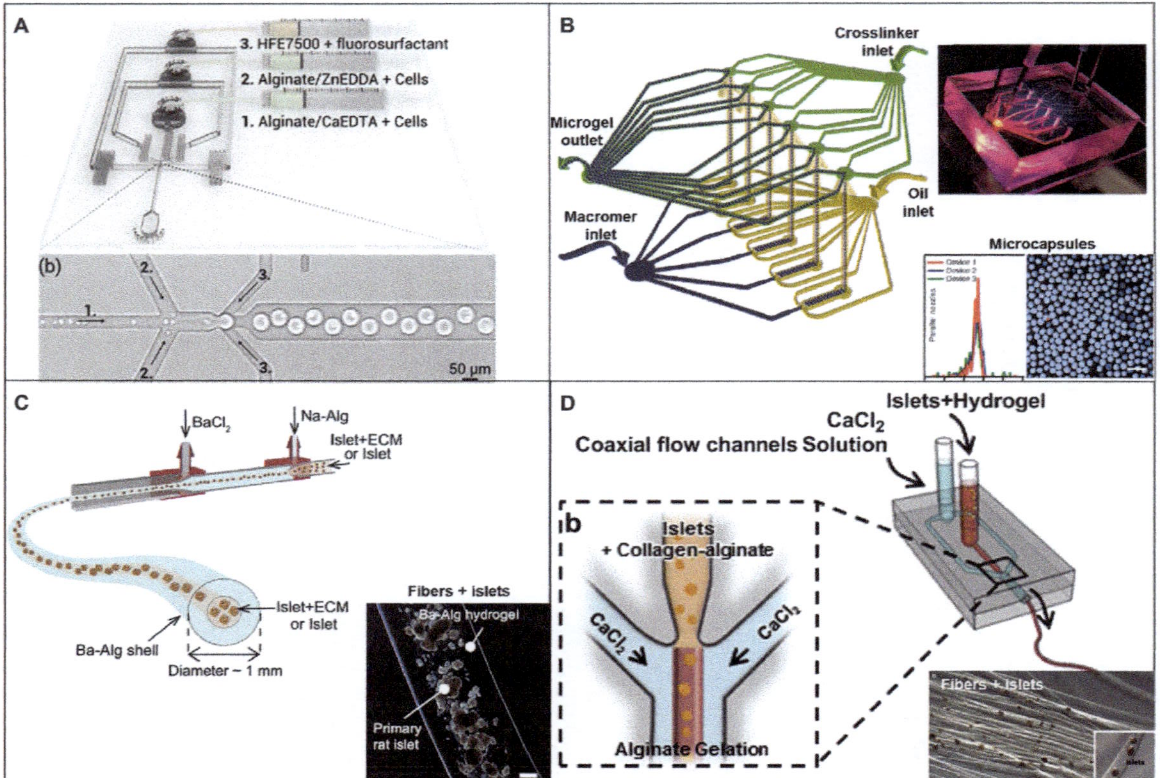

Figure 7.4 Microfluidic platforms for islets/β-cells encapsulation. (A) Microfluidic co-flow device and (B) parallel droplet generators on a two-layer microfluidic device for small-sized monodisperse microcapsule production. (C) Double coaxial microfluidic device and (D) cylindrical and coaxial-flow channels used for islets and β-cells encapsulation in microfibers. (A) Reproduced from ref. 102 with permission from the Royal Society of Chemistry. (B) Reproduced from ref. 99, https://doi.org/10.1038/micronano.2017.76, under the terms of the CC BY 4.0 license https://creativecommons.org/licenses/by/4.0/. (C) Reproduced from ref. 105 with permission from Elsevier, Copyright 2020. (D) Reproduced from ref. 108 with permission from Elsevier, Copyright 2013.

morphological features, and adaptability for cell encapsulation and *in vivo* immunoprotection.[106] A variety of microfluidic designs have been used to encapsulate islets, β-cells and other mammalian cells using different polymer matrices (Figure 7.4C,D).[105,107–112] Using a microfluidic co-axial device (Figure 7.4C), the Takeuchi group developed microfibers encapsulating extracellular matrix (ECM) proteins and islets or β-cell lines within alginate and alginate/polyacrylamide matrices. The microfibers formed were transplanted into diabetic mice and were able to secrete insulin and normalize blood glucose concentrations for a period of up to 100 days.[105,111,113] In another approach, to overcome the problem of pancreatic cell shortages, Jun *et al.* encapsulated hybrid spheroids (hepatocytes/pancreatic cells) into alginate/collagen microfibers using a cylindrical flow channel in a PDMS-based microfluidic chip (Figure 7.4D). They observed effective control of glucose levels by the hybrid microfibres transplanted into the diabetic mice.[109]

7.5 The Pancreas Organ-on-chip Model

Organ-on-chip (OoC) is defined by Bathia and Ingber as a microfluidic device dedicated to living cell cultures. Cells are continuously perfused within microchambers in order to reproduce the behaviours of *in vivo* tissues and organs.[114] Thanks to progress in microfabrication, biomaterials and cell engineering, OoC technology has become a powerful tool for reproducing physiological cell behaviour *in vitro* and replacing the traditional paradigms based on animal experiments and 2D *in vitro* cell culture methods.[114–116] Of the others, OoC technology makes it possible to construct a well-controlled microenvironment and create "physiological-like" situations such as zonation, cell–cell interaction, shear stress and chemical gradients.[114]

Although OoC technology generates a suitable microenvironment that reproduces endocrine function, there is less development of the technology for the pancreas than for other tissues or organs, such as the liver, lung, kidney, gut and heart.[11,117] Pancreas-on-chip applications for diabetes research can be classified in three groups: islet evaluation, drug research and the study of islet physiology and function.[118]

To mimic the *in vivo* physiology and functionality of native islets as much as possible, pancreas-on-chip models are developed using cultures of islets, or cells aggregated into spheroids (pseudo-islets). The pseudo-islet approach makes it possible to engineer uniform, small-sized spheroids (<150 μm), which enhance oxygen and nutrient diffusion, viability and functionality when compared to native islets (heterogeneous size 50–400 μm).[119,120] Pseudo-islets can be engineered by aggregating β-cells, iPSC-derived β-cells or primary β-cells obtained after islet dissociation.[121] To maintain the islets/pseudo-islets under flow inside the microfluidic biochip, trapping microstructures fabricated in the microchambers of the device are used (Figure 7.5). Microwells with different geometries (flat, pyramidal and concave) are the most commonly used designs for islet trapping.[119,122–129] Islets

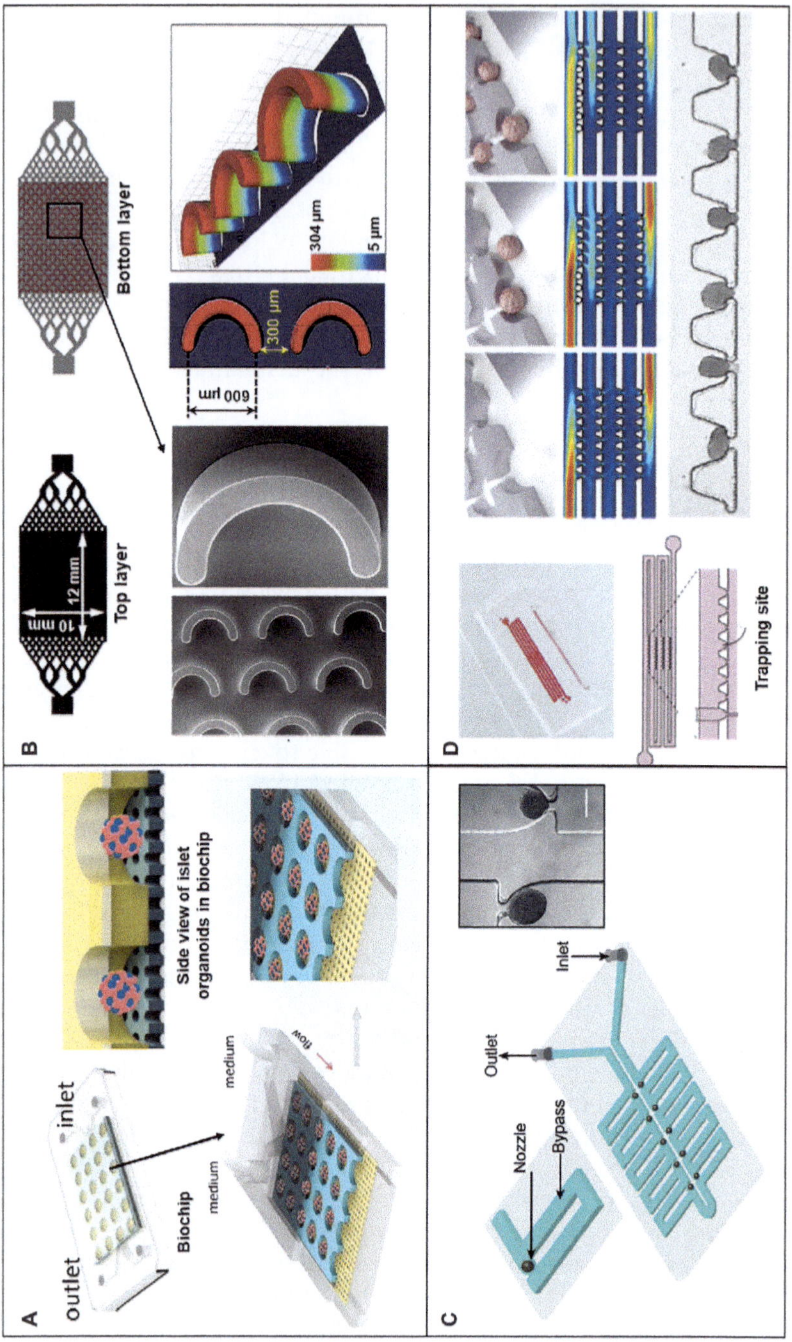
A
outlet
inlet
Biochip
medium
medium
flow
Side view of islet organoids in biochip
B
10 mm
12 mm
Top layer
Bottom layer
600 µm
300 µm
304 µm
5 µm
C
Nozzle
Bypass
Outlet
Inlet
D
Trapping site

can also be immobilized using crescent-shaped structures,[130] mesh systems,[131] nozzle systems, or channel reduction,[132–134] and other constructions involving hydrodynamic trapping principles.[117,135]

7.5.1 Potential Cell Sources for Pancreas-on-chip Models

The success and capacity of *in vitro* pancreatic models to reproduce *in vivo* physiology and functionality depend on the cell types used and their sources.[11] The potential cell sources for *in vitro* models can be classified in four groups: human primary islets of Langerhans or β-cells, animal primary islets of Langerhans or β-cells, cell lines (human or animal: EndoC-βH, MIN6, INS-1, RIN5-F and αTC1.9) and induced pluripotent stem cells (iPSCs).[11,121,136–142] The advantages and limitations of the different cell types are summarized in Figure 7.6.

Primary human β-cells or islets isolated from cadaveric organs are still considered to be the gold standard for transplantation and *in vitro* pancreatic models.[121] Due to their origin, they accurately reflect the physiology and functionality of the organ *in vivo*. However, the shortage of donors, the high costs associated with isolating islets, and the high variability limit their application.[136,143] For animal primary cells, although they are widely used because of their attractivity (availability and stability), there are considerable limitations, including functional differences between animal and human islets, interspecies variability and ethical concerns.[121,143] As an alternative to primary islets, several animal (MIN6, INS-1, RIN5-F and αTC1.9) and human (EndoC-βH1-3) β-cell lines have been investigated.[137–140] Nevertheless, they present several lacunae, such as limited functionality, and differences in the expression of β-cell markers when compared to primary cells.[143,144]

In recent years, β-cells derived from stem cells have emerged as an attractive source of pancreatic cells. Stem cells have the ability to self-renew and differentiate into most of the body's cell types, including pancreatic cells.[143] The use of embryonic stem cells (ESCs) raises ethical problems and is strictly regulated (prohibited in some countries).[145] hiPSCs do not raise any ethical problems and offer the possibility of developing patient-specific models with the genetic background associated with the disease desired.[136,143] The differentiation of hiPSCs into pancreatic β-cells has been widely reported in the literature.[141,142,146–148] However, the high cost and

Figure 7.5 Geometric designs used for islets/pseudo-islets trapping in microfluidic devices under flow: (A) concave microwells; (B) crescent-shaped structures; (C) nozzle system or channel reduction; and (D) geometric constructions using hydrodynamic trapping principle. (A) Adapted from ref. 125 with permission from the Royal Society of Chemistry. (B) Reproduced from ref. 130 with permission from Elsevier, Copyright 2021. (C) Reproduced from ref. 133 with permission from the Royal Society of Chemistry. (D) Reproduced from ref. 117, https://doi.org/10.1016/j.matbio.2019.06.008, under the terms of the CC BY 4.0 license https://creativecommons.org/licenses/by/4.0/.

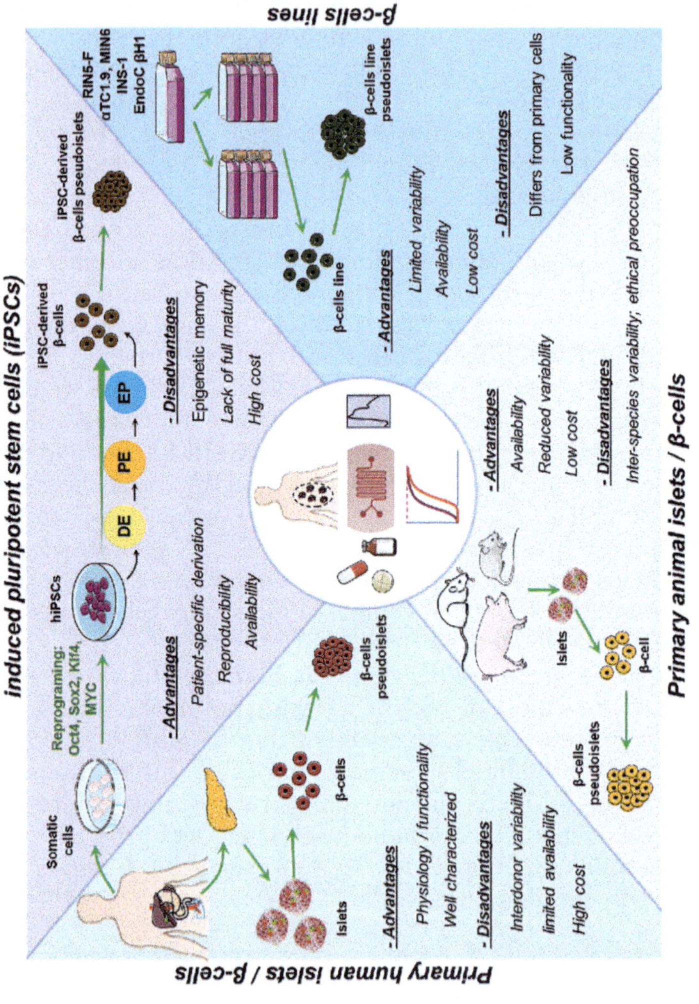

Figure 7.6 Advantages and limitations of the potential cell sources for *in vitro* pancreas-on-chip models (DE: definitive endoderm, PE: pancreatic endoderm, EP: endocrine progenitor).

inability of obtaining fully mature β-cells remain obstacles to the generalization of the use of hiPSCs.[143]

7.5.2 Pancreas-on-chip for Islets Physiology Study

Circumstantial evidence has shown the existence of fluctuations in plasma glucose and insulin levels.[149] Under *in vivo* conditions, islets are subjected to a dynamic environment in which insulin is secreted by β-cells in a biphasic pattern. Pyruvate oxidation by the tricarboxylic acid cycle (TCA) in mitochondria is the signalling pathway involved in insulin release. In brief, glucose catabolism generates ATP through TCA, which leads to an increase in the intracellular ATP/ADP ratio and closure of ATP-sensitive potassium (KATP) channels. As a result, the plasma membrane depolarizes and the Ca^{2+} voltage-dependent channels (VDCCs) open. The rapid influx of glucose-stimulated Ca^{2+} triggers fusion of insulin granules with the cell membrane and subsequent exocytosis of insulin, C-peptide and proinsulin. Thus, rapid stimulation of β-cells with glucose induces biphasic insulin secretion with a first phase corresponding to a rapid increase in the rate of secretion for 4–8 min. This is followed by a decrease in the insulin rate and a second stable or gradually increasing phase which lasts as long as the glucose simulation is applied.[150] Impairment of dynamic insulin release can have a severe influence on glucose homeostasis and related physiological functions. Loss in the first phase, reduction in the second phase, and impairment of the oscillatory pattern of insulin secretion are, for example, characteristic features of T2DM and contribute significantly to its progression. Therefore, studying dynamic insulin secretion is crucial for understanding the pathogenesis and pathophysiology of diabetes, as well as for assessing the pharmacokinetics and potential action mechanisms of antidiabetic medication.[151]

In the past decade, many research groups have specialized in developing a microfluidic perfusion system (MPS) for islet physiology research (Table 7.2).[152–156] Currently, integrating analytical throughput into microfluidic devices is of great interest for studying the pulsatility of pancreatic hormone secretion by detecting and monitoring the electrical, biochemical and ionic signalling activities of β-cells. As an example, Xing *et al.* developed pumpless microfluidic devices including real-time fluorescent imaging and insulin secretion kinetics for both mouse and human islets.[157] Their results showed an increase in intracellular calcium and a typical biphasic profile of insulin secretion, affirming the involvement of calcium oscillations in the phenomenon of pulsatile insulin secretion. In fact, coherent insulin and calcium oscillations are either disturbed or lost in patients with T2DM.[158] To standardize the pancreatic islet model, Misun *et al.* designed a hanging drop microfluidic chip that showed significant potential for enhancing the spatiotemporal resolutions of stimuli, and sampling and observing molecular signals. Reaggregated islets produced physiologically close and reproducible insulin secretion with a pronounced first phase and pulsatile second phase

Table 7.2 Applications and characteristics of main pancreas-on-chip models reported in the literature.

Purpose	Type of cells	Species	Biochips design	Aspect study	References
Islet evaluation	Pancreatic islets	Human Mouse	Chamber with microwells	Cell viability and morphology, GSIS, insulin secretion, mitochondrial potential, intracellular calcium	122
Islet evaluation	Alpha- and beta-cells derived from iPSC	Human	Multilayer device	Cell viability and morphology, expression of islets-specific genes and proteins, GSIS, insulin secretion, intracellular calcium	125
Islet evaluation	Pancreatic islets	Mouse	Chamber with microwells	Islet functionality under glucosegradients (insulin secretion, mitochondrial potential, intracellular calcium)	156
Islet evaluation	Pancreatic islets and adipose tissue	Mouse	Microchamber with 3D-printed PLA template interface	GSIS, glycerol secretion	172
Islets evaluation and preservation	Pancreatic islets and endothelial cells	Mouse	Microchamber	Cell viability and morphology, islets functions under glucose gradients (insulin secretion, intracellular calcium and membrane depolarization)	132
Islets evaluation	Pancreatic islets and stem cell-derived beta cells	Human	Trapping sites	Automated measurement of glucose-stimulated insulin secretion	134
Islets evaluation	Pancreatic islets and adipose tissue	Mouse	Microchamber with 3D-templated tissue culture interface	Dynamic and quantitative measurements of both hormone secretion and nutrient sensing/uptake	173

Islets preservation	Pancreatic islets and bone marrow mesenchymal stem cells	Rat	Microchamber	Cell migration and insulin secretion in response to high-glucose challenge	165
Islets evaluation	Pancreatic islets	Mouse	Microstructured borosilicate glass	Oxygen consumption, NADPH autofluorescence, intracellular calcium, insulin secretion	123
Islets evaluation	EndoC-βH3	Human	Trapping sites	Cell viability, GSIS, insulin secretion, mitochondrial activity	117
Islets preservation and physiology	Pancreatic islets within 3D alginate hydrogel	Human Mouse	Microchamber	*In situ* tracking of viability and calcium signalling, oxygen gradient, oxidative stress under glucolipotoxicity environment, GSIS	161
Islets evaluation and physiology	Pancreatic islets	Human Mouse	Pumpless biochip with trapping sites	Insulin secretion kinetics under glucose stimulation, intracellular calcium oscillation	157
Islets physiology	Pancreatic islets	Mouse	Cellulose-base scaffold	In situ insulin monitoring, GSIS, expression of islets-specific genes	152
Islets physiology	Pancreatic β-cells	Human	Microchamber	Beta cells spheroids formation, oxidative stress under glucolipotoxicity environment, intracellular calcium	129
Islets physiology	Pancreatic islets	Mouse	Symmetric and antiparallel chambers	Islets activity, continuous glucose monitoring	153
Islets physiology	Pancreatic islets	Mouse	Independent channel networks	Continuous monitoring of glucose-stimulated insulin secretion	154

Table 7.2 (*Continued*)

Purpose	Type of cells	Species	Biochips design	Aspect study	References
Islets physiology	Pancreatic islets	Mouse	Independent microchamber	Automated measurement of glucose-stimulated insulin secretion, effect of lipotoxicity environment on intracellular calcium and insulin secretion	155
Islets evaluation, physiology and drug screening	Reaggregated islets	Human	Microfluidichanging-drop biochip	Dynamics of insulin release at high temporal resolution	151
Islets functions and drug screening	Pancreatic islets	Rat	Concave microwells	Cell viability, ultrastructure morphology, expression of islets-specific genes and proteins, GSIS, drugs screening	119
Islets functions and drug screening	hiPSCs derived β-cells	Human	Crest and honeycombs wells	Cell viability, expression of islets-specific genes and proteins, GSIS, insulin and C-peptide secretion, GLP-1 stimulation	130
Islets functions and drug screening	Pancreatic islets	Rat	Flat microwells	Islet viability, expression of islets-specific genes and proteins, GSIS, insulin, glucagon and C-peptide secretion, GLP-1 stimulation	126

throughout, indicating robust cellular communication and synchronization.[151] Other studies have reported a dissociation between oscillations of insulin secretion and glucose, suggesting that other metabolic signals may be involved. However, their efficacy has been demonstrated to be less than that dependent on Ca^{2+} concentration oscillations.[159,160] In another strategy, Patel *et al.* developed an organoid microphysiological platform combining dynamic 3D cultures, optical assessment and functional assays, while also limiting the quantity of sample required.[161] Replacing standard culture dishes with their MPS platform made possible an *in situ*, temporal and live cell assessment of the dynamic impact of glucolipotoxicity on pancreatic islets. Outcomes revealed an increase in Ca^{2+} signalling and a rate of dead cells in the β-cells exposed to high glucose-fatty acid media when compared to β-cells exposed to basal glucose media. These findings are consistent with the theory that hyperlipidaemia and hyperglycaemia contribute to the impaired β-cell function observed in T2DM.[162] Microfluidic biochips are thus a promising tool for studying the mechanisms and restoration of insulin secretion in diabetic islets.

7.5.3 Pancreas-on-chip for Islets Evaluation and Preservation

Microfluidic biochips have been explored extensively to assess the functionality of pancreatic islets (Table 7.2). For instance, the pressure and flow volume of the islet vascular system can be reproduced, making these platforms ideal for *in vitro* islet analysis. This technology could thus make it possible to reduce bioartificial pancreas failure by obtaining precise knowledge of the quality of a donor's pancreatic tissue and accelerating studies in drug discovery and toxicology.

Prior to transplantation, islets are checked through rigorous quality control to assess their viability, morphology, hormone secretion and response to glucose stimulation. The *in vitro* gold standard for evaluating islet quality is static 2D culture using conventional Petri dishes or multi-well plates. Although this model has significantly contributed to medical research, it presents certain limitations. To determine how the culture conditions affect gene expression in islets, Jun *et al.* compared expression levels of islet-specific genes in islet spheroids cultured in microfluidic biochips under static or dynamic conditions and islets cultured in Petri dishes using conventional methods for 7 and 14 days.[119] The results showed that β-cell genes such as insulin, Pdx1 and Glut2 had significantly higher expression in dynamic groups than in intact or static groups. They demonstrated that a microfluidic platform makes long-term islet maintenance possible for up to 1 month. Essaouiba *et al.* developed a microfluidic biochip composed of 600 micro-wells for assessment of rat islets and showed that biochip culture improves the viability and expression of pancreatic genes.[126] In a different strategy, islets were cocultured with mesenchymal stem cells (MSCs) to investigate the viability and preservation of islet functions. MSCs have been shown to secrete several paracrine molecules, which mediate trophic effects on neighbouring cells.[163,164] Lin *et al.* used a coculture microfluidic chip where rat bone marrow mesenchymal stem cells (BM-MSCs) and islets were introduced respectively into two microchambers which could be connected by a traffic tunnel.[165] They reported that the BM-MSCs had the ability to migrate to the microchamber containing murine islets and provide improved GSIS during 3 to 21 days of culture.

In vivo pancreatic islets are heavily vascularized with fenestrated endothelial cells (ECs) to facilitate blood glucose sensing and endocrine hormone secretion. The close proximity of insulin secreting β-cells and ECs plays a major role in modulating the proliferation and survival of both cell types.[166] Transplanting isolated islets causes disruption in their vascular connections, making the islets dependent on the formation of new blood vessels for optimal function. Evidence from experimental islet transplants indicates insufficient revascularization of transplanted islets with subsequent chronically decreased blood perfusion and oxygen tension, which has metabolic consequences within the tissue.[167] It has effectively been established that, during *ex vivo* culture, the characteristic cobblestone morphology of ECs is deteriorated, leading to a rapid decline in cell density (−50% in the first day and total loss after 7 days of culture).[168] The Rocheleau research group postulated that this deterioration occurs in the absence of blood flow due to limited diffusion of media inside the tissue. Thus, they developed a haemodynamic environment using a custom-designed microfluidic device to test the effect of fluid flow on the maintenance of rat islet ECs.[132] Outcomes revealed that isolated pancreatic islets treated with media flow in a microfluidic device achieved wide-reaching diffusion within the tissue of serum albumin, which is an anti-apoptotic signal for ECs.[169] Although they observed a significant reduction in necrosis in flow-treated islets, EC morphology was only partially maintained.[133] To improve previous results, they explored treatments to stimulate endogenous expression of angiogenic factors. The data suggested mild hypoxia as a potential method for slowing the demise of ECs while maintaining islet β-cell function.[170]

Other perfused well platforms were developed by the Easley research group to assess the dynamics of hormone secretion from endocrine tissues. Based on their first report of mouse primary adipocyte cultures within microfluidic systems,[171] they developed a robust method for a temporally resolved examination of endocrine tissues, including pancreatic islets. The concept consists in a macro-to-micro interfacing of microfluidic systems using 3D-printed interface templates.[172] The main results highlighted that glycerol release rates from adipocytes increased after exposure to low glucose and insulin levels (LGLI). In this way, they investigated a fully automated 16-channel microfluidic input/output multiplexer composed of 3D-printed templates to interface both islets and adipose tissue.[173] This system served essentially as a mimic of the circulatory system and upstream endocrine signals, while allowing dynamic and quantitative measurements of both hormone secretion and nutrient sensing/uptake. In addition, the device provided new information on temporally resolved free fatty acid exchange in adipose tissue.

Recently, hiPSCs-derived organoids have emerged as a new class of *in vitro* organ models for disease modelling and regenerative medicine. Tao *et al.* presented a new strategy for creating heterogeneous human islet organoids derived from hiPSCs using organ-on-chip technology.[125] The islet organoids generated displayed suitable tissue morphology and multicellular complexity resembling human pancreatic islets *in vivo*. Moreover, they showed enhanced expression of mature β-cell-associated genes and proteins, insulin secretion levels and Ca^{2+} flux release ability in response to glucose under perfused culture conditions, highlighting the role of biomimetic mechanical cues in enhancing islet organoid function and maturation. Therefore, human islet-on-a-chip provides a proof of concept for synergistic engineering of heterogeneous islet organoids from hiPSCs by combining organ-on-a-chip technology and stem cell development biology.[130]

7.5.4 Pancreas-on-chip for Drug Development and Screening

Insulin secretagogues are antidiabetic medications that aim to increase the insulin output of pancreatic islets. They include sulfonylureas such as tolbutamide, glinides and incretin-related drugs such as dipeptidyl peptidase 4 (DPP-4) inhibitors, and glucagon-like peptide-1 receptor (GLP-1R) agonists which are widely used to treat type 2 diabetes.[174] The process for studying and developing antidiabetic drugs usually depends on *in vivo* animal models and *in vitro* 2D monolayer cell culture models. Although animal models and 2D cell cultures have made significant medical contributions, there is a lack of predictability when results are extrapolated to humans because of anatomical and physiological differences. Ramachandran *et al.* highlighted evidence of these differences in a study where re-aggregated human pancreatic islets were assessed and compared to native human islet and rat islet drug screening.[175] The response of the different islets was significantly different when tested against a variety of compounds such as the calcium channel agonist Bay K 8644, glibenclamide, tolbutamide, caffeine, carbachol and glucagon-like peptide-1, among others. As the authors pointed out, islet reaggregates may represent a more homogeneous model for drug screening as native islets are heterogeneous in terms of both size and composition.

Currently, several works are being carried out on developing *in vitro* models to improve medical prediction. Recent progress in microfabrication, cell engineering and imaging technologies has led organ-on-a-chip to become an innovative technology capable of reproducing physiological cell behaviours and drug screening (Table 7.2). For instance, in order to study compound-stimulation mechanisms for insulin secretion, Misun *et al.* exposed reaggregated islets to tolbutamide and exendin-4, a sulfonylurea-class compound and a glucagon-like-peptide 1 receptor (GLP-1R) agonist, respectively.[151] Continuous tolbutamide treatment resulted in a sharp increase of the first-phase insulin secretion, which could lead to severe hypoglycaemia if treatment timing is not respected. On the other hand, exendin-4 showed a potentiated action on glucose-induced insulin secretion by a most pronounced second-phase with sustained oscillations, which is consistent with a previous study by Ritzel *et al.*[176] Therefore, recent progress in β-cell signalling studies in biochips has not only enhanced our understanding of insulin secretion, but also revealed the mechanisms of insulin secretagogues. Immunosuppressive medication, which is an inevitable part of islet transplants, could also benefit from OoC technologies. Several studies have demonstrated that OoC devices can provide a robust platform for testing the compatibility of the immunosuppressive regimen with the immune system, as well as the effect of immunosuppressive reagents on isolated islet function and viability.[118]

In physiological conditions, cells reside in a three-dimensional (3D) environment and interact with other cells and the extracellular matrix (ECM).[177] These interactions are necessary for the proper differentiation and function of typical cells, and they more precisely mimic physiological pathological conditions. With this in mind, Jun *et al.* performed a drug efficacy test on rat islet spheroids cultured under dynamic conditions.[119] They tested two typical antidiabetic drugs, tolbutamide and glucagon-like peptide-1 (GLP-1), at different concentrations. The results revealed that islet spheroids under dynamic conditions exhibited higher sensitivity to drugs. These outcomes were also demonstrated by Essaouiba *et al.* on both native rat islets and hiPSC-derived islets.[126,130] Moreover, for drug toxicity testing, Jun *et al.* exposed the islet models to rapamycin, which has been used in islet transplantation as an immunosuppressant. Higher toxicity resistance in the dynamic group reflects that *in vitro* toxicity assays may produce significantly different results depending on the culture environment.[119] For drug screening applications, spheroid biochip models can improve assay reproducibility and quality with enhanced response to therapeutics.

7.6 Multi-organ-on-chip for Diabetes Research

T2DM has been linked to different causes, particularly dysfunctions affecting β-cells, growing resistance to insulin, or chronic inflammatory conditions which cause the loss of control of blood glucose levels. It is a multi-organ disease involving interactions between different organs including the pancreas, liver, muscles, nervous system, kidneys, adipose tissue and small intestine (Figure 7.7).[11] The interorgan specificity of T2DM is due to the endocrine and exocrine functionalities of the pancreas. The pancreas contains exocrine glands producing important enzymes for digestion, including trypsin, amylases and lipase which break down, respectively, proteins, carbohydrates and fats in the intestine. This enzyme mix is routed through the pancreatic duct and joins the bile duct to reach the duodenum in the small intestine, its site of action. On the other hand, the endocrine function of the pancreas is centred on islets of Langerhans, which are responsible for releasing important hormones into the bloodstream. The main hormones

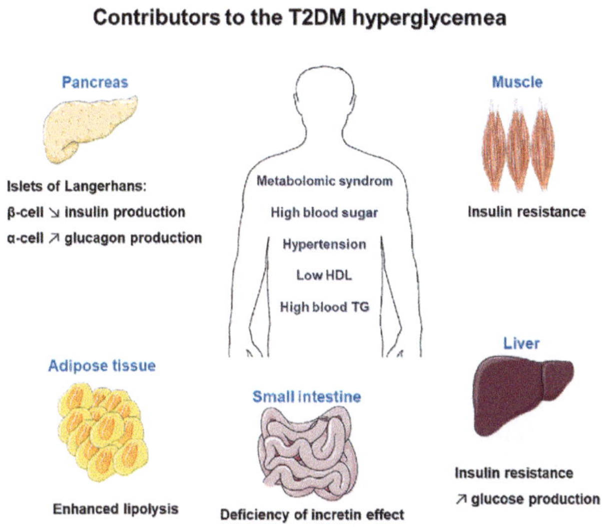

Figure 7.7 Main organs involved in T2DM (adapted from ref. 11).

are insulin and glucagon, which act to control blood sugar. Maintaining sugar homeostasis is imperative for the functioning of the rest of the organs.[178]

To understand the biological processes and develop therapeutic strategies for T2DM, scientists conventionally used *in vivo* models because of the complexity and integrated multi-organ responses they confer. However, these models tended to fail due to the phylogenetic distance between humans and animals, and their use has been limited for ethical reasons.[11,179] An alternative to these models is the use of *in vitro* methods; the reference technique relies on the classic culture of primary cells using *in vitro* culture platforms. These strategies aim to investigate cells' biological responses without controlling the context surrounding them. This kind of simplification results in the model failing to mimic key aspects of the human body.[180] To obtain physiologically meaningful and reproducible data, it is important to generate models that take the different organs involved into account. Different research has focused on this subject and tried to develop culture systems that combine different organs to reproduce T2DM. The most recent innovations used organ-on-chip technology to reproduce the behaviour of an organ or a group of organs. This technology improves the transport of nutrients, oxygen, hormones and metabolic waste to create a "physiological-like" condition. The idea is to combine different cell types in different culture compartments (each compartment mimicking a specific microphysiological condition) and connect them through microfluidic channels to mimic and ensure the crosstalk between two or more organs.[181]

Most of the time, multi-organ-on-chip tries to faithfully emulate the *in vivo* environment and interactions between organs. The implied signalling pathways create synergic effects on cells which enhance their functions compared to monocultured cells.[182] Despite the potential that these approaches offer for reproducing the multi-organ interactions implied in DM, several limitations are encountered in reference to the optimizations needed to ensure the optimal culture of the different cells, vascularization of the organs, which has not yet been mastered, and standardization of the materials and cells in the models used which causes variabilities that could impact results.[183]

7.6.1 Liver Pancreas-on-chip

The liver is the largest organ in the human body and has an indispensable role in overseeing digestion, metabolism and the elimination of toxins from the body. In patients with T2DM, the liver is one of the first organs to be severely affected by insulin resistance and reacts to this by increasing the production of glucose. Many hepatic *in vitro* models and liver-on-chip have been introduced and advanced to reproduce hepatic functions and architecture.[11] Only a few studies have focused on the pancreas and its interactions with the liver to produce a liver–pancreas multi-organ-on-chip.[184] Recent works have demonstrated the potential for OoC technology to ensure pancreatic islet-liver crosstalk. Bauer *et al.* designed a two-organ-chip system to ensure the dynamic culture of human pancreatic islets and liver spheroids composed of the HepaRG cell line and primary human stellate cells[178] (Figure 7.8A). The aim of their study was to develop a reliable human T2DM model. They observed an increase and a conservation of insulin secretion which supported stabilization of the homeostatic state when comparing the coculture condition with the monoculture. In addition, they confirmed that islet microtissues lose their function after being subjected to prolonged hyperglycaemia. In another approach, Essaouiba *et al.* used a perfusion loop with two biochips hosting primary hepatocytes and rat islet to investigate the interaction between the organs compared with monoculture conditions184 (Figure 7.8B). They proved that coculturing pancreatic cells with hepatocytes helped to recover hepatic functions (compared to the hepatic monoculture without insulin) and modified the expression of the genes involved in insulin/glucagon homeostasis.

Integrating multiple organs on a chip has always been a challenging goal because of the optimizations that must be taken into consideration to allow optimal culture of all the cell types involved in the system. Lee *et al.* developed a pancreas–muscle–liver OoC, given the relation that links these three organs.[185] Indeed, *in vivo*, the muscle is responsible for the most important glucose uptake of the body. As a consequence, signalling pathways are activated in the pancreas to produce glucagon, and start glucose production in the liver (gluconeogenesis and glycogenolysis) and the clearance of insulin.[186] In addition to the experimental results obtained from glucose metabolism stimulation, the combination of the multi-organ-on-chip led to the construction of a mathematical model describing time-dependent concentration changes in glucose and insulin.

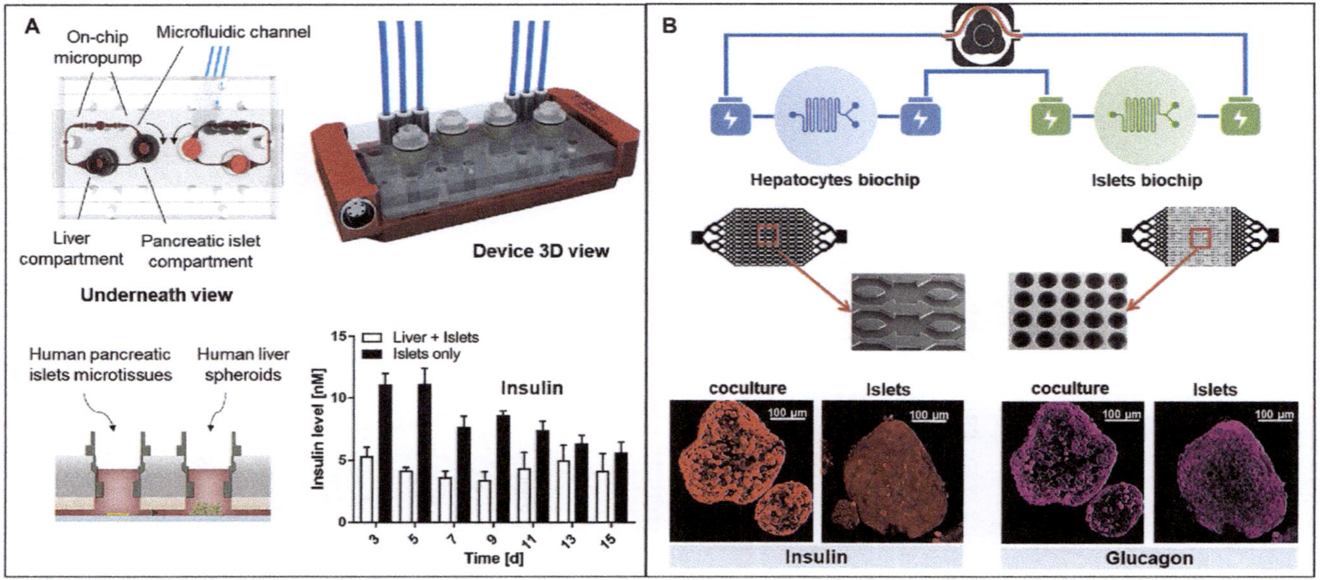

Figure 7.8 Multiorgan-on-chip models for liver/pancreas interaction studies: (A) microphysiological two-organ-chip (2-OC) device developed by TissUse GmbH (Berlin, Germany) for human pancreatic islets/HepaRG spheroids study; (B) setup used by Essaouiba *et al.* to study crosstalk between rat islets/rat primary hepatocytes. (A) Reproduced from ref. 178, https://doi.org/10.1038/s41598-017-14815-w, under the terms of the CC BY 4.0 license https://creativecommons.org/licenses/by/4.0/. (B) Reproduced from ref. 184 with permission from Elsevier, Copyright 2020.

7.6.2 Organ-on-chip Coupling Pancreas and Other Organs

The small intestine and the pancreas crosstalk through a complex endocrine system composed of hormones. In the case of T2DM, endogenous insulin is affected, leading to its deficiency or diminished effectiveness. Nguyen *et al.* focused on this connection and developed an endocrine system on chip reassembling insulin-secreting β-cells and immortalized L-cell lines, which are located in the small intestines that stimulate insulin secretion when exposed to glucose intake from the diet.[139] In normal conditions, the glucose intake sets off the production of glucagon-like peptide-1 (GLP-1) by the L-cells, which in turn activate the secretion of insulin by the β-cells. The authors observed that after 3 days of culture, and after exposure to a glucose stimulus, the multi-organ-on-chip system responded by an increase in, and saturation of, the insulin level. These results were compared to 2D static cultures and monocultures of the two cell lines separately, and the OoC exhibited higher expression of insulin. The results demonstrated the potential of the endocrine system on a chip to screen GLP-1 analogues and stimulants and ensure follow-up of natural insulin production in the treatment of diabetes.[139]

The pancreas interacts also with adipose tissues. Several studies have proved the supportive effects, anti-inflammatory role and immunomodulatory properties of these tissues on the viability, structure and insulin production of β-cells. This has been studied by Lu *et al.*, who developed a coculture system composed of three cell chambers, two holding adipocytes connected to one holding islets.[187] Then to study the interactions between the two cell types, the adipocyte perfusate, containing adipocyte secretions, passes into the islets chamber. GSIS was measured in different conditions to analyse the coculture effect. They observed an increase in insulin production in the coculture condition compared to the monocultured one. These results highlight the endocrine interaction between adipocytes and islets by delivering cytokines and growth factors, which promotes cellular functions.[188]

Although multi-organ-on-chip technology provides a powerful tool for recapitulating several "physiological" situations and studying organ-to-organ crosstalk, further developments are needed to accurately mimic *in vivo* native tissues. The construction of 3D tissue onto the biochip represents a promising approach for reproducing the behaviour of an organ or a group of organs and providing a more appropriate microenvironment for tissue maintenance and development.[189] While pancreas-on-chip models usually use 3D cultures of islets of Langerhans or β-cells spheroids, cells are still mostly cultivated in 2D (monolayer inside the biochip) for the majority of other OoC such as liver, muscle and adipose-on-chip. Recent advances in tissue engineering, biomaterials and microfabrication technology offered the opportunity to create OoC with complex 3D tissues combining multiple cell types.[181,190] Several approaches can be used for 3D cell culture: design of hydrogels and scaffolds, cell aggregation in spheroids, 3D bioprinting, and 3D printing of microfluidic biochip.[190–193] In recent years, many 3D organ-on-chip models have been reported in the literature including liver-,[194–198] muscle-[199–201] and adipose-on-chip.[202,203] These models have been shown to produce the key characteristics and functionalities of *in vivo* tissues/organs. The combination of such models with pancreas-on-chip to build multi-organ-on-chip devices can reproduce accurately the physiological interactions between pancreas and organs related to DM, and help to elucidate the pathophysiological mechanisms in diabetes.

7.7 Conclusions

In this chapter, we have presented the basic concept of microfluidic technologies linked to diabetes. More particularly, we have shown how microfluidics can be used as one of the essential building blocks for various bioengineered solutions in the fight against metabolic syndrome and related disorders, with a focus on diabetes. Of these solutions, the point of care approach can take part in enhancing patient autonomy by providing technologies that can be carried out easily in daily life. Regarding the disease model, the organ-on-chip solution makes it possible to access human models involving complex crosstalk and organ-to-organ interactions. Organ-on-chip coupled with the adequate cell sources, obtained from patient-derived iPSCs, will in the near future contribute to reproducing human variability and the idiosyncrasy encountered in heterogeneous patient populations. Overall, we believe that microfluidics will play a part in generating advanced tools suitable for large-scale, personalized medicine solutions. In addition to conventional animal models, microfluidic-based technology contributes to bridging human *in vitro* models, human cohort data and patient therapy.

List of abbreviations

2D	Two-dimensional
3D	Three-dimensional
ADP	Adenosine diphosphate
ADSCs	Adipose-derived stem cells
ATP	Adenosine triphosphate
AuNPs	Gold nanoparticles
BAP	Bioartificial pancreas
BM-MSCs	Bone marrow mesenchymal stem cells
C_4F_8	Octafluorocyclobutane
Ca^{2+}	Calcium ions
CDD	Charge-coupled device
CF_4	Carbon tetrafluoride
COC	Cyclic olefin copolymer
CRP	C-reactive protein
DM	Diabetes mellitus
DPP-4	Dipeptidyl peptidase 4
ECM	Extracellular matrix

ECs	Endothelial cells
ESCs	Embryonic stem cells
GDM	Gestational diabetes mellitus
GLP-1	Glucagon-like peptide-1
Glut2	Glucose transporter 2
GSIS	Glucose-stimulated insulin secretion
H_2O_2	Hydrogen peroxide
HbA1C	Glycosylated haemoglobin
hiPSCs	Human-induced pluripotent stem cells
HSA	Human serum albumin
IDF	International Diabetes Foundation
iPSCs	Induced pluripotent stem cells
KATP	ATP-sensitive potassium
LADA	Latent autoimmune diabetes in adults
LGIL	Low glucose and insulin levels
LOD	Limit of detection
MEMS	Micro-electro-mechanical systems
Min	Minutes
mm	Millimetre
MODY	Maturity-onset diabetes of the young
MS	Mass spectrometry
MSCs	Mesenchymal stem cells
nM	Nanomolar
NOA81	Norland Optical Adhesive 81
OoC	Organ-on-chip
PC	Polycarbonate
PDMS	Polydimethylsiloxane
Pdx1	Pancreatic and duodenal homeobox 1
PEG	Polyethylene glycol
PFA	Perfluoroalkoxy alkane
PFPE	Perfluoropolyether
PMMA	Poly(methyl methacrylate)
POC	Point-of-care
PS	Polystyrene
PTFE	Polytetrafluoroethylene
PU	Polyurethane
PVC	Polyvinyl chloride
RBC	Red blood cells
T1DM	Type 1 diabetes mellitus
T2DM	Type 2 diabetes mellitus
TCA	Tricarboxylic acid cycle
TPE	Thermoset polyester
USD	United States dollars
VDCCs	Voltage-dependent channels
WHO	World Health Organization
μPAD	Micro paper-based analytical device

References

1. N. Jouvet and J. L. Estall, *Exp. Cell Res.*, 2017, **360**, 19–23.
2. P. V. Röder, B. Wu, Y. Liu and W. Han, *Exp. Mol. Med.*, 2016, **48**, e219.
3. R. Jellali, A. Essaouiba, E. Leclerc and C. Legallais, in *Current Trends and Future Developments on (Bio-) Membranes*, Elsevier, 2020, pp. 77–108.
4. E. Maillard and S. Sigrist, in *Biomaterials for Organ and Tissue Regeneration*, Elsevier, 2020, pp. 299–333.
5. World Health Oorganization (WHO) Official Web Site.
6. IDF Diabetes Atlas, 9th edn, 2019, https://www.diabetesatlas.org/en/, (accessed 22 May 2021).
7. P. Saeedi, I. Petersohn, P. Salpea, B. Malanda, S. Karuranga, N. Unwin, S. Colagiuri, L. Guariguata, A. A. Motala, K. Ogurtsova, J. E. Shaw, D. Bright and R. Williams, *Diabetes Res. Clin. Pract.*, 2019, **157**, 107843.
8. P. Saeedi, P. Salpea, S. Karuranga, I. Petersohn, B. Malanda, E. W. Gregg, N. Unwin, S. H. Wild and R. Williams, *Diabetes Res. Clin. Pract.*, 2020, **162**, 108086.
9. K. Papatheodorou, M. Banach, E. Bekiari, M. Rizzo and M. Edmonds, *J. Diabetes Res.*, 2018, **2018**, 3086167.
10. J. L. Harding, M. E. Pavkov, D. J. Magliano, J. E. Shaw and E. W. Gregg, *Diabetologia*, 2019, **62**, 3–16.
11. J. Rogal, A. Zbinden, K. Schenke-Layland and P. Loskill, *Adv. Drug Deliv. Rev.*, 2019, **140**, 101–128.
12. Z. Punthakee, R. Goldenberg and P. Katz, *Can. J. Diabetes*, 2018, **42**, S10–S15.
13. S. Szunerits, S. Melinte, A. Barras, Q. Pagneux, A. Voronova, A. Abderrahmani and R. Boukherroub, *Chem. Soc. Rev.*, 2021, **50**, 2102–2146.
14. B. Kepsutlu, C. Nazli, T. B. Bal and S. Kizilel, *Curr. Pharm. Biotechnol.*, 2014, **15**, 590–608.
15. R. A. DeFronzo, E. Ferrannini, L. Groop, R. R. Henry, W. H. Herman, J. J. Holst, F. B. Hu, C. R. Kahn, I. Raz, G. I. Shulman, D. C. Simonson, M. A. Testa and R. Weiss, *Nat. Rev. Dis. Prim.*, 2015, **1**, 1–22.
16. D. Tripathy and A. O. Chavez, *Curr. Diab. Rep.*, 2010, **10**, 184–191.
17. A. D. Association, *Diabetes Care*, 2019, **42**, S13–S28.
18. J. F. Plows, J. L. Stanley, P. N. Baker, C. M. Reynolds and M. H. Vickers, *Int. J. Mol. Sci.*, 2018, **19**, 3342.
19. H. D. McIntyre, P. Catalano, C. Zhang, G. Desoye, E. R. Mathiesen and P. Damm, *Nat. Rev. Dis. Prim.*, 2019, **5**, 1–19.
20. A. T. Kharroubi, *World J. Diabetes*, 2015, **6**, 850.
21. G. M. Whitesides, *Nature*, 2006, **442**, 368–373.
22. M. I. Khot, M. A. Levenstein, G. N. de Boer, G. Armstrong, T. Maisey, H. S. Svavarsdottir, H. Andrew, S. L. Perry, N. Kapur and D. G. Jayne, *Sci. Rep.*, 2020, **10**, 1–13.

23. P. S. Dittrich and A. Manz, *Nat. Rev. Drug Discovery*, 2006, **5**, 210–218.
24. K. Ren, J. Zhou and H. Wu, *Acc. Chem. Res.*, 2013, **46**, 2396–2406.
25. S. F. Berlanda, M. Breitfeld, C. L. Dietsche and P. S. Dittrich, *Anal. Chem.*, 2021, **93**, 311–331.
26. A. G. Niculescu, C. Chircov, A. C. Bîrcă and A. M. Grumezescu, *Int. J. Mol. Sci.*, 2021, **22**, 1–26.
27. L. J. Pan, J. W. Tu, H. T. Ma, Y. J. Yang, Z. Q. Tian, D. W. Pang and Z. L. Zhang, *Lab Chip*, 2018, **18**, 41–56.
28. X. Zhang and S. J. Haswell, *MRS Bull.*, 2006, **31**, 95–99.
29. D. Sticker, R. Geczy, U. O. Häfeli and J. P. Kutter, *ACS Appl. Mater. Interfaces*, 2020, **12**, 10080–10095.
30. X. Hou, Y. S. Zhang, G. T. De Santiago, M. M. Alvarez, J. Ribas, S. J. Jonas, P. S. Weiss, A. M. Andrews, J. Aizenberg and A. Khademhosseini, *Nat. Rev. Mater.*, 2017, **2**, 1–15.
31. H. Gao, C. Yan, W. Wu and J. Li, *Sensors*, 2020, **20**, 1792.
32. P. N. Nge, C. I. Rogers and A. T. Woolley, *Chem. Rev.*, 2013, **113**, 2550–2583.
33. J. B. Nielsen, R. L. Hanson, H. M. Almughamsi, C. Pang, T. R. Fish and A. T. Woolley, *Anal. Chem.*, 2020, **92**, 150–168.
34. A. Cignarelli, V. A. Genchi, S. Perrini, A. Natalicchio, L. Laviola and F. Giorgino, *Int. J. Mol. Sci.*, 2019, **20**, 1–20.
35. H. Li and A. J. Steckl, *Anal. Chem.*, 2019, **91**, 352–371.
36. J. P. Rolland, R. M. Van Dam, D. A. Schorzman, S. R. Quake and J. M. DeSimone, *J. Am. Chem. Soc.*, 2004, **126**, 2322–2323.
37. K. W. Bong, J. Lee and P. S. Doyle, *Lab Chip*, 2014, **14**, 4680–4687.
38. R. Jellali, P. Paullier, M. J. Fleury and E. Leclerc, *Sens. Actuators, B*, 2016, **229**, 396–407.
39. M. Tonin, N. Descharmes and R. Houdré, *Lab Chip*, 2016, **16**, 465–470.
40. C. W. Chang, Y. J. Cheng, M. Tu, Y. H. Chen, C. C. Peng, W. H. Liao and Y. C. Tung, *Lab Chip*, 2014, **14**, 3762–3772.
41. H. J. Chiang, S. L. Yeh, C. C. Peng, W. H. Liao and Y. C. Tung, *J. Vis. Exp*, 2017, **2017**, 55292.
42. Y. Gao, G. Stybayeva and A. Revzin, *Lab Chip*, 2019, **19**, 306–315.
43. A. Waldbaur, H. Rapp, K. Länge and B. E. Rapp, *Anal. Methods*, 2011, **3**, 2681–2716.
44. C. W. Tsao, *Micromachines*, 2016, **7**, 225.
45. C. Iliescu, H. Taylor, M. Avram, J. Miao and S. Franssila, *Biomicrofluidics*, 2012, **6**, 016505.
46. T. Wang, J. Chen, T. Zhou and L. Song, *Micromachines*, 2018, **9**, 269.
47. Y. Lin, C. Gao, D. Gritsenko, R. Zhou and J. Xu, *Microfluid. Nanofluidics*, 2018, **22**, 97.
48. B. K. Gale, A. R. Jafek, C. J. Lambert, B. L. Goenner, H. Moghimifam, U. C. Nze and S. K. Kamarapu, *Inventions*, 2018, **3**, 60.
49. T. Kilic, F. Navaee, F. Stradolini, P. Renaud and S. Carrara, *Microphysiol. Syst.*, 2018, **2**, 1–32.

50. N. Azizipour, R. Avazpour, D. H. Rosenzweig, M. Sawan and A. Ajji, *Micromachines*, 2020, **11**, 1–15.
51. A. Aziz, C. Geng, M. Fu, X. Yu, K. Qin and B. Liu, *Bioengineering*, 2017, **4**, 39.
52. N. Wongkaew, M. Simsek, C. Griesche and A. J. Baeumner, *Chem. Rev.*, 2019, **119**, 120–194.
53. S. K. Metkar and K. Girigoswami, *Biocatal. Agric. Biotechnol.*, 2019, **17**, 271–283.
54. G. A. Clarke, B. X. Hartse, A. E. N. Asli, M. Taghavimehr, N. Hashemi, M. A. Shirsavar, R. Montazami, N. Alimoradi, V. Nasirian, L. J. Ouedraogo and N. N. Hashemi, *Sensors*, 2021, **21**, 1–44.
55. S. Patel, R. Nanda, S. Sahoo and E. Mohapatra, *Biochem. Res. Int.*, 2016, **2016**, 3130469.
56. A. B. González-Guerrero, J. Maldonado, S. Herranz and L. M. Lechuga, *Anal. Methods*, 2016, **8**, 8380–8394.
57. J. Wu, M. Dong, C. Rigatto, Y. Liu and F. Lin, *npj Digit. Med.*, 2018, **1**, 7.
58. F. Fontana, J. P. Martins, G. Torrieri and H. A. Santos, *Adv. Mater. Technol.*, 2019, **4**, 1970034.
59. N. Pørksen, B. Nyholm, J. D. Veldhuis, P. C. Butler and O. Schmitz, *Am. J. Physiol.*, 1997, **273**, E908.
60. P. Yao, S. Tung, Z. Zhan, J. Hua and Z. Dong, in *2011 IEEE International Conference on Cyber Technology in Automation, Control, and Intelligent Systems*, CYBER 2011, 2011, pp. 107–111.
61. P. Yao, S. Tung, Z. Zhan, J. Hua and Z. Dong, *Trans. Inst. Meas. Control*, 2013, **35**, 893–900.
62. A. Avoundjian, M. Jalali-Heravi and F. A. Gomez, *Anal. Bioanal. Chem.*, 2017, **409**, 2697–2703.
63. N. Moodley, U. Ngxamngxa, M. J. Turzyniecka and T. S. Pillay, *J. Clin. Pathol.*, 2015, **68**, 258–264.
64. P. Mao and D. Wang, *J. Proteome Res.*, 2014, **13**, 1560–1569.
65. P. Yao, Z. Liu, B. Liu, L. Liu, N. Jiao, Z. Dong and S. Tung, in *2013 IEEE International Conference on Cyber Technology in Automation, Control and Intelligent Systems*, IEEE-CYBER 2013, 2013, 149–152.
66. H. Wan, L. Zhuang, Y. Pan, F. Gao, J. Tu, B. Zhang and P. Wang, in *Biomedical Information Technology*, Elsevier, 2020, pp. 51–79.
67. G. C. Ilacas, A. Basa, K. J. Nelms, J. D. Sosa, Y. Liu and F. A. Gomez, *Anal. Chim. Acta*, 2019, **1055**, 74–80.
68. S. Oyola-Reynoso, A. P. Heim, J. Halbertsma-Black, C. Zhao, I. D. Tevis, S. Çınar, R. Cademartiri, X. Liu, J.-F. Bloch and M. M. Thuo, *Talanta*, 2015, **144**, 289–293.
69. P. De Tarso Garcia, T. M. Garcia Cardoso, C. D. Garcia, E. Carrilho and W. K. Tomazelli Coltro, *RSC Adv.*, 2014, **4**, 37637–37644.
70. T. Lam, J. P. Devadhasan, R. Howse and J. Kim, *Sci. Rep.*, 2017, **7**, 1188.
71. E. Evans, E. F. Moreira Gabriel, T. E. Benavidez, W. K. Tomazelli Coltro and C. D. Garcia, *Analyst*, 2014, **139**, 5560–5567.

72. D. Sechi, B. Greer, J. Johnson and N. Hashemi, *Anal. Chem.*, 2013, **85**, 10733–10737.
73. G. Palazzo, L. Facchini and A. Mallardi, *Sens. Actuators, B*, 2012, **161**, 366–371.
74. A. Kugimiya and E. Matsuzaki, *Appl. Biochem. Biotechnol.*, 2014, **174**, 2527–2536.
75. H. Zhang, E. Smith, W. Zhang and A. Zhou, *Biomed. Microdevices*, 2019, **21**, 48.
76. S. S. Nadar, P. D. Patil, M. S. Tiwari and D. J. Ahirrao, *Crit. Rev. Biotechnol.*, 2021, **41**, 1046–1080.
77. S. Liu, W. Su and X. Ding, *Sensors*, 2016, **16**, 1–17.
78. Y. A. Yang and C. H. Lin, *Biomicrofluidics*, 2015, **9**, 022402.
79. C. C. Wu, H. I. Lin, K. W. Chang, J. D. Mai, S. C. Shiesh and G. B. Lee, *Microfluid. Nanofluidics*, 2015, **18**, 613–621.
80. E. A. Redman, M. Ramos-Payan, J. S. Mellors and J. M. Ramsey, *Anal. Chem.*, 2016, **88**, 5324–5330.
81. J. Li, K. W. Chang, C. H. Wang, C. H. Yang, S. C. Shiesh and G. Bin Lee, *Biosens. Bioelectron.*, 2016, **79**, 887–893.
82. K. W. Chang, J. Li, C. H. Yang, S. C. Shiesh and G. Bin Lee, *Biosens. Bioelectron.*, 2015, **68**, 397–403.
83. B. Zhang, R. B. Kumar, H. Dai and B. J. Feldman, *Nat. Med.*, 2014, **20**, 948–953.
84. Y. Zhan, D. N. Loufakis, N. Bao and C. Lu, *Lab Chip*, 2012, **12**, 5063–5068.
85. S. Cha, T. Shin, S. S. Lee, W. Shim, G. Lee, S. J. Lee, Y. Kim and J. M. Kim, *Anal. Chem.*, 2012, **84**, 10471–10477.
86. C. H. D. Tsai, S. Sakuma, F. Arai and M. Kaneko, *IEEE Trans. Biomed. Eng.*, 2014, **61**, 1187–1195.
87. D. Bento, R. O. Rodrigues, V. Faustino, D. Pinho, C. S. Fernandes, A. I. Pereira, V. Garcia, J. M. Miranda and R. Lima, *Micromachines*, 2018, **9**, 151.
88. Z. Pu, C. Zou, R. Wang, X. Lai, H. Yu, K. Xu and D. Li, *Biomicrofluidics*, 2016, **10**, 011910.
89. M. Yin, B. Huang, S. Gao, A. P. Zhang and X. Ye, *Biomed. Opt. Express*, 2016, **7**, 2067.
90. A. M. Smink, K. Skrzypek, J. Visser, R. Kuwabara, B. J. de Haan, P. de Vos and D. Stamatialis, *Biomed. Mater.*, 2021, **16**, 035036.
91. V. Iacovacci, L. Ricotti, A. Menciassi and P. Dario, *Biochem. Pharmacol.*, 2016, **100**, 12–27.
92. A. A. Tomei, C. Villa and C. Ricordi, *Expert Opin. Biol. Ther.*, 2015, **15**, 1321–1336.
93. D. Velasco, E. Tumarkin and E. Kumacheva, *Small*, 2012, **8**, 1633–1642.
94. T. Alkayyali, T. Cameron, B. Haltli, R. G. Kerr and A. Ahmadi, *Anal. Chim. Acta*, 2019, **1053**, 1–21.
95. A. A. Tomei, V. Manzoli, C. A. Fraker, J. Giraldo, D. Velluto, M. Najjar, A. Pileggi, R. D. Molano, C. Ricordi, C. L. Stabler and J. A. Hubbell, *Proc. Natl. Acad. Sci. U. S. A.*, 2014, **111**, 10514–10519.

96. U. Barkai, A. Rotem and P. de Vos, *World J. Transplant.*, 2016, **6**, 69.
97. Š. Selimović, J. Oh, H. Bae, M. Dokmeci and A. Khademhosseini, *Polymers*, 2012, **4**, 1554–1579.
98. D. M. Headen, G. Aubry, H. Lu and A. J. García, *Adv. Mater.*, 2014, **26**, 3003–3008.
99. D. M. Headen, J. R. García and A. J. García, *Microsyst. Nanoeng.*, 2018, **4**, 1–9.
100. J. D. Weaver, D. M. Headen, M. M. Coronel, M. D. Hunckler, H. Shirwan and A. J. García, *Am. J. Transplant.*, 2019, **19**, 1315–1327.
101. S. Akbari and T. Pirbodaghi, *Microfluid. Nanofluidics*, 2014, **16**, 773–777.
102. A. G. Håti, D. C. Bassett, J. M. Ribe, P. Sikorski, D. A. Weitz and B. T. Stokke, *Lab Chip*, 2016, **16**, 3718–3727.
103. H. Liu, Y. Wang, H. Wang, M. Zhao, T. Tao, X. Zhang and J. Qin, *Adv. Sci.*, 2020, **7**, 1903739.
104. D. An, A. Chiu, J. A. Flanders, W. Song, D. Shou, Y. C. Lu, L. G. Grunnet, L. Winkel, C. Ingvorsen, N. S. Christophersen, J. J. Fels, F. W. Sand, Y. Ji, L. Qi, Y. Pardo, D. Luo, M. Silberstein, J. Fan and M. Ma, *Proc. Natl. Acad. Sci. U. S. A.*, 2017, **115**, E263–E272.
105. T. Watanabe, T. Okitsu, F. Ozawa, S. Nagata, H. Matsunari, H. Nagashima, M. Nagaya, H. Teramae and S. Takeuchi, *Biomaterials*, 2020, **255**, 120162.
106. J. Cheng, Y. Jun, J. Qin and S. H. Lee, *Biomaterials*, 2017, **114**, 121–143.
107. K. H. Lee, S. J. Shin, Y. Park and S. H. Lee, *Small*, 2009, **5**, 1264–1268.
108. Y. Jun, M. J. Kim, Y. H. Hwang, E. A. Jeon, A. R. Kang, S. H. Lee and D. Y. Lee, *Biomaterials*, 2013, **34**, 8122–8130.
109. Y. Jun, A. R. Kang, J. S. Lee, G. S. Jeong, J. Ju, D. Y. Lee and S. H. Lee, *Biomaterials*, 2013, **34**, 3784–3794.
110. Y. Jun, A. R. Kang, J. S. Lee, S. J. Park, D. Y. Lee, S. H. Moon and S. H. Lee, *Biomaterials*, 2014, **35**, 4815–4826.
111. H. Onoe, T. Okitsu, A. Itou, M. Kato-Negishi, R. Gojo, D. Kiriya, K. Sato, S. Miura, S. Iwanaga, K. Kuribayashi-Shigetomi, Y. T. Matsunaga, Y. Shimoyama and S. Takeuchi, *Nat. Mater.*, 2013, **12**, 584–590.
112. A. Soltani, M. Soleimani, M. A. Ghiass, S. E. Enderami, S. Rabbani, A. Jafarian and A. Allameh, *Life Sci.*, 2021, **274**, 119338.
113. F. Ozawa, T. Okitsu and S. Takeuchi, *ACS Biomater. Sci. Eng.*, 2017, **3**, 392–398.
114. S. N. Bhatia and D. E. Ingber, *Nat. Biotechnol.*, 2014, **32**, 760–772.
115. A. Polini, L. Prodanov, N. S. Bhise, V. Manoharan, M. R. Dokmeci and A. Khademhosseini, *Expert Opin. Drug Discovery*, 2014, **9**, 335–352.
116. H. Kimura, Y. Sakai and T. Fujii, *Drug Metab. Pharmacokinet.*, 2018, **33**, 43–48.
117. A. Zbinden, J. Marzi, K. Schlünder, C. Probst, M. Urbanczyk, S. Black, E. M. Brauchle, S. L. Layland, U. Kraushaar, G. Duffy, K. Schenke-Layland and P. Loskill, *Matrix Biol.*, 2020, **85–86**, 205–220.

118. S. Abadpour, A. Aizenshtadt, P. A. Olsen, K. Shoji, S. R. Wilson, S. Krauss and H. Scholz, *Curr. Diabetes Rep.*, 2020, **20**, 1–13.
119. Y. Jun, J. S. Lee, S. Choi, J. H. Yang, M. Sander, S. Chung and S. H. Lee, *Sci. Adv.*, 2019, **5**, 4520–4547.
120. Y. Yu, A. Gamble, R. Pawlick, A. R. Pepper, B. Salama, D. Toms, G. Razian, C. Ellis, A. Bruni, B. Gala-Lopez, J. L. Lu, H. Vovko, C. Chiu, S. Abdo, T. Kin, G. Korbutt, A. M. J. Shapiro and M. Ungrin, *Diabetologia*, 2018, **61**, 2016–2029.
121. P. L. Lewis and J. M. Wells, *Stem Cells*, 2021, **39**, 522–535.
122. J. S. Mohammed, Y. Wang, T. A. Harvat, J. Oberholzer and D. T. Eddington, *Lab Chip*, 2009, **9**, 97–106.
123. T. Schulze, K. Mattern, E. Früh, L. Hecht, I. Rustenbeck and A. Dietzel, *Biomed. Microdevices*, 2017, **19**, 1–11.
124. T. Schulze, K. Mattern, P. Erfle, D. Brüning, S. Scherneck, A. Dietzel and I. Rustenbeck, *Front. Bioeng. Biotechnol.*, 2021, **9**, 158.
125. T. Tao, Y. Wang, W. Chen, Z. Li, W. Su, Y. Guo, P. Deng and J. Qin, *Lab Chip*, 2019, **19**, 948–958.
126. A. Essaouiba, T. Okitsu, R. Jellali, M. Shinohara, M. Danoy, Y. Tauran, C. Legallais, Y. Sakai and E. Leclerc, *Mol. Cell. Endocrinol.*, 2020, **514**, 110892.
127. K. Hirano, S. Konagaya, A. Turner, Y. Noda, S. Kitamura, H. Kotera and H. Iwata, *Biochem. Biophys. Res. Commun.*, 2017, **487**, 344–350.
128. L. J. Y. Ong, L. H. Chong, L. Jin, P. K. Singh, P. S. Lee, H. Yu, A. Ananthanarayanan, H. L. Leo and Y. C. Toh, *Biotechnol. Bioeng.*, 2017, **114**, 2360–2370.
129. S. H. Lee, S. G. Hong, J. Song, B. Cho, E. J. Han, S. Kondapavulur, D. Kim and L. P. Lee, *Adv. Healthc. Mater.*, 2018, **7**, 1701111.
130. A. Essaouiba, R. Jellali, M. Shinohara, B. Scheidecker, C. Legallais, Y. Sakai and E. Leclerc, *J. Biotechnol.*, 2021, **330**, 45–56.
131. T. Hori, K. Yamane, T. Anazawa, O. Kurosawa and H. Iwata, *Biomed. Microdevices*, 2019, **21**, 1–9.
132. K. S. Sankar, B. J. Green, A. R. Crocker, J. E. Verity, S. M. Altamentova and J. V. Rocheleau, *PLoS One*, 2011, **6**, e24904.
133. P. N. Silva, B. J. Green, S. M. Altamentova and J. V. Rocheleau, *Lab Chip*, 2013, **13**, 4374–4384.
134. A. L. Glieberman, B. D. Pope, J. F. Zimmerman, Q. Liu, J. P. Ferrier, J. H. R. Kenty, A. M. Schrell, N. Mukhitov, K. L. Shores, A. B. Tepole, D. A. Melton, M. G. Roper and K. K. Parker, *Lab Chip*, 2019, **19**, 2993–3010.
135. M. Nourmohammadzadeh, Y. Xing, J. W. Lee, M. A. Bochenek, J. E. Mendoza-Elias, J. J. McGarrigle, E. Marchese, Y. Chun-Chieh, D. T. Eddington, J. Oberholzer and Y. Wang, *Lab Chip*, 2016, **16**, 1466–1472.
136. D. Balboa, J. Saarimäki-Vire and T. Otonkoski, *Stem Cells*, 2019, **37**, 33–41.
137. R. Scharfmann, W. Staels and O. Albagli, *J. Clin. Invest.*, 2019, **129**, 3511–3520.

138. A. D. Green, S. Vasu, R. C. Moffett and P. R. Flatt, *Biochimie*, 2016, **125**, 119–125.
139. D. T. T. Nguyen, D. Van Noort, I. K. Jeong and S. Park, *Biofabrication*, 2017, **9**, 015021.
140. J. Y. Kim, H. W. Kim, S. J. Bae, D. J. Joo, K. H. Huh, Y. H. Fang, Y. Cho, J. H. Jeong, Y. S. Kim and J. I. Lee, in *Transplantation Proceedings*, Elsevier, 2012, 44, 1095–1098.
141. S. G. Yabe, S. Fukuda, F. Takeda, K. Nashiro, M. Shimoda and H. Okochi, *J. Diabetes*, 2017, **9**, 168–179.
142. S. G. Yabe, S. Fukuda, J. Nishida, F. Takeda, K. Nashiro and H. Okochi, *Regen. Ther.*, 2019, **10**, 69–76.
143. N. S. Amirruddin, B. S. J. Low, K. O. Lee, E. S. Tai and A. K. K. Teo, *Semin. Cell Dev. Biol.*, 2020, **103**, 31–40.
144. M. Bakhti, A. Böttcher and H. Lickert, *Nat. Rev. Endocrinol.*, 2019, **15**, 155–171.
145. V. Volarevic, B. S. Markovic, M. Gazdic, A. Volarevic, N. Jovicic, N. Arsenijevic, L. Armstrong, V. Djonov, M. Lako and M. Stojkovic, *Int. J. Med. Sci.*, 2018, **15**, 36–45.
146. S. Zhu, H. A. Russ, X. Wang, M. Zhang, T. Ma, T. Xu, S. Tang, M. Hebrok and S. Ding, *Nat. Commun.*, 2016, **7**, 1–13.
147. S. Kahraman, E. R. Okawa and R. N. Kulkarni, *Curr. Diabetes Rep.*, 2016, **16**, 1–8.
148. M. Hohwieler, M. Müller, P. O. Frappart and S. Heller, *Stem Cells Int.*, 2019, **2019**, 9301382.
149. K. S. Polonsky, B. D. Given and E. Van Cauter, *J. Clin. Invest.*, 1988, **81**, 442–448.
150. Y. Wang, J. F. Lo, J. E. Mendoza-Elias, A. F. Adewola, T. A. Harvat, K. P. Kinzer, D. Lee, M. Qi, D. T. Eddington and J. Oberholzer, *Bioanalysis*, 2010, **2**, 1729–1744.
151. P. M. Misun, B. Yesildag, F. Forschler, A. Neelakandhan, N. Rousset, A. Biernath, A. Hierlemann and O. Frey, *Adv. Biosyst.*, 2020, **4**, 1900291.
152. M. A. Ortega, J. Rodríguez-Comas, O. Yavas, F. Velasco-Mallorquí, J. Balaguer-Trias, V. Parra, A. Novials, J. M. Servitja, R. Quidant and J. Ramón-Azcón, *Biosensors*, 2021, **11**, 138.
153. R. Perrier, A. Pirog, M. Jaffredo, J. Gaitan, B. Catargi, S. Renaud, M. Raoux and J. Lang, *Biosens. Bioelectron.*, 2018, **117**, 253–259.
154. J. F. Dishinger and R. T. Kennedy, *Anal. Chem.*, 2007, **79**, 947–954.
155. J. F. Dishinger, K. R. Reid and R. T. Kennedy, *Anal. Chem.*, 2009, **81**, 3119–3127.
156. A. F. Adewola, Y. Wang, T. Harvat, D. T. Eddington, D. Lee and J. Oberholzer, *J. Visualized Exp.*, 2010, **26**, 1649.
157. Y. Xing, M. Nourmohammadzadeh, J. E. M. Elias, M. Chan, Z. Chen, J. J. McGarrigle, J. Oberholzer and Y. Wang, *Biomed. Microdevices*, 2016, **18**, 1–9.
158. L. S. Satin, P. C. Butler, J. Ha and A. S. Sherman, *Mol. Aspects Med.*, 2015, **42**, 61–77.

159. M. C. Beauvois, C. Merezak, J. C. Jonas, M. A. Ravier, J. C. Henquin and P. Gilon, *Am. J. Physiol. - Cell Physiol.*, 2006, **290**, 1503–1511.
160. J. A. Lees, M. Messa, E. W. Sun, H. Wheeler, F. Torta, M. R. Wenk, P. De Camilli and K. M. Reinisch, *Science*, 2017, **355**, 6171.
161. S. N. Patel, M. Ishahak, D. Chaimov, A. Velraj, D. LaShoto, D. W. Hagan, P. Buchwald, E. A. Phelps, A. Agarwal and C. L. Stabler, *Sci. Adv.*, 2021, **7**, 5515–5527.
162. R. P. Robertson, J. Harmon, P. O. T. Tran and V. Poitout, *Diabetes*, 2004, **53**, S119–S124.
163. K. S. Park, Y. S. Kim, J. H. Kim, B. K. Choi, S. H. Kim, S. H. Oh, Y. R. Ahn, M. S. Lee, M. K. Lee, J. B. Park, C. H. Kwon, J. W. Joh, K. W. Kim and S. J. Kim, *Transplant. Proc.*, 2009, **41**, 3813–3818.
164. J. Chen, Y. F. Ye, L. M. Liao, J. Q. Cai, L. H. Huang, S. L. Yang, Y. J. Ma, Y. F. Fu, X. M. Xu and J. M. Tan, *CellR4*, 2013, **1**, e382.
165. P. Lin, L. Chen, D. Li, N. Yang, Y. Sun and Y. Xu, *Neuroendocrinol. Lett.*, 2009, **30**, 204–208.
166. M. Brissova and A. C. Powers, *Diabetes*, 2008, **57**, 2269–2271.
167. L. Jansson and P. O. Carlsson, *Diabetologia*, 2002, **45**, 749–763.
168. D. Nyqvist, M. Köhler, H. Wahlstedt and P. O. Berggren, *Diabetes*, 2005, **54**, 2287–2293.
169. W. Hong, S. Lin, M. Zippi, W. Geng, S. Stock, Z. Basharat, B. Cheng, J. Pan and M. Zhou, *Can. J. Gastroenterol. Hepatol.*, 2017, **2017**, 5297143.
170. K. S. Sankar, S. M. Altamentova and J. V. Rocheleau, *PLoS One*, 2019, **14**, e0222424.
171. L. A. Godwin, J. C. Brooks, L. D. Hoepfner, D. Wanders, R. L. Judd and C. J. Easley, *Analyst*, 2015, **140**, 1019–1025.
172. J. C. Brooks, K. I. Ford, D. H. Holder, M. D. Holtan and C. J. Easley, *Analyst*, 2016, **141**, 5714–5721.
173. X. Li, J. C. Brooks, J. Hu, K. I. Ford and C. J. Easley, *Lab Chip*, 2017, **17**, 341–349.
174. S. Seino, K. Sugawara, N. Yokoi and H. Takahashi, *Diabetes, Obes. Metab.*, 2017, **19**, 22–29.
175. K. Ramachandran, S. J. Williams, H. H. Huang, L. Novikova and L. Stehno-Bittel, *Tissue Eng., Part A*, 2013, **19**, 604–612.
176. R. Ritzel, M. Schulte, N. Pørksen, M. S. Nauck, J. J. Holst, C. Juhl, W. März, O. Schmitz, W. H. Schmiegel and M. A. Nauck, *Diabetes*, 2001, **50**, 776–784.
177. N. Gupta, J. R. Liu, B. Patel, D. E. Solomon, B. Vaidya and V. Gupta, *Bioeng. Transl. Med.*, 2016, **1**, 63–81.
178. S. Bauer, C. Wennberg Huldt, K. P. Kanebratt, I. Durieux, D. Gunne, S. Andersson, L. Ewart, W. G. Haynes, I. Maschmeyer, A. Winter, C. Ämmälä, U. Marx and T. B. Andersson, *Sci. Rep.*, 2017, 7, 1–11.
179. J. Bailey, M. Thew and M. Balls, *ATLA Altern. Lab. Anim.*, 2013, **41**, 335–350.

180. N. Gaio, B. van Meer, W. Q. Solano, L. Bergers, A. van de Stolpe, C. Mummery, P. M. Sarro and R. Dekker, *Micromachines*, 2016, 7, 120.
181. N. Picollet-D'hahan, A. Zuchowska, I. Lemeunier and S. Le Gac, *Trends Biotechnol.*, 2021, DOI: 10.1016/j.tibtech.2020.11.014.
182. Y. Zhao, R. K. Kankala, S. Bin Wang and A. Z. Chen, *Molecules*, 2019, **24**, 675.
183. Q. Wu, J. Liu, X. Wang, L. Feng, J. Wu, X. Zhu, W. Wen and X. Gong, *Biomed. Eng. Online*, 2020, **19**, 1–19.
184. A. Essaouiba, T. Okitsu, R. Kinoshita, R. Jellali, M. Shinohara, M. Danoy, C. Legallais, Y. Sakai and E. Leclerc, *Biochem. Eng. J.*, 2020, **164**, 107783.
185. D. W. Lee, S. H. Lee, N. Choi and J. H. Sung, *Biotechnol. Bioeng.*, 2019, **116**, 3433–3445.
186. O. Pivovarova, W. Bernigau, T. Bobbert, F. Isken, M. Möhlig, J. Spranger, M. O. Weickert, M. Osterhoff, A. F. H. Pfeiffer and N. Rudovich, *Diabetes Care*, 2013, **36**, 3779–3785.
187. S. Lu, C. E. Dugan and R. T. Kennedy, *Anal. Chem.*, 2018, **90**, 5171–5178.
188. Y. Ohmura, M. Tanemura, N. Kawaguchi, T. MacHida, T. Tanida, T. Deguchi, H. Wada, S. Kobayashi, S. Marubashi, H. Eguchi, Y. Takeda, N. Matsuura, T. Ito, H. Nagano, Y. Doki and M. Mori, *Transplantation*, 2010, **90**, 1366–1373.
189. O. Yesil-Celiktas, O. Hassan, A. K. Miri, S. Maharjan, R. Al-kharboosh, A. Quiñones-Hinojosa and Y. S. Zhang, *Adv. Biosys.*, 2018, **2**, 1800109.
190. Y. Liu, E. Gill and Y. Y. Shery Huang, *Future Sci. OA*, 2017, **3**, FSO173.
191. M. Ruoß, M. Vosough, A. Königsrainer, S. Nadalin, S. Wagner, S. Sajadian, D. Huber, Z. Heydari, S. Ehnert, J. G. Hengstler and A. K. Nussler, *Food Chem. Toxicol.*, 2020, **138**, 111188.
192. H. G. Yi, H. Lee and D. W. Cho, *Bioengineering*, 2017, **4**, 10.
193. K. Duval, H. Grover, L. H. Han, Y. Mou, A. F. Pegoraro, J. Fredberg and Z. Chen, *Physiology*, 2017, **32**, 266–277.
194. N. S. Bhise, V. Manoharan, S. Massa, A. Tamayol, M. Ghaderi, M. Miscuglio, Q. Lang, Y. Shrike Zhang, S. R. Shin, G. Calzone, N. Annabi, T. D. Shupe, C. E. Bishop, A. Atala, M. R. Dokmeci and A. Khademhosseini, *Biofabrication*, 2016, **8**, 014101.
195. H. Lee, S. Chae, J. Y. Kim, W. Han, J. Kim, Y. Choi and D. W. Cho, *Biofabrication*, 2019, **11**, 025001.
196. Y. Zhang, N. Yang, L. Xie, F. Shu, Q. Shi and N. Shaheen, *Micromachines*, 2020, **11**, 1118.
197. Q. Meng, Y. Wang, Y. Li and C. Shen, *Biotechnol. Bioeng.*, 2021, **118**, 612–621.
198. L. Boulais, R. Jellali, U. Pereira, E. Leclerc, S. Bencherif and C. Legallais, *ACS Appl. Bio Mater.*, 2021, **4**, 5617–5626.
199. G. Agrawal, A. Aung and S. Varghese, *Lab Chip*, 2017, **17**, 3447–3461.

200. K. Shimizu, H. Araki, K. Sakata, W. Tonomura, M. Hashida and S. Konishi, *J. Biosci. Bioeng.*, 2015, **119**, 212–216.
201. C. G. Anene-Nzelu, K. Y. Peh, A. Fraiszudeen, Y. H. Kuan, S. H. Ng, Y. C. Toh, H. L. Leo and H. Yu, *Lab Chip*, 2013, **13**, 4124–4133.
202. F. Yang, A. Carmona, K. Stojkova, E. I. Garcia Huitron, A. Goddi, A. Bhushan, R. N. Cohen and E. M. Brey, *Lab Chip*, 2021, **21**, 435–446.
203. J. Rogal, C. Binder, E. Kromidas, J. Roosz, C. Probst, S. Schneider, K. Schenke-Layland and P. Loskill, *Sci. Rep.*, 2020, **10**, 6666.

CHAPTER 8

3D Bioprinting of Islets

SARAH DUIN,[a] ANJA LODE,[a] BARBARA LUDWIG[b,c] AND MICHAEL GELINSKY*[a]

[a] Centre for Translational Bone, Joint and Soft Tissue Research, University Hospital Carl Gustav Carus and Faculty of Medicine of Technische Universität Dresden, Fetscherstr. 74, 01307 Dresden, Germany; [b] Paul Langerhans Institute Dresden of Helmholtz Centre, Munich, at University Hospital Carl Gustav Carus of Technische Universität Dresden, and German Centre for Diabetes Research, Dresden, Tatzberg 47-49, 01307 Dresden, Germany; [c] Department of Medicine III, University Hospital Carl Gustav Carus, Technische Universität Dresden, Germany, Fetscherstraße 74, 01307 Dresden, Germany
*Email: michael.gelinsky@tu-dresden.de

Diabetes mellitus type 1 (T1D) is a mostly autoimmune-mediated disease which results in the destruction of the β-cells in the pancreas, leading to a complete lack of the peptide-hormone insulin. Insulin is vital for the uptake of glucose by somatic cells and thereby the regulation of blood-glucose levels. The conventional treatment for patients with insulin-dependent diabetes is the replacement of insulin *via* regular injections 30–60 min before meals, calculated according to its carbohydrate content. Nowadays, with the availability of ultra-fast or ultra-long-acting insulin analogues through adapted pharmacokinetics in combination with continuous glucose monitoring, metabolic control in many patients has improved by far, indicated by lower levels of glycated haemoglobin (HbA1c) and less (nocturnal) hypoglycaemia.[1–3] Even with these improvements the administration it

Nanoscience & Nanotechnology Series No. 59
Nanotechnology for Diabetes Management
Edited by Amar Abderrahmani, Sabine Szunerits, Rabah Boukerroub and Abdelfattah El Ouaamari

Published by the Royal Society of Chemistry, www.rsc.org

remains an invasive procedure, which could possibly be alleviated by the use of nanotechnology.

In general, the term "nanotechnology" refers to techniques for the fabrication or manipulation of objects in the range of up to 100 nm, although oftentimes structures of up to 1000 nm are also described as nanomaterials.[4–6] Furthermore, hydrogels such as sodium alginate show characteristics at the nanoscale level, which affect their macroscale properties (reviewed in Ernst *et al.*[7]), for example, the mesh size of cross-linked alginate gels is in the range of only a few nanometres.[8,9] Nanoparticles and hydrogels with nanoscale features are a highly promising tool for the delivery of therapeutic molecules through intravenous or oral administration, or the encapsulation of therapeutic cells. In preclinical studies for the therapy of diabetes mellitus with rats, nanoparticles have been used to successfully prevent digestion of orally administered insulin through protection within alginate/chitosan particles,[10] and to facilitate the intestinal uptake of the oral antidiabetic metformin by encapsulation into alginate nanogels.[11] Metformin is used in the therapy of diabetes mellitus type 2, a disease that is characterized by a reduced sensitivity of somatic cells for insulin and thereby a reduced uptake of glucose.

Even with promising studies such as those mentioned, which could prevent the necessity for continual injections, and the clinical application of advanced monitoring systems for blood glucose levels as well as improved pharmacokinetics of insulin, only a minority of patients reach HbA1c levels comparable to healthy individuals.[12] Furthermore, episodes of severe hypoglycaemia and diabetic coma cannot be ruled out completely,[13] and the overall life expectancy of individuals with long-standing T1D is reduced by 8–13 years.[14,15]

8.1 Transplantation of Insulin-producing Cells

To circumvent such complications, the only truly curative treatment to date remains the replacement of functional β-cells, which can currently only be achieved *via* whole pancreas or pancreatic islet transplantation, whereby the transplantation of islets is far less invasive than that of whole pancreas. For allogenic and, in cases of diseases of the exocrine pancreas, also autologous islet transplantation, pancreatic islets are isolated from the whole pancreas and transplanted into the portal vein of the recipient by gravity infusion,[16] whereas whole pancreas transplantation requires a large surgery. Both procedures increase the glycaemic control of patients significantly, but for a long time, patients receiving whole pancreas transplantation had a higher likelihood of reaching insulin independence, while the vast majority of patients receiving islet transplantations remained dependent on external insulin but achieved improved glycaemic control. However, with growing experience and improved protocols in islet transplantation centres around the world, clinical outcomes between the procedures are levelling out.[16,17]

With refined protocols for increased safety and efficacy, islet transplantation has been declared a standard therapy for selected patients in a number of North American and European countries since 2001.[18] After infusion into the portal vein, ideally the islets settle into small capillaries and are revascularized over a period of 10–14 days. Infusion into the portal vein had shown early success and is now an established method so that today up to 70% of recipients achieve insulin-independence initially.[16] It has been shown over the years however, that the liver is not an ideal transplantation site for islets anatomically, physiologically or immunologically. Mainly due to early hypoxia, coagulation and an instant blood-mediated inflammatory reaction (IBMIR), a considerable number of islets are lost shortly after transplantation.[19,20] This means that in many cases islets from two donors are required to achieve euglycaemia in the recipient.[21] IBMIR is a result of the direct non-physiological contact of the islets with blood in the hepatic capillaries, which initiates complement activation and clotting. This, in turn, can result in thrombosis and ischemia of hepatic tissue due to blocking of capillary ends. A high loss of transplanted β-cell-mass is critical, since the transplantation outcome mainly depends on the number of functional islets in the recipient,[16] yet the number of transplanted islets cannot be increased without restriction to prevent an inner-hepatic rise of blood pressure and an increased clogging of capillaries.[22] Furthermore, the placement within the liver chronically exposes the transplanted islets to unphysiologically high levels of nutrients but also to many toxic substances and a high level of immunosuppressive drugs.[22]

Alternative transplantation sites such as the gastric submucosa, skeletal muscle, omentum or bone marrow have been tested in preclinical studies, some with promising results.[22] Despite this, the liver remains the sole transplantation site at which insulin independence has been achieved clinically so far,[20] likely because of prolonged hypoxia at alternative sites.[23]

As with any transplantation, immunosuppression is mandatory in the case of allogeneic islet transplantation to prevent allo-rejection even with a high level of major histocompatibility complex/human leucocyte antigen (MHC/HLA) compatibility. Immunosuppressive regimens for early islet transplantations in the 1990s relied on the use of corticosteroids which are effective immunosuppressants but toxic to islets. Transplantations were only sporadically successful until the publication of the so-called Edmonton protocol which introduced a corticoid-free immunosuppressive regimen in 2000.[24–26] Since then, protocols have been refined further so that success rates for islet transplantation are now comparable to those of whole pancreas transplantation, with 50%–70% of patients becoming insulin-independent initially.[16] However, long-term immunosuppression remains challenging since many immunosuppressive agents, especially calcineurin inhibitors which block interleukin 2 (IL-2)-mediated T-cell activation, leading to a reduction in insulin secretion, can interfere with angiogenesis and β-cell proliferation, and are directly toxic to the kidneys which are already affected by continual hyperglycaemia in T1D.[16,22] In addition, long-term immunosuppression results in a strongly increased risk for infectious diseases and the development of tumours.[3,16,24]

Even in the case of successful engraftment and initial insulin independence, in a not insignificant number of patients, there is a progressive loss of function of transplanted islets resulting in recurring dependence on exogenous insulin.[18] In light of these disadvantages, transplantations for T1D patients are only performed in the case of pancreas resection, concurrent or prior kidney transplantations due to diabetes-induced nephropathy which require immunosuppression in any case, or for patients with severely labile glucose control.[3]

However, even with a return to insulin dependence over time, the vast majority of recipients still retain endogenous secretion of C-peptide. As a direct result, these patients still exhibit far greater glycaemic control, even 10 years after transplantation and fewer complications than patients treated with improved cell-free sensor/pump systems.[16,20] It would therefore be preferable to extend this therapeutic option to a far wider range of patients, which necessitates alleviation of the disadvantages (transplantation site, immunosuppression, loss-of-function) and the procurement of alternative sources of insulin-producing cells, as even with the present select group of recipients, there is a severe shortage of donors. Potential unlimited sources of insulin-producing cells are embryonic stem cells (ESCs) or induced pluripotent stem cells (iPSCs) that can be differentiated into β-cells, and xenogeneic pancreatic islets. However, even in the case of differentiated (autologous) stem cells (SCs), the fact remains that T1D is an autoimmune disease that selectively destroys β-cells. The use of xenogeneic islets, on the other hand, almost inevitably results in hyperacute rejection due to xenoreactive components of the innate immune system, the intensity of which increases the less related the species are.[27,28]

8.2 Carrier Systems for Pancreatic Islet Transplantation

To alleviate the disadvantages present in islet transplantation, different strategies have been researched over the years: the use of an extrahepatic transplantation site generally requires a carrier or scaffolding system for the islets to avoid fusion and thereby avoid the development of large necrotic centres; the omission of immunosuppressants requires the use of semipermeable encapsulation materials; and to be able to easily exchange islets in case of a loss of function over time, refillable "pouches" have been the focus of research.

The two general principles underlying approaches for scaffolding systems are based on either providing transient stability while promoting vascularization until the islets are fully integrated into the vascular network, or providing permanent stability in combination with immunoprotection. Approaches aiming at vascular and tissue integration typically use materials that are highly bioactive and quickly biodegradable, such as components of the extracellular matrix. With these approaches, exposure of islets to hypoxic conditions is minimized, but they are disadvantaged by requiring immunosuppression and cannot be used with xenogeneic islets even with immunosuppression. Approaches aiming at immunoprotection, on the other hand, generally use biologically inert materials that are largely non-degradable in the human body, such as alginate. Approaches for immunoprotection can be further categorized into either micro- or macro-encapsulation (see Figure 8.1), whereby both approaches are naturally based on semi-permeable materials that are permissive for small molecules such as oxygen, nutrients and insulin, but form a barrier for immune cells and preferably also for antibodies and inflammatory cytokines.[29–31] Microencapsulation (see Figure 8.1A) describes capsules of 200–800 μm in diameter, generally only containing a single islet (reviewed in Scharp and Marchetti[32]). Macroencapsulation (see Figure 8.1B) is based on collectively encasing several thousand islets in structures of several cm in diameter. With these approaches, administration of immunosuppressants can be omitted and especially macroencapsulation approaches are practical for extrahepatic transplantation sites. Macrocapsules are however disadvantaged by increased diffusion distances limiting supply and glucose sensing of the islets due to a reduced surface-to-volume ratio.[33,34]

In light of the impact of diffusion distance on the oxygenation of islets and insulin release, the main foci of macroencapsulation research have been the maximization of the surface-to-volume ratio, enhanced vascularization of the outer capsule surface, or an external supply of oxygen. An example for alginate-based macroencapsulation approaches which combines a high surface-to-volume ratio with sufficient stability is the TRAFFIC device, an alginate covered nylon fibre presented by An *et al.*[35] A successfully applied approach for an external supply with oxygen is the "beta-Air" device that has also been shown to provide full immunoprotection.[36–38] In contrast to microencapsulation approaches, macroscopic devices have already been commercialized in some cases, such as the "beta-Air" and the so-called "islet sheet",[39] which are both alginate-based, but also others such as the "TheraCyte"[40] and the "Cellpouch" device,[41] the exact compositions of which are confidential.

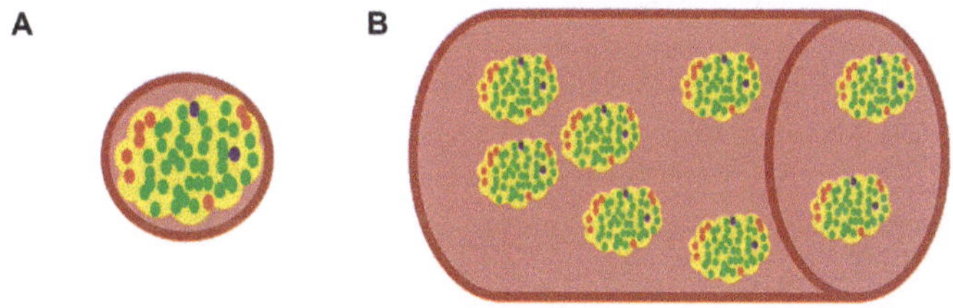

Figure 8.1 Schematic depiction of encapsulation approaches. (A) Microencapsulation: each individual islet is covered by a thin layer of hydrogel. (B) Macroencapsulation: a large number of islets is collectively encased in a macroscopic hydrogel-based device. Reproduced from ref. 42, https://nbn-resolving.org/urn:nbn:de:bsz:14-qucosa2-733889, under the terms of the CC BY 4.0 license https://creativecommons.org/licenses/by/4.0/.

In general, macroencapsulation approaches have a number of advantages for the transplantation of pancreatic islets, yet the upscaling to clinically relevant sizes and especially to clinically relevant islet densities remains a major challenge, mainly due to increasing diffusion distances.[33,43]

Independent of whether the structural support is only required transiently or permanently, in macrostructures the islets benefit from a controlled porous environment, homogeneous distribution and the precise spatial distribution of biologically active molecules. The development of macrocapsules with defined macropores to increase the surface-to-volume ratio therefore seems to be a promising strategy, which can be realized by employing 3D bioprinting.

8.3 3D (Bio)printing in Medical Research

Over the last 30 years, 3D printing, or computer-aided additive manufacturing, developed into a broad area of research within the field of regenerative medicine, and now encompasses a variety of different technologies. Partially, those technologies are purely material-based such as fused deposition modelling (FDM), selective laser sintering (SLS) and stereolithography (SLA).[44,45] In these cases, the resulting constructs, termed scaffolds, can only be seeded with cells retrospectively. Other techniques however even allow for the incorporation of cells during the printing process such as inkjet-printing, laser-induced forward-transfer (LIFT) and extrusion-based printing.[46–48] In the case of cell incorporation, the process is termed bioprinting and the cell-containing material is referred to as bioink.

The spectrum of medical applications that could benefit from the different 3D printing techniques covers a broad range, spanning from the cell-free – such as the generation of patient-specific implants, supportive structures or patient-individual models for the planning of surgeries and prosthetics[49–51] – to the cell-containing, that is research in tissue engineering (TE). Bioprinting for TE aims at the generation of tissue models for drug screening,[52] disease or cancer tissue models for basic research[53] or the generation of patient-individual tissue equivalents containing live cells, which can also be supplied with bioactive drugs or proteins such as for the induction of vascularization or the prevention of inflammation.[54] If live cells are to be used, the material needs a viscosity that does not impair cell survival through shear stress yet allows for shape fidelity until cross-linking under physiological conditions.[55] Further requirements are cytocompatibility, the support of cellular functions, and, depending on the application, either biodegradability or stability *in vivo*.[56]

Additive manufacturing techniques in general are used to create 3D structures in a layer-by-layer fashion. The most promising of these for the generation of volumetric cell-containing constructs is extrusion-based 3D bioprinting, often also referred to as bioprinting or plotting, whereby hydrogels containing cells are extruded through a needle tip and deposited strand by strand in horizontal layers (see Figure 8.2A). This technique

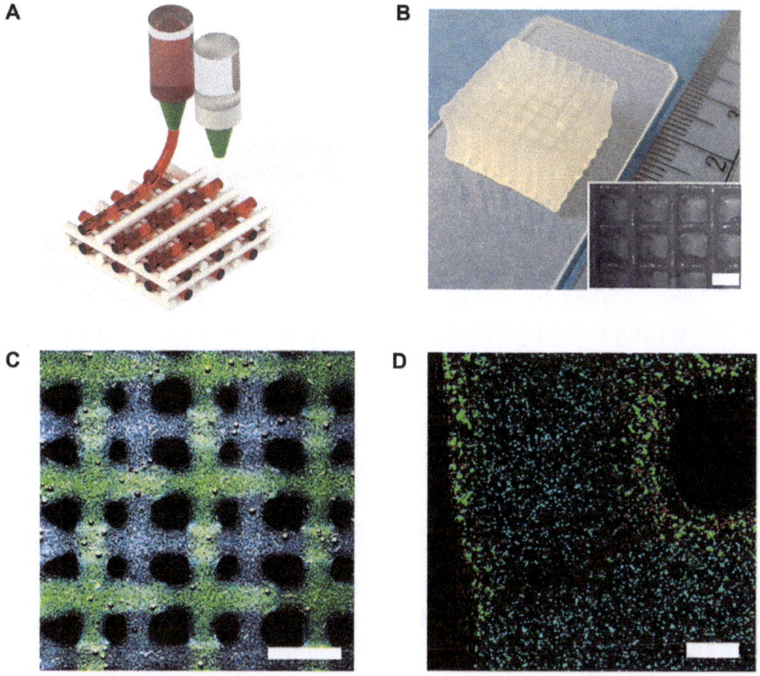

Figure 8.2 3D bioprinting, allowing for the co-cultivation of different cell types within one scaffold. (A) Schematic depiction of extrusion-based printing using two different materials. (B) Plotted 3D scaffold consisting of 3% alginate and 9% methylcellulose (algMC); scale bar = 1 mm. Reproduced from ref. 63 with permission from John Wiley and Sons, Copyright © 2015 John Wiley, Ltd. (C) Scaffold created by multichannel extrusion printing with alternating strands of algMC containing green microalgae (*Chlamydomonas reinhardtii*), and strands of algMC containing SaOS-2, cells from a human bone osteosarcoma cell line, appearing in blue; scale bar = 3 mm. Reproduced from ref. 64 with permission from John Wiley and Sons, Copyright © 2015 WILEY-VCH Verlag GmbH & Co. KGaA, Weinheim. (D) Core–shell hydrogel strand created by coaxial extrusion printing with encapsulated DiD-labelled HepG2, immortalized cells from a human liver carcinoma (red), in the algMC shell and DiI-labelled NIH 3T3 murine fibroblasts (cyan) embedded in the algMC core. Viable and active cell clusters depicted in green with calcein AM; scale bar = 500 μm. Reproduced from ref. 65, https://doi.org/10.1038/s41598-021-84384-6, under the terms of the CC BY 4.0 license https://creativecommons.org/licenses/by/4.0/.

enables the generation of macroporous scaffolds in clinically relevant sizes (see Figure 8.2B) and a variety of shapes from highly viscous materials with high cell densities. Extrusion-based bioprinting has so far been used successfully with a variety of different cell types such as fibroblasts, chondrocytes, hepatocytes and mesenchymal stem cells (MSCs) as also stated in recent reviews.[48,56] Furthermore, with extrusion-based multichannel bioprinting, different materials and cell types can be used simultaneously and with precise spatial distribution within a scaffold. This has been used successfully for the co-encapsulation of mammalian cells and microalgae in separate strands of material (see Figure 8.2C), and for the co-encapsulation of hepatocytes and fibroblasts combined within one strand with the so-called core–shell technique (see Figure 8.2D).

In addition to direct administration, nanoparticles can also be used in so-called multicomponent hydrogels,[57] whereby the particles are incorporated into macroscopic structures, *i.e.*, to achieve a local and sustained delivery of factors. With respect to diabetes research, this concept has been applied by Liu *et al.* who achieved a sustained release of insulin from nanoparticles embedded in bulk gels of poly(vinyl alcohol) (PVA) in mice.[4] In another study, chitosan nanoparticles loaded with the anti-inflammatory compound curcumin and embedded in alginate/collagen scaffolds resulted in improved wound healing in diabetic rats.[58] Such multicomponent hydrogels have also been investigated in combination with 3D bioprinting, for example to achieve an increase in shape fidelity during printing or to enable the monitoring of chemical parameters: Through the inclusion of modified silica nanoparticles, Lee *et al.* managed to increase the viscosity of an alginate/gellan gum hydrogel blend and were able to print stable ear structures with a size of several cm.[59] To measure oxygen distribution within bioprinted structures, Trampe *et al.* incorporated luminescent optical sensor nanoparticles into an alginate/methylcellulose hydrogel blend.[60] With this system it is possible to observe diffusion of oxygen through printed structures and thereby analyse the oxygen supply and oxygen consumption (as an indicator of metabolic activity) of encapsulated cells. For a more thorough overview of nanoparticles and nanomaterials in 3D printing beyond the examples presented here, the reader is also directed to several recent reviews.[61,62]

8.4 3D (Bio)printing Approaches for Pancreatic Islets

As mentioned above, all carrier systems for pancreatic islets are based on providing stability and a cell-supportive environment either with the aim of vascularization, or with the aim of immunoprotection, whereby only the second strategy could be applicable for xenotransplantation. Despite the advantages (*i.e.*, the rapid preparation of macroporous constructs in clinically relevant sizes), so far relatively few studies on applying 3D printing techniques to prepare such carrier systems for islets have been published and the majority of these concern 3D-printed structures which are retrospectively filled with insulin-producing cells or pancreatic islets, hence not bioprinted constructs. As the requirements and therefore the materials and techniques applicable for 3D-printed structures, often filled with bulk gels, and for bioprinted hydrogels containing cells already during the printing process, are quite different, the two approaches are discussed in separate sections below, whereby only materials applicable for bioprinting will be elucidated in further detail.

8.4.1 3D Printing of Cell-free Matrices for Islet Transplantation

3D-printed cell-free matrices are generally comprised of materials which are either not degradable within the human body, or show very slow biodegradation rates.

Approaches, where only the printed structure, without an additional hydrogel component, was used as cellular support have been introduced by Bernard *et al.* in 2012 and by Burzava *et al.* in 2019, both of whom used stereolithography to create microwell arrays. Bernard *et al.* used polyethyleneglycol (PEG) to create a grid-like pattern in which cells of the insulin-producing mouse insulinoma line 6 (MIN6) formed spheroids of defined sizes.[66] Burzava *et al.* created bucket-like well-plate inserts from polydimethylsiloxane (PDMS) which were successfully coated with endothelial cells before human islets were seeded on top.[67]

A further approach used by a number of groups is the creation of 3D-printed structures by fused deposition modelling, which are then filled with a soft, generally quickly biodegradable, hydrogel for support and positioning of cells. One of the first studies employing this concept was presented by Daoud *et al.* in 2011, in which the cylindrical scaffolds of poly(lactic-*co*-glycolic acid) (PLGA) were filled with islets in extracellular matrix (ECM)-based gels, mainly collagen and fibronectin, or in plain medium, whereby the presence of the gels led to increased functionality of islets after a culture time of 10 days *in vitro*.[68] All other studies using FDM employed capsules made of the slowly biodegradable polylactic acid (PLA) but varied in the internal hydrogel used for cell support. Hereby Farina *et al.* filled the capsule with platelet lysate enriched with vascular endothelial growth factor (VEGF) to induce vascularization *in vivo*. They had additionally incorporated a loading port within the capsule for retrospective loading of the device with human islets and could show functionality of these islets over 22 weeks. A large advantage of this system is the possibility of refilling with fresh islets in case of loss of function over time.[69] As more widely available cell sources than human islets, Song *et al.* and Soltanian *et al.* used stem cell-derived insulin producing cells and pancreatic organoids, respectively. In the study by Song and co-workers, the device was filled with a fibrin gel and showed functionality for 12 weeks after subcutaneous implantation.[70] Soltanian *et al.* instead soaked the device in a hydrogel consisting of collagen and Matrigel and could observe vascularization as well as enhanced maturation of β-cells followed by enhanced functionality after transplantation into immunocompromised mice.[71]

Instead of preparing a durable, mechanically stable capsule, Akkouch *et al.* prepared a completely scaffold-free tissue of fibroblasts and insulin-producing βTC3 cells by infusing 3D-printed alginate hollow strands, prepared by extrusion-based printing, with these cells and could show high viability and the presence of insulin and C-peptide *in vitro*.[72]

While the outer structure is prepared from materials stable over the long term *in vivo*, all the approaches described above aim at quick vascularization with the use of either non-encapsulated cells, or cells encapsulated in hydrogels that are quickly biodegradable and oftentimes pro-angiogenic.

As mentioned above (see Section 8.2), approaches aiming at immunoprotection instead necessarily employ biocompatible but inert materials the human body cannot degrade, such as the polysaccharide alginate. Combinations of a cell-free 3D-printed structure with cells encased in alginate were recently introduced by Espona-Noguera *et al.*,[73] Marchioli *et al.*[74] and Mridha *et al.*[75] Hereby, the former used laser sintering to create non-degradable microporous polyamide disks with a loading port, and the latter two both employed polycaprolactone (PCL), creating honeycomb-like ring structures by extrusion-based printing, and boxed as well as tubular devices by melt-electro-writing (MEW), respectively. Espona-Noguera *et al.* and Marchioli *et al.* both filled the devices with cells in alginate bulk gels, whereas Mridha *et al.* microencapsulated the cells before insertion into the device. The polyamide disk has so far only been tested *in vitro*, but the encased INS1E cells, insulin-producing cells from a rat insulinoma, were shown to form clusters of increasing size and cell number over time, and after 1 week of culture insulin release could be observed.[73] The honeycomb structure by Marchioli *et al.* was functionalized with heparin to bind the pro-angiogenic VEGF, leading to rapid vascularization in a chicken chorioallantoic membrane model and a functional response of human islets after 7 days.[74] The scaffolds prepared by MEW were filled with pseudo-islets prepared from MIN6 cells or primary mouse islets, both microencapsulated in alginate. When the tubular structure was pre-vascularized *in vivo* and retrospectively filled with islets, normalization of blood glucose levels could be observed within 48 h after implantation into isogenic mice. However, this normalization was only transient, the transplanted islets showed functionality for 60 days, but after 105 days all animals had relapsed into hyperglycaemia due to a fibrotic response towards the scaffolds, which spread to the microencapsulated islets.[75]

8.4.2 Materials used in 3D Bioprinting Approaches for Pancreatic Islets

In contrast to the 3D printing of cell-free matrices, whereby the only limitations in regards to the choice of materials are printability and the biocompatibility of the finished construct, materials and techniques for bioprinting have to meet additional requirements. Specifically, the materials need to be soft enough to not damage the cells during incorporation, should offer a hydrated environment of appropriate stiffness in the finished scaffold, and the entire post-processing regime including cross-linking or curing must be possible under cell-compatible conditions. With respect to the techniques, they should allow for islets, cell clusters of several hundred μm in diameter, to be bioprinted without damaging them, and be capable of producing cell-containing scaffolds of clinically relevant sizes. In light of these requirements, generally biopolymer-based hydrogels are chosen for bioprinting with extrusion-based printing systems. The biopolymers already investigated for the bioprinting of pancreatic islets either alone or in polymer blends are briefly discussed in the following, before the specific islet bioprinting approaches that have been published so far are introduced.

8.4.2.1 Alginate

Independent of whether the approach is micro- or macroencapsulation, a bulk gel within a stable printed structure, or a bioprinted scaffold, alginate is a promising candidate for the immunoprotective encapsulation of pancreatic islets, as it creates a barrier which is impermeable for cells of the immune system, yet barely impacts the diffusion of glucose and insulin whereby the functional response of islets can be retained.[43,76,77] Indeed, using alginate for microencapsulation, many promising studies in rodents have been conducted showing biocompatibility and stability of the capsule materials *in vivo* as well as survival and function of the transplanted islets for extended periods of time. For example, alginate capsules were shown to be immunoprotective and containing functional islets for more than 6 months.[78–80]

Alginate is a linear anionic polysaccharide, found in brown algae and some bacterial species, which is comprised of (1,4)-linked β-D-mannuronate (M) and α-L-guluronate (G) residues arranged in varying MM, GG, or MG blocks (see Figure 8.3A). Alginates for research are usually available as water-soluble alginic acid sodium salts in varying grades of molecular weight (Mw).

Alginate can easily be cross-linked at room temperature and neutral pH by a number of different multivalent cations resulting in different gel characteristics, although only Ca^{2+}, Sr^{2+} and low concentrations of Ba^{2+} are not toxic to cells.[81] While there is evidence that Ba^{2+} can also bind M residues, the divalent cations mainly bind to blocks of G residues. This results in a characteristic structure of the cross-linked gel chains, termed "egg-box" model (see Figure 8.3B) by Grant *et al.* in 1973.[82] The general mechanism of cross-linking is the ionic interaction between the divalent cations and two carboxy as well as two hydroxy groups of the alginate.[83] The carboxy groups especially are in closer vicinity in the "buckle-shaped" GG- than in MM-blocks, which results in a higher density of charges and therefore a stronger interaction with the cross-linking ions.[84] Furthermore, in MM-blocks, one of the oxygens involved in the binding of cross-linking ions is a ring oxygen instead of a hydroxy group, which results in a weaker interaction.[84] Cross-linking can occur between, but also within, alginate chains, whereby the interchain binding between GG-blocks of different alginate chains is responsible for the actual gel formation. This formation of junctions increases with increasing affinity between ion and alginate,[85] likely because the length of blocks required for sufficient binding of ions decreases

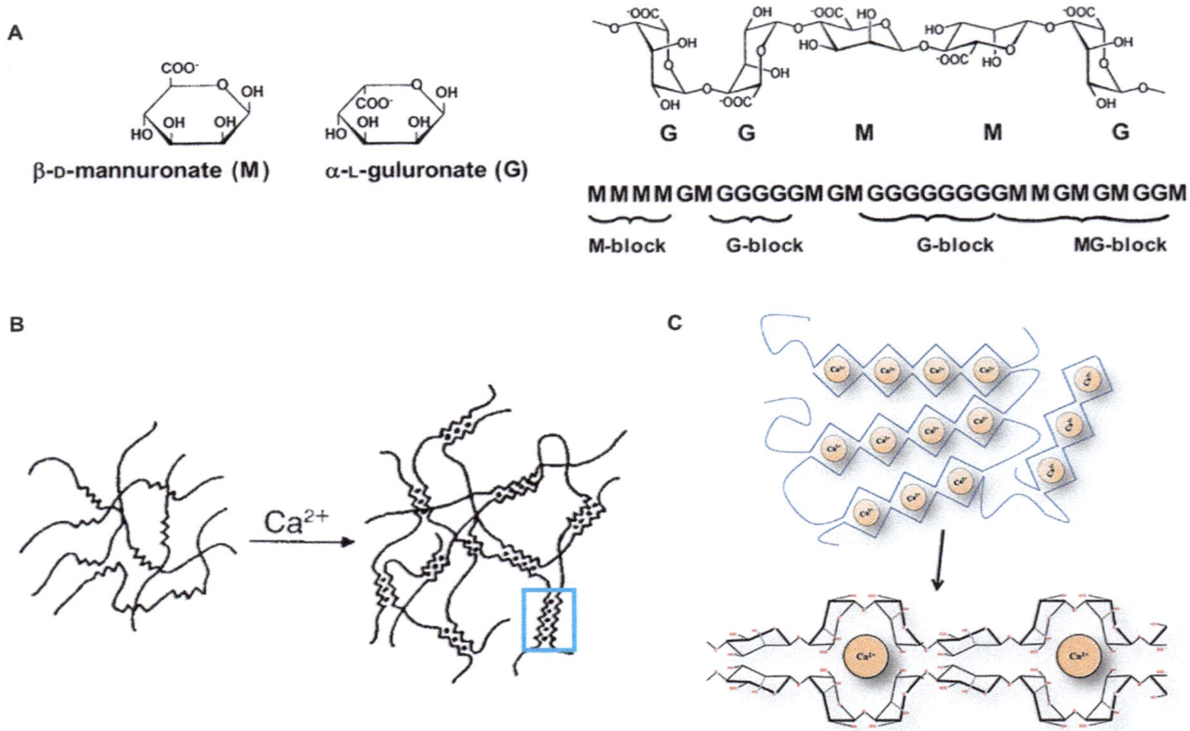

Figure 8.3 Alginate. (A) Structure of alginate with β-D-mannuronate (M) and α-L-guluronate (G) on the left, linked molecules in the middle and a schematic depiction of M, G and MG block on the right. Adapted from ref. 103 with permission from Elsevier, Copyright 2011. (B) Cross-linking of alginate with divalent cations binding to the G residues resulting in a characteristic "egg-box" structure, indicated by the blue box. Reproduced from ref. 81 with permission from Elsevier, Copyright 1990. (C) Detailed depiction of the egg-box model for the cross-linking of alginate with multivalent cations, specifically Ca^{2+}. Adapted from ref. 104, https://doi.org/10.3389/fbioe.2014.00026, under the terms of the CC BY 3.0 license https://creativecommons.org/licenses/by/3.0/.

with increasing affinity[86] and the number of junctions that can be expected to form between alginate chains increases accordingly.[87] Another reason for increased gel stability when cross-linking is achieved with Sr^{2+} or Ba^{2+} instead of Ca^{2+} is that Ca^{2+} has a stronger affinity to chelating agents such as phosphate than to alginate.[88] In addition to that, mainly due to its low affinity, there is a constant exchange between Ca^{2+} and monovalent non-gelling ions such as Na^{+} in saline solutions,[89] which results in osmotic swelling. Ba^{2+} is the strongest cross-linking ion which is not cytotoxic at low concentrations, but has been reported to lead to the inhibition of K^{+}-channels, as reviewed by Santos *et al.*[88] While release from alginate capsules cross-linked with 50 mM $BaCl_2$ remained below this threshold and did not induce a response in mice,[90] it has also been reported that Ba^{2+} ions leaking from alginate capsules cross-linked with only 20 mM $BaCl_2$ have been found to accumulate in the femur and blood of mice after transplantation.[87] Sr^{2+} ions cross-link alginate in a strength between those of Ca^{2+} and Ba^{2+} ions[91] and a concentration of 70 mM has been successfully used for the encapsulation of murine islets in alginate.[92] In general, not much information about the effect of Sr^{2+} on islets can be found: Sr^{2+} ions can be a substitute for Ca^{2+} ions in glucose-stimulated insulin release[93,94] although with a slightly lower affinity,[95] but other effects on islets, if they were investigated, have not been published. While little is known for the direct effect of Sr^{2+} on pancreatic islets, there is plenty of research on its overall biological effects, which was reviewed by S. Nielsen in 2004.[96] In general, it can replace Ca^{2+} in many biological processes even though the result tends to be slightly weaker.[96] Furthermore, it has been described to stimulate angiogenesis *in vivo*[97] and to have immune-modulatory effects through downregulating the expression of pro-inflammatory cytokines[98] as well as modulating macrophage phenotypes towards the anti-inflammatory M2 type.[98,99] There is evidence of cytotoxicity of Sr^{2+} on MSC at concentrations above 5 mM for continuous exposure in direct cell culture.[100] On the other hand, when used for the cross-linking of alginate, which is generally performed for 10 minutes before transfer to cell culture medium, 70 mM $SrCl_2$ had no impact on the survival of pancreatic islets[37,92] nor adrenocortical cells.[101] Indeed, the use of 100 mM $SrCl_2$ for cross-linking even supported the long-term survival and proliferation of chondrocytes encapsulated in alginate.[102] This discrepancy can likely be attributed to the vast majority of Sr^{2+} ions being bound by the alginate matrix and therefore likely not accessible to the embedded cells.

For soft TE applications aiming at immunoprotection, alginate hydrogels are ideally suited not only because of their cytocompatibility, but also to a large extent because mammalian species lack enzymes to cleave the alginate chains and thereby degrade the gels.[105] Highly relevant for TE applications is the immunogenicity of a given material, especially concerning the degree of fibrotic overgrowth which strongly impacts diffusion through the membrane and thereby cell survival.[106] While generally biocompatible, in the past, there has often been a lack of reproducibility in biocompatibility studies with highly differing amounts of immune reaction and fibrotic overgrowth for alginate. This has in part been attributed to the use of M-rich alginates which might leak and provoke an immune response, but more likely is a result of impurities in the alginate preparations.[103,107–109] Since alginates are isolated from brown algae, many commercially available preparations contain various impurities.[110] The contaminants can leak out from the cross-linked hydrogels and elicit an immune response through their sole presence, but could also affect the morphology and biocompatibility of alginate coatings by reducing the intrinsic wettability of the gel and interfere with the interaction between single chains during cross-linking.[111] Thus, a large focus of research had been on the development of purification protocols: Proper purification of the alginate preparations has been shown to greatly reduce fibrotic overgrowth,[112,113] influence survival of pancreatic islets *in vitro*[114] or even to completely prevent an immune response.[105,115]

In summary, due to their high capacity for binding water, high biocompatibility, permeability for oxygen and nutrients, and gelation under mild cross-linking conditions, alginate hydrogels are suitable for cell encapsulation and have been successfully used with a range of different cell types including hepatocytes, MSCs and pancreatic islets, as excellently reviewed by Lee and Mooney in 2012.[105] Furthermore, important in regards to clinical applicability, alginate has been approved for medical use by the US Food and Drug Administration, *e.g.*, as a wound covering.[116]

8.4.2.2 Alginate/methylcellulose (algMC)

At low concentrations, the matrix-forming alginate supports the survival of different cell types. However, as reviewed by Malda *et al.*, alginate solutions with a low polymer content have very low viscosity, whereas successful printing requires a highly viscous shear-thinning hydrogel (blend) to achieve shape-fidelity during the printing process.[55] The addition of thickening biopolymers such as methylcellulose (MC) to an alginate solution creates a highly viscous hydrogel blend (algMC) ideal for 3D (bio)printing which to the best of the authors' knowledge was first published by our group in 2015.[63]

The fundamental properties of MC are well-known and have been summarized in a number of reviews over the years.[117–119] In general terms, MC is a water-soluble molecule, prepared from cellulose through the substitution of hydroxy groups with methyl groups, which allows for the incorporation of water molecules otherwise hindered by the crystalline structure of cellulose. The extent by which hydroxy groups are substituted is described by the degree of substitution (DS). MC is an inversely thermo-gelling polymer with high solubility at low temperatures and gelation at high temperatures. Hereby the gelation temperature is lowered by an increase in the Mw, concentration, and DS of MC, whereas the presence of salts can lower or increase it.[120] Orally ingested MC is considered non-toxic for humans[121] and it is known for its use as a thickening agent in the food industry.[122] While it has not yet been approved for transplantation, it has been reported to be biocompatible after injection into the brains of rats[123] and muscle of mice.[124]

With the increased viscosity caused by the addition of MC to alginate, it was possible to create scaffolds up to a height of at least 50 layers, corresponding to a height of several mm, with lasting shape fidelity during the printing process prior to cross-linking (see Figure 8.2B). Within the blend, MC provides stability during the printing process and increases the shear-thinning properties, thereby enabling the fabrication of macroporous scaffolds. Furthermore, when MSCs were bioprinted within this blend they showed high viability and were able to differentiate into adipocytes.[63] The same blend was later used by Li *et al.* in 2017 for printing of mouse fibroblasts. In this case, the authors did not incorporate the cells into the blend prior to printing, but printed material and cell layers alternately.[125] In a later study by our group, a blend consisting of 3% alginate and 9% methylcellulose (algMC) was further characterized in terms of the influence of sterilization on MC. In a comparison between different sterilization methods, highly energetic γ-irradiation was shown to impact the M_w of MC and thereby the printability of the blend, whereas the viscosity was adequate for printing when MC had been autoclaved.[126]

In contrast to alginate, MC cannot directly be ionically cross-linked and, as recently discussed by the author S. Duin, in all likelihood remains at least partially non-gelled within algMC scaffolds stored under cell culture conditions,[42] theoretically enabling its release from the scaffolds over time. That MC is indeed partially released from printed algMC scaffolds over time has been strongly indicated by previous results from our lab,[63,126] and this release seems to be directly correlated to the Mw of MC, as the shorter chains after exposure to γ-irradiation were released almost completely, whereas the other sterilization methods resulted in longer chains and only partial release.[126]

Furthermore, the simple presence of MC chains during alginate cross-linking, but likely also their release from the cross-linked alginate network, create microporosity within the scaffold strands.[63,126] It has also been shown that this microporosity can at least partially be tailored by the amount of MC used and could influence cell viability.[127]

8.4.2.3 Gelatine/GelMA

Another biopolymer frequently used in TE, though not applicable for immunoprotective approaches due to its fast biodegradability, is gelatine, including in its methacrylated form (GelMA). Gelatine is an amphoteric polypeptide produced through the partial acidic or alkaline hydrolytic degradation of collagen, its specific composition of amino acids is therefore dependent on the respective collagen used as a source material and the hydrolysis conditions.[128,129] Collagen, as an abundant component of the mammalian ECM, contains a high amount of cell attachment sequences such as the RGD (arginine–glycine–aspartic acid)-motif, which are still present in gelatine; in contrast to fibrillar collagen however, gelatine is soluble in water. Further advantages of gelatine are its low cost, wide availability, high biocompatibility, low immunogenicity and biodegradability.[129,130] With its sol–gel transition temperature of approximately 30 °C, gelatine gels in a very cell-friendly manner, but quickly solubilizes at 37 °C *in vivo*. To increase its stability, gelatine can be further synthesized into GelMA by reaction with methacrylic anhydride in phosphate buffer, which can be cross-linked with UV light in the presence of a photoinitiator.[131,132] Similarly to collagen and gelatine, GelMA is still subject to proteolytic breakdown *in vivo* but retains its stability for a longer time than gelatine does.[133] Gelatine and GelMA can easily be combined with other hydrogels such as alginate for bioprinting, which has been shown by Zehnder *et al.* and Giuseppe *et al.* in recent years. Zehnder and co-workers successfully used a blend of gelatine and alginate-dialdehyde with MG-63 osteoblast-like cells, whereby the gelatine was partially released from the cross-linked gels over time and the cells were shown to migrate to the surface of the scaffold strands and spread there.[134] In the study by Giuseppe *et al.* the scaffolds comprised of gelatine and alginate showed high shape fidelity (*i.e.*, high accuracy) as well as a high viability of incorporated MSC, though the scaffolds degraded within 3 days in cell culture conditions.[135] The same group recently introduced a blend using GelMA instead of gelatine, which resulted in a strongly reduced degradation of only 20% over 14 days while retaining its printing characteristics and its support of encapsulated cells.[136]

8.4.2.4 Decellularized ECM (dECM)

In vivo, all cells are surrounded by a tissue-specific extracellular matrix, which not only offers a host of cell-attachment sites but also plays a crucial role in regulating cellular function, proliferation and differentiation. The ECM is composed of a vast number of molecules, among them different proteoglycans and proteins, whereby collagen is one of the most abundant, and the specific architecture, while important for cell signalling, is near impossible to fully mimic. The only known way to closely mimic the composition is to extract ECM from tissues *via* decellularization, generally with a combination of physical, chemical and enzymatic treatments such as sonication, organic solvents and proteases.[137] Successful decellularization has been shown for the majority of organs from different species, *i.e.*, adipose[138,139] and pancreatic tissue[140,141] from both humans and pigs, to mention just a small selection. As most components of the ECM are highly similar between species, properly decellularized ECM does not elicit an immune response in immune-competent transplant recipients,[139,141] but the tissue-specific cues could lead to the ectopic formation of new tissue.[139] Further aspects complicating the broad use of dECM for TE are the high cost and labour expenditures necessary for production, the low reproducibility due to high batch variability and the impossibility of scaling up the process. On the other hand, dECM offers the possibility of a tissue-specific environment for cell guidance and generally results in high cell viability and functionality.[142] dECM can be further processed into a bioink *via* lyophilization, crushing into powder, digestion and dissolution followed by neutralizing the pH as, partially due to the high collagen content, an acidic environment is necessary for successful dissolution which would be harmful to the encapsulated cells.143 dECM bioinks are in the sol state at room temperature and gelate at 37 °C.[141,143] Overall, their rheological properties depend on the tissue they are sourced from and the total protein content, exemplarily shown by the higher storage modulus of dECM from cartilage than from adipose tissue.[143] Nevertheless, while it is possible to create layered macroporous constructs with 3D bioprinting of dECM bioinks, due to their low viscosity the shape-fidelity of the constructs is oftentimes sub-par.[143,144] This could be alleviated by creating a blend with other more viscous materials,[142] but this impacts the composition and specific architecture of the dECM.

8.4.2.5 *Pectin*

Pectin describes a family of water-soluble oligo- and polysaccharides that are abundant in plant cell walls and are used as gelling agents and stabilizers in the food industry.[145,146] All pectins are rich in galacturonic acid, but the exact composition varies depending on the source and the extraction process. Galacturonic acid is structurally highly similar to the guluronic acid found in alginate, and can also be cross-linked with different multivalent ions. Ionic cross-linking of pectins results in a similar, but not identical, egg-box-like structure to ionically cross-linked alginate[147,148] and due to the similar cross-linking mechanism, pectins can interact synergistically with alginate.[149] Pectins are known to be cytocompatible, gelate in a cell-friendly manner and as a hydrogel have a high water content and flexibility. Due to these advantages, they have not only been the focus of research in drug delivery and TE, but are also commercially available as hydrocolloid wound dressings.[148] Furthermore, they have been shown to be non-immunogenic in a study investigating them as part of a nucleus pulposus, *i.e.*, intervertebral disc, replacement material,[150] and Hu *et al.* were recently able to observe a protective effect of different pectins in pancreatic islets and β-cells exposed to oxidative and inflammatory stress.[151] Similar to other polymers such as alginate, it is also possible to create hydrogel blends for bioprinting with pectin as has been exemplarily shown in a blend of pectin and nanocellulose by Cernecu *et al.* recently.[152]

8.4.3 3D Bioprinting of Pancreatic Islets

In contrast to 3D printing approaches with retrospective seeding of scaffold devices, 3D bioprinting is a more complex issue, which is reflected in the low number of studies published so far. In addition to the requirements for normal encapsulation, *i.e.*, immune protection and support of functionality, it is also required that the hydrogels are viscous enough to remain stable until cross-linking, which often interferes with the diffusion of glucose and insulin. The first publication reporting on bioprinted islets was that by Marchioli *et al.* who used a hydrogel blend consisting of 4% alginate and 5% gelatine to print structures of up to 17 layers (roughly 1.5 mm in height) using a printing needle with an outlet diameter of 410 μm. Analysis after bioprinting revealed that encapsulated human islets retained their round-shaped morphology, and both INS1E cells and human islets showed a high viability, but neither of them reacted to glucose stimulation within the scaffolds. As the human islets recovered functionality after removal from the bioprinted constructs and the diffusion of glucose was shown to be slower than reported for 4% plain alginate, it is likely that the dense meshwork of the scaffold interfered with the functionality through impaired diffusion of glucose and insulin.[153]

Improved functionality of human islets and SC-derived insulin-producing cells in porcine pancreatic tissue-derived dECM gels was shown in a study by Kim *et al.*: In a concentration of 2%–3%, the dECM bioink was printable for up to five layers with a needle outlet diameter of 800 μm. The group reported functionality of human islets in the bulk gel, which was further improved by a co-culture with human umbilical vein endothelial cells (HUVEC) in addition to decreased central necrosis. Furthermore, if the blend was bioprinted instead of being used as a bulk gel, the incorporated islets retained their morphology and showed high viability (see Figure 8.4A). However, functionality testing was only performed on bulk gels, the authors did not repeat the analysis explicitly for printed scaffolds.[154] With this approach, the dECM-based material, especially in combination with HUVEC, is likely to result in faster revascularization of transplanted islets, but the scaffolds lack immunoprotective properties due to their fast biodegradability and expected direct vascularization.

Transient immunoprotection through the use of 2% alginate with 7.5% GelMA and the co-printing of a second cell type was presented by Liu *et al.*, though partial degradation of the scaffold material *in vivo* is to be expected as GelMA, though not alginate, is biodegradable. This group employed the core–shell extrusion printing technique to bioprint murine islets in the protected core, surrounded by a shell containing endothelial progenitor cells, whereby the material blend itself did not vary between core and shell. The needle outlet diameter is reported as 450 μm for the core and 1045 μm for the shell, resulting in an effective shell material thickness of 300 μm and the analysed constructs consisted of a single layer. The islets encased in the printed core retained a high viability and their round-shaped morphology, but again no functional response to glucose stimulation could be observed, which was also attributed to a reduced diffusion of glucose.[155] Very recently, the same group showed that a slightly adapted hydrogel blend consisting of 2% alginate and 7.5% GelMA with an additional 3.5% of gelatine can also be used for the co-printing of islets and regulatory T-cells (Tregs) in scaffolds of up to 10 layers. Again, the islets were encased in the core, the Tregs were incorporated into the shell material. The presence of Tregs is thought to eventually induce localized immunomodulation at the transplant site and

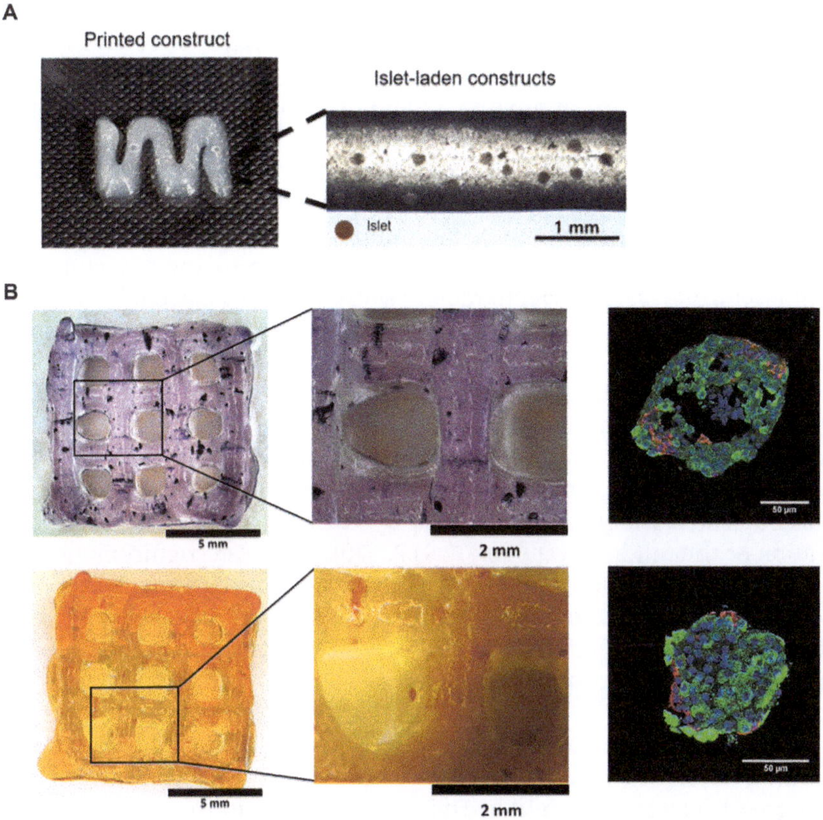

Figure 8.4 Islet bioprinting. Core–shell structure containing murine islets on the left, scale bar = 1 mm, viability of bioprinted murine islets within core–shell printed constructs 1 day and 5 days after bioprinting on the right, scale bars = 100 µm. Reproduced from ref. 154 with permission from the Royal Society of Chemistry. (B) Bioprinted murine islets in algMC stained for metabolic activity with MTT, day 1 after bioprinting (top row) and stained for presence of insulin with DTZ, day 7 after bioprinting (bottom row), scale bars = 5 mm (left) and 2 mm (right). Reproduced from ref. 158 with permission from John Wiley and Sons, Copyright © 2019 WILEY-VCH Verlag GmbH & Co. KGaA, Weinheim. (C) Representative images of murine control islets (top) and islets bioprinted in algMC (bottom) immunofluorescently stained for nuclei (DAPI), insulin (green) and glucagon (red), day 4 after bioprinting, scale bars = 50 µm. Adapted from ref. 158 with permission from John Wiley and Sons, Copyright © 2019 WILEY-VCH Verlag GmbH & Co. KGaA, Weinheim.

has been shown to protect the viability of printed murine islets during exposure to peripheral blood mononuclear cells (PBMC) *in vitro* in this study.[156] Functionality of islets in the hydrogel blend was not analysed though – it is possible that the biodegradability of gelatine and GelMA results in high microporosity over time, which could improve the diffusion characteristics of the scaffolds. This could, however, in turn result in reduced immunoprotection, and both these aspects have to be clarified in future studies.

An approach directly aiming at immunomodulation was recently presented by the de Vos lab who developed a material blend of 6% alginate, 2% pectin and 8% Pluronic F127, whereby the alginate is the matrix-forming component, the Pluronic is added to increase the initial viscosity to allow for bioprinting, and the pectin acts as an immunomodulatory component. The scaffolds were printed with a strand diameter of 1.5 mm and a height of up to four layers, and were shown to remain stable for at least two weeks *in vitro* and four weeks *in vivo*. Concerning biocompatibility, the group successfully bioprinted MIN6 cells with high viability over 7 days *in vitro*. Compared to pectin-free scaffolds, the presence of pectin led to reduced cell death after exposure to pro-inflammatory cytokines for 24 h *in vitro*, and to reduced fibrotic overgrowth *in vivo* after subcutaneous implantation of the scaffolds into immunocompetent mice. That pectin not only increases the viability of β-cells, but also has an immunomodulatory function, was shown *via* reduced activation of the toll-like receptor (TLR) 2. This pattern-recognition receptor is involved in the innate immune response and expressed on the reporter cell line THP-1, and showed reduced activity in the presence of pectin-containing, but not pectin-free, scaffolds. This approach is promising concerning immunoprotection, but the group is yet to show functionality of encapsulated cells.[157]

To the best of our knowledge, the only study reporting functionality of bioprinted islets in scaffolds which are likely to offer lasting immunoprotection was presented by Duin *et al.* in 2019. In this study, we used the algMC hydrogel blend described previously in this chapter and islets were isolated from adult rats; scaffolds of four layers were printed with a needle outlet diameter of 840 μm. Using this approach, islet morphology was slightly impacted by the incorporation into the highly viscous hydrogel blend but not by the printing process, and the islets retained a high viability for more than 7 days. Despite the high viscosity, it was possible to achieve an adequately homogeneous distribution of islets within the blend and finally the printed scaffolds (see Figure 8.4B) and all observed islets retained their insulin and glucagon production as shown by immunostainings (see Figure 8.4C). Furthermore, *in vitro* the printed islets showed a reaction to glucose stimulation which was comparable to control islets in suspension culture 4 and 7 days after printing.[158] Overall, the polymer content of this hydrogel blend is in the same range as that used in the other islet bioprinting studies, and in general, a high polymer content is known to decrease the permeability of gels. In contrast to the previously described studies, in the algMC blend, the diffusion of glucose is not slowed down compared to 3% plain alginate although the diffusion of insulin was later reported to be slightly impaired.[42] Nevertheless, the results from the study indicate that with this approach it is possible to create functional macroporous macrocapsules.

In summary, the bioprinting of islets offers the possibility to create macroporous constructs of clinically relevant size to decrease the diffusion distance in macroencapsulation devices and additionally print islet supportive cell types in close vicinity with precise spatial distribution. While some macrocapsules based on plain alginate have successfully been tested not only pre-clinically, but also in human trials,[36,38] most 3D-bioprinted macrocapsules are yet to be tested *in vivo* to further analyse them in terms of long-term stability and immunoprotection in combination with long-term functionality. Currently, the main limitations of bioprinting approaches for pancreatic islets are finding the balance between a printable gel, which requires high viscosity usually through a high polymer content, and adequate permeability to allow for the diffusion of oxygen, nutrients and, of course, glucose and insulin, which is generally decreased by a higher polymer content. Nevertheless, many of the presented approaches are promising, indicating that 3D bioprinting is a technique that could eventually be used for the generation of islet-containing constructs for clinical transplantation.

References

1. M. A. Atkinson, G. S. Eisenbarth and A. W. Michels, *Lancet*, 2014, **383**, 69–82.
2. M. A. Atkinson and G. S. Eisenbarth, *Lancet*, 2001, **358**, 221–229.
3. J. T. Warshauer, J. A. Bluestone and M. S. Anderson, *Cell Metab.*, 2020, **31**, 46–61.
4. J. Liu, S. M. Zhang, P. P. Chen, L. Cheng, W. Zhou, W. X. Tang, Z. W. Chen and C. M. Ke, *J. Mater. Sci.: Mater. Med.*, 2007, **18**, 2205–2210.
5. D. H. Abdelkader, M. M. Tambuwala, C. A. Mitchell, M. A. Osman, S. A. El-Gizawy, A. M. Faheem, M. El-Tanani and P. A. McCarron, *Drug Delivery Transl. Res.*, 2018, **8**, 1053–1065.
6. I. P. S. Fernando, W. W. Lee, E. J. Han and G. Ahn, *Chem. Eng. J.*, 2020, **391**, 123823.
7. A. U. Ernst, D. T. Bowers, L.-H. Wang, K. Shariati, M. D. Plesser, N. K. Brown, T. Mehrabyan and M. Ma, *Adv. Drug Delivery Rev.*, 2019, **139**, 116–138.
8. M. Grassi, C. Sandolo, D. Perin, T. Coviello, R. Lapasin and G. Grassi, *Molecules*, 2009, **14**, 3003–3017.
9. J. Klein, J. Stock and K. D. Vorlop, *Eur. J. Appl. Microbiol. Biotechnol.*, 1983, **18**, 86–91.
10. B. Sarmento, A. Ribeiro, F. Veiga, P. Sampaio, R. Neufeld and D. Ferreira, *Pharm. Res.*, 2007, **24**, 2198–2206.
11. S. Kumar, G. Bhanjana, R. K. Verma, D. Dhingra, N. Dilbaghi and K. H. Kim, *J. Pharm. Pharmacol.*, 2017, **69**, 143–150.
12. N. C. Foster, R. W. Beck, K. M. Miller, M. A. Clements, M. R. Rickels, L. A. Dimeglio, D. M. Maahs, W. V. Tamborlane, R. Bergenstal,

E. Smith, B. A. Olson and S. K. Garg, *Diabetes Technol. Ther.*, 2019, **21**, 66–72.
13. K. M. Miller, N. C. Foster, R. W. Beck, R. M. Bergensta, S. N. DuBose, L. A. DiMeglio, D. M. Maahs and W. V. Tamborlane, *Diabetes Care*, 2015, **38**, 971–978.
14. L. A. DiMeglio, C. Evans-Molina and R. A. Oram, *Lancet*, 2018, **391**, 2449–2462.
15. S. J. Livingstone, D. Levin, H. C. Looker, R. S. Lindsay, S. H. Wild, N. Joss, G. Leese, P. Leslie, R. J. McCrimmon, W. Metcalfe, J. A. McKnight, A. D. Morris, D. W. M. Pearson, J. R. Petrie, S. Philip, N. A. Sattar, J. P. Traynor and H. M. Colhoun, *JAMA, J. Am. Med. Assoc.*, 2015, **313**, 37–44.
16. A. M. J. Shapiro, M. Pokrywczynska and C. Ricordi, *Nat. Rev. Endocrinol.*, 2017, **13**, 268–277.
17. B. Ludwig, S. Ludwig, A. Steffen, H. D. Saeger and S. R. Bornstein, *Curr. Diabetes Rep.*, 2010, **10**, 506–511.
18. C. A. Welsch, W. L. Rust and M. Csete, *Stem Cells Transl. Med.*, 2019, **8**, 209–214.
19. D. J. Van Der Windt, R. Bottino, A. Casu, N. Campanile and D. K. C. Cooper, *Xenotransplantation*, 2007, **14**, 288–297.
20. A. R. Pepper, A. Bruni and A. M. J. Shapiro, *Curr. Opin. Organ Transplant.*, 2018, **23**, 428–439.
21. J. A. Emamaullee, A. M. J. Shapiro, R. V. Rajotte, G. Korbutt and J. F. Elliott, *Transplantation*, 2006, **82**, 945–952.
22. E. Cantarelli and L. Piemonti, *Curr. Diabetes Rep.*, 2011, **11**, 364–374.
23. M. De Groot, T. A. Schuurs and R. Van Schilfgaarde, *J. Surg. Res.*, 2004, **121**, 141–150.
24. M. R. Rickels and R. P. Robertson, *Endocr. Rev.*, 2019, **40**, 631–668.
25. A. Bruni, M. McCall and A. M. J. Shapiro, in *Endocrine Surgery in Children*, ed. D. J. Ledbetter and P. R. V. Johnson, Springer, Berlin, 2017, pp. 181–196.
26. A. M. J. Shapiro, J. R. T. Lakey, E. A. Ryan, G. S. Korbutt, E. Toth, G. L. Warnock, N. M. Kneteman and R. V. Rajotte, *N. Engl. J. Med.*, 2000, **343**, 230–238.
27. D. J. Van Der Windt, M. Marigliano, J. He, T. V. Votyakova, G. J. Echeverri, B. Ekser, D. Ayares, F. G. Lakkis, D. K. C. Cooper, M. Trucco and R. Bottino, *Cell Transplant.*, 2012, **21**, 1791–1802.
28. D. K. C. Cooper, M. B. Ezzelarab, H. Hara, H. Iwase, W. Lee, M. Wijkstrom and R. Bottino, *Xenotransplantation*, 2016, **23**, 83–105.
29. A. C. Farney, D. E. R. Sutherland and E. C. Opara, *Pancreas*, 2016, **45**, 8–20.
30. R. Calafiore, G. Basta, G. Luca, A. Lemmi, L. Racanicchi, F. Mancuso, M. P. Montanucci and P. Brunetti, *Transplant. Proc.*, 2006, **38**, 1156–1157.
31. P. De Vos and P. Marchetti, *Trends Mol. Med.*, 2002, **8**, 363–366.
32. D. W. Scharp and P. Marchetti, *Adv. Drug Delivery Rev.*, 2014, **67–68**, 35–73.

33. A. J. Hwa and G. C. Weir, *Curr. Diab. Rep.*, 2018, **18**, 50.
34. O. Korsgren, *Diabetes*, 2017, **66**, 1748–1754.
35. D. An, A. Chiu, J. A. Flanders, W. Song, D. Shou, Y. C. Lu, L. G. Grunnet, L. Winkel, C. Ingvorsen, N. S. Christophersen, J. J. Fels, F. W. Sand, Y. Ji, L. Qi, Y. Pardo, D. Luo, M. Silberstein, J. Fan and M. Ma, *Proc. Natl. Acad. Sci. U. S. A.*, 2017, **115**, E263–E272.
36. B. Ludwig, A. Reichel, A. Steffen, B. Zimerman, A. V. Schally, N. L. Block, C. K. Colton, S. Ludwig, S. Kersting, E. Bonifacio, M. Solimena, Z. Gendler, A. Rotem, U. Barkai and S. R. Bornstein, *Proc. Natl. Acad. Sci. U. S. A.*, 2013, **110**, 19054–19058.
37. U. Barkai, G. C. Weir, C. K. Colton, B. Ludwig, S. R. Bornstein, M. D. Brendel, T. Neufeld, C. Bremer, A. Leon, Y. Evron, K. Yavriyants, D. Azarov, B. Zimermann, S. Maimon, N. Shabtay, M. Balyura, T. Rozenshtein, P. Vardi, K. Bloch, P. De Vos and A. Rotem, *Cell Transplant.*, 2013, **22**, 1463–1476.
38. P. O. Carlsson, D. Espes, A. Sedigh, A. Rotem, B. Zimerman, H. Grinberg, T. Goldman, U. Barkai, Y. Avni, G. T. Westermark, L. Carlbom, H. Ahlström, O. Eriksson, J. Olerud and O. Korsgren, *Am. J. Transplant.*, 2018, **18**, 1735–1744.
39. M. Lamb, R. Storrs, S. Li, O. Liang, K. Laugenour, R. Dorian, D. Chapman, H. Ichii, D. Imagawa, C. Foster, S. King and J. R. T. Lakey, *Transplant. Proc.*, 2011, **43**, 3265–3266.
40. M. Kumagai-Braesch, S. Jacobson, H. Mori, X. Jia, T. Takahashi, A. Wernerson, M. Flodström-Tullberg and A. Tibell, *Cell Transplant.*, 2013, **22**, 1137–1146.
41. A. R. Pepper, R. Pawlick, B. Gala-Lopez, A. MacGillivary, D. M. Mazzuca, D. J. G. White, P. M. Toleikis and A. M. James Shapiro, *Transplantation*, 2015, **99**, 2294–2300.
42. S. Duin, PhD thesis, Technische Universität Dresden, 2020.
43. M. A. Bochenek, O. Veiseh, A. J. Vegas, J. James, M. Qi, E. Marchese, M. Omami, J. C. Doloff, J. Mendoza-elias, M. Nourmohammadzadeh, A. Khan, C. Yeh, Y. Xing, D. Isa, S. Ghani, J. Li, C. Landry, A. R. Bader, K. Olejnik, M. Chen, J. Hollister-Lock, Y. Wang, D. L. Greiner, G. C. Weir, B. L. Strand, A. Mari, A. Rokstad, I. Lacik, R. Langer and D. G. Anderson, *Nat. Biomed. Eng.*, 2019, **2**, 810–821.
44. M. Jiménez, L. Romero, I. A. Domínguez, M. D. M. Espinosa and M. Domínguez, *Complexity*, 2019, **2019**, 1–30.
45. Z. Xie, M. Gao, A. O. Lobo and T. J. Webster, *Polymers*, 2020, **12**, 1717.
46. J. Li, M. Chen, X. Fan and H. Zhou, *J. Transl. Med.*, 2016, **14**, 271.
47. D. Kilian, T. Ahlfeld, A. R. Akkineni, A. Lode and M. Gelinsky, *MRS Bull.*, 2017, **42**, 585–592.
48. I. Matai, G. Kaur, A. Seyedsalehi, A. McClinton and C. T. Laurencin, *Biomaterials*, 2020, **226**, 119536.
49. A. J. Bauermeister, A. Zuriarrain and M. I. Newman, *Ann. Plast. Surg.*, 2016, **77**, 569–576.

50. P. Tack, J. Victor, P. Gemmel and L. Annemans, *Biomed. Eng. Online*, 2016, **15**, 1–21.
51. Q. Yan, H. Dong, J. Su, J. Han, B. Song, Q. Wei and Y. Shi, *Engineering*, 2018, **4**, 729–742.
52. J. Kim, J. S. Kong, W. Han, B. S. Kim and D.-W. Cho, *Int. J. Mol. Sci.*, 2020, **21**, 7757.
53. F. Meng, C. M. Meyer, D. Joung, D. A. Vallera, M. C. McAlpine and A. Panoskaltsis-Mortari, *Adv. Mater.*, 2019, **31**, 1–10.
54. T. Billiet, M. Vandenhaute, J. Schelfhout, S. Van Vlierberghe and P. Dubruel, *Biomaterials*, 2012, **33**, 6020–6041.
55. J. Malda, J. Visser, F. P. Melchels, T. Jüngst, W. E. Hennink, W. J. A. Dhert, J. Groll and D. W. Hutmacher, *Adv. Mater.*, 2013, **25**, 5011–5028.
56. R. Badwaik, *J. Clin. Diagnostic Res.*, 2019, **13**, FE01–FE08.
57. S. J. Buwalda, T. Vermonden and W. E. Hennink, *Biomacromolecules*, 2017, **18**, 316–330.
58. V. V. S. R. Karri, G. Kuppusamy, S. V. Talluri, S. S. Mannemala, R. Kollipara, A. D. Wadhwani, S. Mulukutla, K. R. S. Raju and R. Malayandi, *Int. J. Biol. Macromol.*, 2016, **93**, 1519–1529.
59. M. Lee, K. Bae, P. Guillon, J. Chang, Ø. Arlov and M. Zenobi-Wong, *ACS Appl. Mater. Interfaces*, 2018, **10**, 37820–37828.
60. E. Trampe, K. Koren, A. R. Akkineni, C. Senwitz, F. Krujatz, A. Lode, M. Gelinsky and M. Kühl, *Adv. Funct. Mater.*, 2018, **28**, 1–11.
61. A. Bhattacharyya, G. Janarthanan and I. Noh, *Addit. Manuf.*, 2021, **37**, 101639.
62. N. Di Marzio, D. Eglin, T. Serra and L. Moroni, *Front. Bioeng. Biotechnol.*, 2020, **8**, 326.
63. K. Schütz, A. M. Placht, B. Paul, S. Brüggemeier, M. Gelinsky and A. Lode, *J. Tissue Eng. Regener. Med.*, 2017, **11**, 1574–1587.
64. A. Lode, F. Krujatz, S. Brüggemeier, M. Quade, K. Schütz, S. Knaack, J. Weber, T. Bley and M. Gelinsky, *Eng. Life Sci.*, 2015, **15**, 177–183.
65. R. Taymour, D. Kilian, T. Ahlfeld, M. Gelinsky and A. Lode, *Sci. Rep.*, 2021, **11**, 5130.
66. A. B. Bernard, C.-C. Lin and K. S. Anseth, *Tissue Eng., Part C*, 2012, **18**, 583–592.
67. A. L. S. Burzava, A. Forget, F. J. Harding, M. P. Cockshell, D. Penko, C. Rouzaud, V. Ahmadi, P. F. Marina, D. Rojas-Canales, C. S. Bonder, P. T. H. Coates, M. Waibel, H. E. Thomas, T. W. Kay, T. Loudovaris, A. Blencowe and N. H. Voelcker, *Biointerphases*, 2019, **14**, 011002.
68. J. T. Daoud, M. S. Petropavlovskaia, J. M. Patapas, C. E. Degrandpré, R. W. DiRaddo, L. Rosenberg and M. Tabrizian, *Biomaterials*, 2011, **32**, 1536–1542.
69. M. Farina, A. Ballerini, D. W. Fraga, E. Nicolov, M. Hogan, D. Demarchi, F. Scaglione, O. M. Sabek, P. Horner, U. Thekkedath, O. A. Gaber and A. Grattoni, *Biotechnol. J.*, 2017, **12**, 1–5.
70. J. Song and J. R. Millman, *Biofabrication*, 2016, **1**, 015002.

71. A. Soltanian, Z. Ghezelayagh, Z. Mazidi, M. Halvaei, S. Mardpour, M. K. Ashtiani, E. Hajizadeh-Saffar, Y. Tahamtani and H. Baharvand, *J. Cell. Physiol.*, 2019, **234**, 9564–9576.
72. A. Akkouch, Y. Yu and I. T. Ozbolat, *Biofabrication*, 2015, **7**, 031002.
73. A. Espona-Noguera, J. Ciriza, A. Cañibano-Hernández, R. Villa, L. Saenz del Burgo, M. Alvarez and J. Luis Pedraz, *Int. J. Pharm.*, 2019, **566**, 604–614.
74. G. Marchioli, A. Di Luca, E. de Koning, M. Engelse, C. A. Van Blitterswijk, M. Karperien, A. A. Van Apeldoorn and L. Moroni, *Adv. Healthc. Mater.*, 2016, **5**, 1606–1616.
75. A. R. Mridha, T. R. Dargaville, P. D. Dalton, L. Carroll, M. B. Morris, V. Vaithilingam and B. E. Tuch, *Tissue Eng., Part A*, 2022, **28**, 212–224.
76. G. R. Rayat, R. V. Rajotte, Z. Ao and G. S. Korbutt, *Transplantation*, 2000, **69**, 1084–1090.
77. A. Omer, V. F. Duvivier-Kali, N. Trivedi, K. Wilmot, S. Bonner-Weir and G. C. Weir, *Diabetes*, 2003, **52**, 69–75.
78. V. F. Duvivier-Kali, A. Omer, R. J. Parent, J. J. O'Neil and G. C. Weir, *Diabetes*, 2001, **50**, 1698–1705.
79. O. Veiseh, J. C. Doloff, M. Ma, A. J. Vegas, H. H. Tam, A. R. Bader, J. Li, E. Langan, J. Wyckoff, W. S. Loo, S. Jhunjhunwala, A. Chiu, S. Siebert, K. Tang, J. Hollister-Lock, S. Aresta-Dasilva, M. Bochenek, J. Mendoza-Elias, Y. Wang, M. Qi, D. M. Lavin, M. Chen, N. Dholakia, R. Thakrar, I. Lacík, G. C. Weir, J. Oberholzer, D. L. Greiner, R. Langer and D. G. Anderson, *Nat. Mater.*, 2015, **14**, 643–651.
80. S. Schneider, P. J. Feilen, F. Brunnenmeier, T. Minnemann, H. Zimmermann, U. Zimmermann and M. M. Weber, *Diabetes*, 2005, **54**, 687–693.
81. O. Smidsrød and G. Skjåk-Braek, *Trends Biotechnol.*, 1990, **8**, 71–78.
82. G. T. Grant, E. R. Morris, D. A. Rees, P. J. C. Smith and D. Thom, *FEBS Lett.*, 1973, **32**, 195–198.
83. R. G. Schweiger, *J. Org. Chem.*, 1962, **27**, 1789–1791.
84. C. M. DeRamos, A. E. Irwin, J. L. Nauss and B. E. Stout, *Inorg. Chim. Acta*, 1997, **256**, 69–75.
85. O. Smidsrød, *Faraday Discuss. Chem. Soc.*, 1974, **57**, 263–274.
86. B. T. Stokke, O. Smidsrød, P. Bruheim and G. Skjåk-Bræk, *Macromolecules*, 1991, **24**, 4637–4645.
87. Ý. A. Mørch, I. Donati, B. L. Strand and G. Skjåk-Bræk, *Biomacromolecules*, 2006, **7**, 1471–1480.
88. E. Santos, J. Zarate, G. Orive, R. M. Hernández and J. L. Pedraz, in *Therapeutic Applications of Cell Microencapsulation*, ed. J. L. Pedraz and G. Orive, Springer, LLC Landes Bioscience, New York, 2010, vol. 640, pp. 5–21.
89. A. Martinsen, G. Skjåk-Braek and O. Smidsrød, *Biotechnol. Bioeng.*, 1989, **33**, 79–89.
90. P. Montanucci, S. Terenzi, C. Santi, I. Pennoni, V. Bini, T. Pescara, G. Basta and R. Calafiore, *Biomed Res. Int.*, 2015, **2015**, 1–11.

91. O. Smidsrød and G. Skjåk-Braek, *Trends Biotechnol.*, 1990, **8**, 71–78.
92. B. Ludwig, A. Rotem, J. Schmid, G. C. Weir, C. K. Colton, M. D. Brendel, T. Neufeld, N. L. Block, K. Yavriyants, A. Steffen, S. Ludwig, T. Chavakis, A. Reichel, D. Azarov, B. Zimermann, S. Maimon, M. Balyura, T. Rozenshtein, N. Shabtay, P. Vardi, K. Bloch, P. De Vos, A. V. Schally, S. R. Bornstein and U. Barkai, *Proc. Natl. Acad. Sci. U. S. A.*, 2012, **109**, 5022–5027.
93. J. C. Henquin, *Pflügers Arch. Eur. J. Physiol.*, 1980, **383**, 123–129.
94. B. Hellman, E. Gylfe, P. Bergsten, E. Grapengiesser, A. Berts, Y. J. Liu, A. Tengholm and J. Westerlund, *Endocrinology*, 1997, **138**, 3161–3165.
95. B. Ribalet and P. M. Beigelman, *Am. J. Physiol.*, 1981, **241**, C59–C67.
96. S. P. Nielsen, *Bone*, 2004, **35**, 583–588.
97. F. Zhao, B. Lei, X. Li, Y. Mo, R. Wang, D. Chen and X. Chen, *Biomaterials*, 2018, **178**, 36–47.
98. L. Wei, Y. Jiang, W. Zhou, S. Liu, Y. Liu, X. Rausch-Fan and Z. Liu, *J. Periodontal Res.*, 2018, **53**, 999–1008.
99. F. Zhao, B. Lei, X. Li, Y. Mo, R. Wang, D. Chen and X. Chen, *Biomaterials*, 2018, **178**, 36–47.
100. M. Schumacher, A. Lode, A. Helth and M. Gelinsky, *Acta Biomater.*, 2013, **9**, 9547–9557.
101. M. Balyura, E. Gelfgat, M. Ehrhart-Bornstein, B. Ludwig, Z. Gendlerc, U. Barkai, B. Zimerman, A. Rotem, N. L. Block, A. V. Schally and S. R. Bornstein, *Proc. Natl. Acad. Sci. U. S. A.*, 2015, **112**, 2527–2532.
102. S. A. Abbah, W. W. Lu, S. L. Peng, D. M. K. Aladin, Z. Y. Li, W. K. Tam, K. M. C. Cheung, K. D. K. Luk and G. Q. Zhou, *Cell Transplant.*, 2008, **17**, 1181–1192.
103. K. I. Draget and C. Taylor, *Food Hydrocoll.*, 2011, **25**, 251–256.
104. G. A. Paredes-Juárez, M. Spasojevic, M. M. Faas and P. de Vos, *Front. Bioeng. Biotechnol.*, 2014, **2**, 26.
105. K. Y. Lee and D. J. Mooney, *Prog. Polym. Sci.*, 2012, **37**, 106–126.
106. J. Dusseault, S. K. Tam, M. Ménard, S. Polizu, G. Jourdan, L. Yahia and J. P. Hallé, *J. Biomed. Mater. Res., Part A*, 2006, **76**, 243–251.
107. P. De Vos, B. J. De Haan and R. Van Schilfgaarde, *Biomaterials*, 1997, **18**, 273–278.
108. H. Ertesvåg and S. Valla, *Polym. Degrad. Stab.*, 1998, **59**, 85–91.
109. G. Orive, S. Ponce, R. M. Hernández, A. R. Gascón, M. Igartua and J. L. Pedraz, *Biomaterials*, 2002, **23**, 3825–3831.
110. G. Klöck, H. Frank, R. Houben, T. Zekorn, A. Horcher, U. Siebers, M. Wöhrle, K. Federlin and U. Zimmermann, *Appl. Microbiol. Biotechnol.*, 1994, **40**, 638–643.
111. S. K. Tam, J. Dusseault, S. Polizu, M. Ménard, J. P. Hallé and L. Yahia, *Biomaterials*, 2006, **27**, 1296–1305.
112. P. De Vos, B. J. De Haan, G. H. J. Wolters, J. H. Strubbe and R. Van Schilfgaarde, *Diabetologia*, 1997, **40**, 262–270.
113. A. G. Mallett and G. S. Korbutt, *Tissue Eng., Part A*, 2009, **15**, 1301–1309.

114. G. Langlois, J. Dusseault, S. Bilodeau, S. K. Tam, D. Magassouba and J. P. Hallé, *Acta Biomater.*, 2009, **5**, 3433–3440.
115. A. M. Rokstad, O. L. Brekke, B. Steinkjer, L. Ryan, G. Kolláriková, B. L. Strand, G. Skjåk-Bræk, I. Lacík, T. Espevik and T. E. Mollnes, *Acta Biomater.*, 2011, 7, 2566–2578.
116. N. C. Hunt, R. M. Shelton and L. M. Grover, *Biotechnol. J.*, 2009, **4**, 730–737.
117. A. Sannino, C. Demitri and M. Madaghiele, *Materials*, 2009, 2, 353–373.
118. P. L. Nasatto, F. Pignon, J. L. M. Silveira, M. E. R. Duarte, M. D. Noseda and M. Rinaudo, *Polymers*, 2015, **7**, 777–803.
119. T. Ahlfeld, V. Guduric, S. Duin, A. R. Akkineni, K. Schütz, D. Kilian, J. Emmermacher, N. Cubo-Mateo, S. Dani, M. von Witzleben, J. Spangenberg, R. Abdelgaber, R. F. Richter, A. Lode and M. Gelinsky, *Biomater. Sci.*, 2020, **8**, 2102–2110.
120. E. Heymann, *Trans. Faraday Soc.*, 1935, **31**, 846–864.
121. U.S. Food and Drug Administration, Inactive Ingredient Search for Approved Drug Products, https://www.accessdata.fda.gov/scripts/cder/iig/index.Cfm?event=BasicSearch.page (accessed 9 July 2021).
122. D. G. Coffey, D. A. Bell and A. Henderson, in *Food Polysaccharides and Their Applications*, ed. A. M. Stephen, G. O. Phillips and P. A. Williams, CRC Press, Taylor & Francis Group, Boca Raton, 2nd edn, 2006, pp. 159–163.
123. M. Tate, *Biomaterials*, 2001, **22**, 1113–1123.
124. S. Davoudi, C. Y. Chin, M. J. Cooke, R. Y. Tam, M. S. Shoichet and P. M. Gilbert, *Biomaterials*, 2018, **173**, 34–46.
125. H. Li, Y. J. Tan, K. F. Leong and L. Li, *ACS Appl. Mater. Interfaces*, 2017, **9**, 20086–20097.
126. E. Hodder, S. Duin, D. Kilian, T. Ahlfeld, J. Seidel, C. Nachtigall, P. Bush, D. Covill, M. Gelinsky and A. Lode, *J. Mater. Sci.: Mater. Med.*, 2019, **30**, 10.
127. T. Gonzalez-Fernandez, S. Rathan, C. Hobbs, P. Pitacco, F. E. Freeman, G. M. Cunniffe, N. J. Dunne, H. O. McCarthy, V. Nicolosi, F. J. O'Brien and D. J. Kelly, *J. Controlled Release*, 2019, **301**, 13–27.
128. X. Wang, Q. Ao, X. Tian, J. Fan, H. Tong, W. Hou and S. Bai, *Polymers*, 2017, **9**, 401.
129. K. Su and C. Wang, *Biotechnol. Lett.*, 2015, **37**, 2139–2145.
130. M. E. Hoque, T. Nuge, T. K. Yeow, N. Nordin and R. G. S. V. Prasad, *Polym. Res. J.*, 2015, **9**, 15.
131. K. Yue, G. Trujillo-de Santiago, M. M. Alvarez, A. Tamayol, N. Annabi and A. Khademhosseini, *Biomaterials*, 2015, **73**, 254–271.
132. S. Van Vlierberghe, P. Dubruel and E. Schacht, *Biomacromolecules*, 2011, **12**, 1387–1408.
133. Y.-C. Chen, R.-Z. Lin, H. Qi, Y. Yang, H. Bae, J. M. Melero-Martin and A. Khademhosseini, *Adv. Funct. Mater.*, 2012, **22**, 2027–2039.
134. T. Zehnder, B. Sarker, A. R. Boccaccini and R. Detsch, *Biofabrication*, 2015, 7, 025001.

135. M. Di Giuseppe, N. Law, B. Webb, R. A. Macrae, L. J. Liew, T. B. Sercombe, R. J. Dilley and B. J. Doyle, *J. Mech. Behav. Biomed. Mater.*, 2018, **79**, 150–157.
136. A. A. Aldana, F. Valente, R. Dilley and B. Doyle, *Bioprinting*, 2021, **21**, e00105.
137. T. Hoshiba, H. Lu, N. Kawazoe and G. Chen, *Expert Opin. Biol. Ther.*, 2010, **10**, 1717–1728.
138. J. S. Choi, B. S. Kim, J. Y. Kim, J. D. Kim, Y. C. Choi, H.-J. Yang, K. Park, H. Y. Lee and Y. W. Cho, *J. Biomed. Mater. Res., Part A*, 2011, **97A**, 292–299.
139. Y. C. Choi, J. S. Choi, B. S. Kim, J. D. Kim, H. I. Yoon and Y. W. Cho, *Tissue Eng., Part C*, 2012, **18**, 866–876.
140. R. Gaetani, S. Aude, L. L. DeMaddalena, H. Strassle, M. Dzieciatkowska, M. Wortham, R. H. F. Bender, K.-V. Nguyen-Ngoc, G. W. Schmid-Schöenbein, S. C. George, C. C. W. Hughes, M. Sander, K. C. Hansen and K. L. Christman, *Tissue Eng., Part C*, 2018, **24**, 697–708.
141. S. D. Sackett, D. M. Tremmel, F. Ma, A. K. Feeney, R. M. Maguire, M. E. Brown, Y. Zhou, X. Li, C. O'Brien, L. Li, W. J. Burlingham and J. S. Odorico, *Sci. Rep.*, 2018, **8**, 10452.
142. N. Khoshnood and A. Zamanian, *Bioprinting*, 2020, **20**, e00095.
143. F. Pati, J. Jang, D.-H. Ha, S. Won Kim, J.-W. Rhie, J.-H. Shim, D.-H. Kim and D.-W. Cho, *Nat. Commun.*, 2014, **5**, 3935.
144. H. Lee, W. Han, H. Kim, D.-H. Ha, J. Jang, B. S. Kim and D.-W. Cho, *Biomacromolecules*, 2017, **18**, 1229–1237.
145. W. G. Willats, J. P. Knox and J. D. Mikkelsen, *Trends Food Sci. Technol.*, 2006, **17**, 97–104.
146. D. Mohnen, *Curr. Opin. Plant Biol.*, 2008, **11**, 266–277.
147. I. Braccini and S. Pérez, *Biomacromolecules*, 2001, **2**, 1089–1096.
148. F. Munarin, M. C. Tanzi and P. Petrini, *Int. J. Biol. Macromol.*, 2012, **51**, 681–689.
149. P. Walkenström, S. Kidman, A.-M. Hermansson, P. B. Rasmussen and L. Hoegh, *Food Hydrocolloids*, 2003, **17**, 593–603.
150. N. Yao, C. Huang and D. Jin, *Orthop. Surg.*, 2009, **1**, 231–237.
151. S. Hu, R. Kuwabara, M. Beukema, M. Ferrari, B. J. de Haan, M. T. C. Walvoort, P. de Vos and A. M. Smink, *Carbohydr. Polym.*, 2020, **249**, 116863.
152. A. I. Cernencu, A. Lungu, I.-C. Stancu, A. Serafim, E. Heggset, K. Syverud and H. Iovu, *Carbohydr. Polym.*, 2019, **220**, 12–21.
153. G. Marchioli, L. Van Gurp, P. P. Van Krieken, D. Stamatialis, M. Engelse, C. A. Van Blitterswijk, M. B. J. Karperien, E. De Koning, J. Alblas, L. Moroni and A. A. Van Apeldoorn, *Biofabrication*, 2015, 7, 025009.
154. J. Kim, I. K. Shim, D. G. Hwang, Y. N. Lee, M. Kim, H. Kim, S. W. Kim, S. Lee, S. C. Kim, D. W. Cho and J. Jang, *J. Mater. Chem. B*, 2019, **7**, 1773–1781.
155. X. Liu, S. S. D. Carter, M. J. Renes, J. Kim, D. M. Rojas-Canales, D. Penko, C. Angus, S. Beirne, C. J. Drogemuller, Z. Yue, P. T. Coates and G. G. Wallace, *Adv. Healthcare Mater.*, 2019, **8**, 1–12.

156. J. Kim, C. M. Hope, N. Gantumur, G. B. Perkins, S. O. Stead, Z. Yue, X. Liu, A. U. Asua, F. D. Kette, D. Penko, C. J. Drogemuller, R. P. Carroll, S. C. Barry, G. G. Wallace and P. T. Coates, *Adv. Funct. Mater.*, 2020, **30**, 2000544.
157. S. Hu, F. D. Martinez-Garcia, B. N. Moeun, J. K. Burgess, M. C. Harmsen, C. Hoesli and P. de Vos, *Mater. Sci. Eng., C*, 2021, **123**, 112009.
158. S. Duin, K. Schütz, T. Ahlfeld, S. Lehmann, A. Lode, B. Ludwig and M. Gelinsky, *Adv. Healthcare Mater.*, 2019, **8**, 1801631.

CHAPTER 9

In Vivo Imaging of the Pancreas and Gut Hormone Receptors

OLOF ERIKSSON,*[a,b] GRY HULSART-BILLSTRÖM,[a] BOGDAN MITRAN,[b] EMMI PUUVUORI[a] AND IRINA VELIKYAN[a]

[a] Science for Life Laboratory, Department of Medicinal Chemistry, Uppsala University, Uppsala, Sweden; [b] Antaros Medical AB, Mölndal, Sweden
*Email: olof.eriksson@ilk.uu.se

9.1 Introduction: Digital Biopsies of the Pancreas

The pancreas, which is located deep in the abdomen, is a small organ, weighing around 100 g. However, its impact on the metabolism of the human body is significant. Despite the importance of the pancreas in, among other things, glucose homeostasis and the development of diabetes, it can be considered "understudied". This is primarily due to the difficulties in obtaining repeatable and accurate biopsies of the pancreas. In fact, biopsy of the pancreas is associated with high risk,[1] in part due to the proteolytic enzymes present in the exocrine compartment.

Thus, most pancreatic biopsies available in biobanks for scientific analysis have been taken post-mortem – introducing uncertainty and – bias to the material, difficulty in matching pancreatic biopsies between subjects with and without diabetes, as well as other differences in quality (such as warm ischaemia time). Consequently, most biopsy analyses are performed at a group level, while longitudinal analyses in individual subjects are rare or non-existing. For example, recent studies on the beta cell mass (BCM; the

Nanoscience & Nanotechnology Series No. 59
Nanotechnology for Diabetes Management
Edited by Amar Abderrahmani, Sabine Szunerits, Rabah Boukerroub and Abdelfattah El Ouaamari

Published by the Royal Society of Chemistry, www.rsc.org

number of insulin-producing cells) in the pancreas have demonstrated a large variability also among non-diabetic subjects (Figure 9.1).[2] Furthermore, there is an overlap in BCM in non-diabetic subjects and subjects diagnosed with type 2 diabetes (T2D).[3] This large variability raises interesting questions, such as whether healthy persons with low BCM are at risk of developing type 1 diabetes (T1D) or T2D. Additionally, at different stages of conditions like T1D, an individual may have both functional and non-functional beta cells (Figure 9.2), and recent reports suggest that beta cells may de-differentiate to alpha cell-like endocrine cells in T2D.[4] However, due to our inability to take safe and reproducible biopsies of the pancreas, it is currently difficult or impossible to directly and longitudinally study these

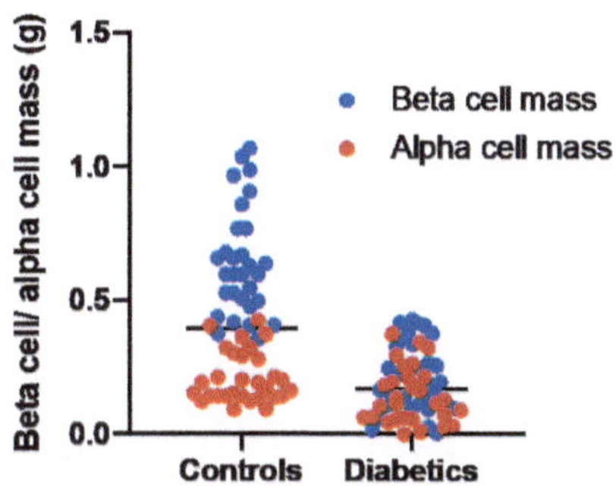

Figure 9.1 Beta and alpha cell mass in non-diabetic subjects and individuals with diabetes, assessed through *ex vivo* morphometry of pancreatic post-mortem biopsies. Data taken from ref. 2.

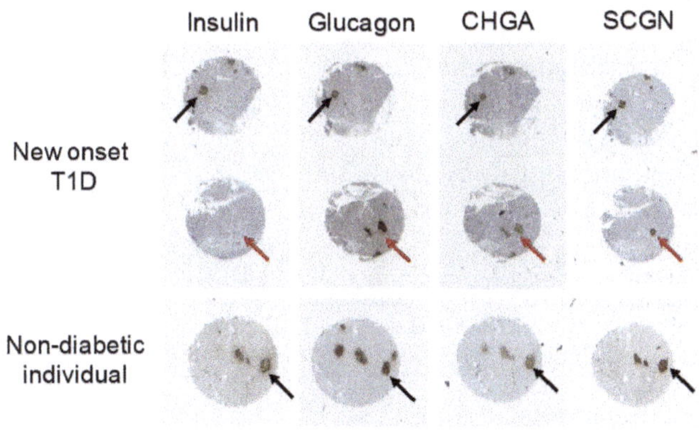

Figure 9.2 In the pancreas of non-diabetic individuals (bottom row), a vast majority of islets are positive for insulin staining (*e.g.*, contain functional beta cells, black arrows). However, although most islets are insulin-negative in T1D (top rows, red arrows), some insulin-positive islets can also be found, especially in T1D of recent onset (black arrows). CHGA: Chromogranin A; SCGN: Secretagogin.

important matters. Clearly, there is a need for the possibility to visualise these processes in the human pancreas in a non-invasive manner – *i.e.*, for new techniques for taking "digital biopsies".

Biomedical imaging techniques such as positron emission tomography (PET), single-photon emission computed tomography (SPECT), magnetic resonance imaging (MRI) and, last but not least, computed tomography (CT) have been implemented into the clinical diagnostic routine in many areas, such as oncology, neurology, and cardiology. However, their use in the clinical management of diabetes is not as advanced or widespread.

Recent decades have seen progress in novel techniques for clinical molecular imaging of receptors and cell subtypes, including those crucial for pancreatic function in both health and disease. This includes radiopharmaceuticals for studying, *e.g.*, the insulin-producing beta cells, as well as the most common incretin receptors. These topics are the main focus of this chapter. Also, seeing the pancreas in unprecedent anatomical detail is now possible thanks to optimised MRI and CT protocols. Importantly, these novel combined techniques are in or under clinical translation and approval, suggesting that they may become part of the routine assessment in diabetes diagnosis and management in coming years.

9.2 Biomedical Imaging Techniques – An Overview

Clinically implemented biomedical imaging techniques can be broadly divided into anatomical and molecular imaging, though there is a large overlap (Table 9.1). The first category includes, *e.g.*, CT and MRI, but certain applications of these techniques could be considered to be molecular imaging (such as contrast agent enhancement). There are many other important imaging techniques (*e.g.*, ultrasound), but these are outside of the scope of this chapter due to limited applicability in the molecular imaging of diabetes.

This section gives a brief overview of PET, CT, and MRI, with a focus on their implementation in the clinical setting and their use in the field of pancreas imaging.

Table 9.1 Overview of imaging modalities discussed in this chapter.

Modality	Abbreviation	Mode of detection	Resolution (approximate)	Sensitivity (approximate)	Ionising radiation
Computed tomography	CT	Tissue density	μm	mM	Yes
Magnetic resonance imaging	MRI	Movement of hydrogen atoms	μm	mM	No
Positron emission tomography	PET	Injected radioactivity	mm	fM	Yes

9.2.1 Positron Emission Tomography

Hybrid PET imaging combines PET with CT or MRI to obtain detailed information about the function and form of a target organ's cells and anatomy. Before a scan, the patient is injected with a small amount of a radiotracer. This is a protein, peptide, or small molecule labelled with a radioactive isotope. The radiotracer will accumulate in certain organs and tissues by either binding to a specific receptor or getting metabolically trapped. After this, the patient is placed in the scanner. Radiotracer decay will result in positron emission, generating two equal energy photons which can be detected when they emanate from the body, primarily from the organ of interest. An image of the physiological uptake of the tracer is then reconstructed and can be fused with CT or MRI images (Figure 9.3).

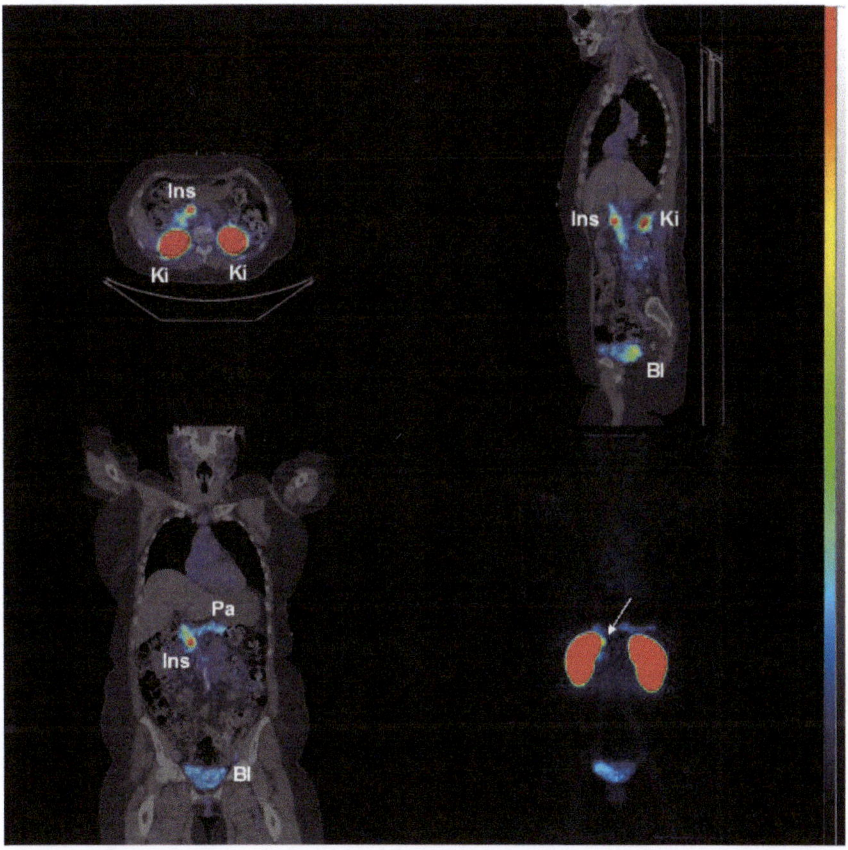

Figure 9.3 Clockwise from bottom left panel: Coronal, transaxial, sagittal, and 3D projections of a ^{68}Ga-NODAGA-exendin-4 PET/CT examination. A GLP-1R-positive pancreatic neuroendocrine tumour (Ins) is located in the pancreatic head.

The benefits of hybrid PET imaging, beyond it being a non-disruptive method, include no recovery or downtime for patients. Most importantly, it may detect biological changes in a tissue indicating the onset of a disease process, even before anatomical changes occur. By combining two different imaging techniques, we can obtain high diagnostic clarity and sufficient sensitivity to visualise the endocrine part of the pancreas.

The limitations of PET include its spatial resolution, which is in the range of several millimetres for clinical scanners. Additionally, PET radiopharmaceuticals carry a radiation dose to the individual, and care must be taken to reduce the dose to sensitive tissue. However, this is often not a problem with the short-lived radionuclides used in clinical routines, such as gallium-68, carbon-11, and fluorine-18. Another challenge that limits the array of molecules that can be reliably and reproducibly radiolabelled is the labelling chemistry. Here, great advances have been made in the last decade, especially for peptides or proteins; this is discussed further below.

9.2.2 Magnetic Resonance Imaging

MRI uses the particular properties of the hydrogen atom to generate images of the human body. Crucially, an MRI examination in its simplest form is non-invasive, in the sense that no contrast agent is administered and no blood sampling is necessary. The MRI scan requires the presence of a strong magnetic field (often 1.5–3 T for clinical scanners) which polarises paramagnetic nuclei (with hydrogen being the most common in the human body). By manipulating the polarisation, using *e.g.*, external radio waves or magnetic gradients, it is possible to extract three-dimensional images with high spatial and temporal resolution. The details of the physics behind the MRI technique have been described many times previously and new applications are constantly being developed. One example is magnetic resonance spectroscopy.

Water and fat contain large amounts of hydrogen. Therefore, MRI is often used to map water and fat in the body. MRI is thus an excellent technique for imaging soft tissue, potentially yielding high contrast between otherwise similar and adjacent tissues and organs, including the pancreas (Figure 9.4). Furthermore, special MRI sequences can be used to track water movement in tissue [apparent diffusion coefficient (ADC) or diffusion-weighted (DW) MRI] or to measure the fat content in tissues and organs (proton density fat fraction, PDFF), including the pancreas – again completely non-invasively.

Although non-invasive, MRI can cause discomfort to the patient being scanned. For example, claustrophobia is a common exclusion criterion for studies involving MRI, as long sequences (>1 h) and the loud noise associated with certain protocols may cause significant discomfort and drop-outs, especially among subjects classified as claustrophobic.

As MRI examinations do not require ionising radiation for image generation, they are normally classified as safe also in vulnerable populations, such as children and pregnant women.

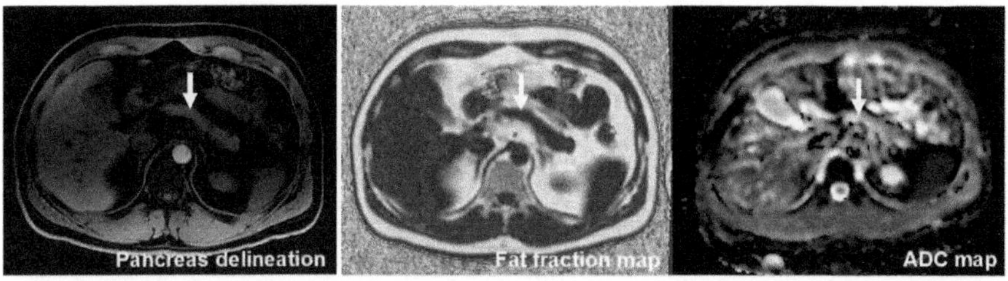

Figure 9.4 MRI of the pancreas, showcasing the versatility of the technique, enabling assessment of pancreas volume, fat content, and diffusion/perfusion in the same scanning session. The corpus of the pancreas is indicated by a white arrow.

Lastly, MRI can be enhanced by the administration of a contrast agent, for example small molecules incorporating strongly paramagnetic nuclides such as gadolinium or manganese. Such agents are administered intravenously and can increase the contrast by accumulating in target tissues. However, since these techniques have relatively poor sensitivity, quite large amounts of a contrast agent must be administered (millimolar range). This increases the risk of adverse events. Gadolinium is contraindicated in subjects with impaired renal function, and manganese has been associated with neurological impairment after chronic exposure.

9.2.3 Computed Tomography

Röntgen discovered X-rays in 1895. Less than a hundred years later, Cormak and Hounsfield revolutionised the world of imaging by developing computed tomography (CT). They all received the Nobel Prize for their discoveries, Röntgen in 1901 and Cormak and Hounsfield in 1979. The discovery of CT made it possible to build cross-sections by back-projection from a series of projections, which in turn allowed detailed imaging of internal structures. This technique quickly advanced to three-dimensional imaging and high resolution on the microscale. In its simplest form, radiography has similarities with photography, where photons are registered by a camera – but instead of such visible light, X-rays, which are high-energy photons, are registered. The subject is placed between an X-ray source and the camera, with the high-energy photons attenuating when passing through the subject. The camera will count the number of hits of photons per camera pixel and this number is used to estimate the density of the sample. In short, radiography is sensitive to the density of the subject. X-rays are attenuated in proportion to the atomic number Z of an atom, *i.e.*, the number of protons in its core. The partial absorption of X-rays is used to build a density map of the subject and can be adjusted by changing the energy of the photons by altering the voltage and current to the imaging machine. This is done on a daily basis in clinical practice, where CT scanners automatically adjust the energy of the X-rays used, to optimise the scanning settings for the specific

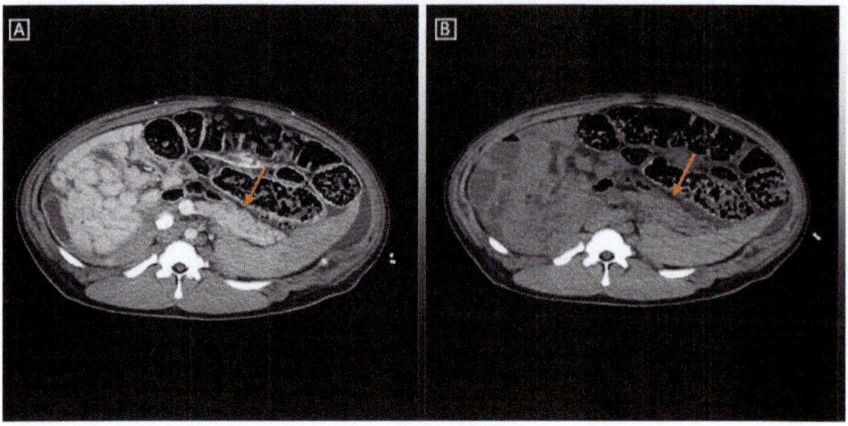

Figure 9.5 Transaxial images of contrast-enhanced (A) and native CT (B) of pig pancreas. The pancreas (orange arrow) is difficult to visualise on CT without the support of contrast media.

task and patient. Low-energy photons are attenuated more than higher energy photons. The camera and X-ray source will rotate around a subject, taking a number of projections. Each of these projections will be used to calculate cross-sectional images through a method called filtered back-projection, which produces a dataset that can be used for 3D analysis. The scans can be improved by changing the scanning parameters or by administration of a contrast agent. CT is non-invasive, but the radiation dose needs to be considered for each scan.

CT is used on a regular basis to diagnose and monitor diseases of the pancreas, as it is especially useful in determination of the parenchyma of the pancreas, which is responsible for the pancreatic exocrine function. It can also easily visualise any calcifications of the pancreas.[5] The administration of contrast agent can significantly enhance the image and facilitate delineation of the pancreas (Figure 9.5). The pancreatic volume is a potential marker of disease; for example, T1D and T2D are associated with reduced volume, whereas an enlarged pancreatic volume can be associated with obesity.[5,6] Additionally, CT is often used together with other imaging modalities to describe the anatomy of a patient.

9.3 Imaging in Diabetes

Metabolic disease and diabetes are conditions with complex aetiology and resolution. A trend of improved understanding of metabolic disease can be considered to have led to progressively improved definition of sub-types of, *e.g.*, diabetes over time. Pathologies resulting in increased blood glucose have been known since antiquity, which is seen in the observation that "diabetes mellitus" is derived from the Greek expression for "sweet urine" It was only in the early 20th century that understanding was gained of there being sub-types of diabetes with distinct pathophysiology and requiring different treatments – what we today call T1D and T2D. Recent progress in genomic, proteomic, and metabolomic profiling now further challenges our understanding of metabolic disease, and additional sub-divisions have been proposed.[7] It is not unreasonable to expect further developments in this area as the paradigm of precision medicine enters the clinical routine.

Imaging has revolutionised the management and treatment of oncological disease, and is currently being widely adopted for diagnostics also in neurology and cardiovascular disease. Given the developments of novel PET techniques designed to study disease processes and biomarkers in diabetes, it is possible to imagine a similar future role for imaging in metabolic disease as well.

For example, anatomical imaging techniques can already be used to non-invasively assess pancreas volume. MRI is probably the most sensitive technique for this measurement, and is particularly useful in children, as it does not require exposure to ionising radiation. Recently, pancreatic volume has been associated with severity of T1D, but it is as yet unclear if the correlation is causative.[8] Pancreas size could potentially provide additional information on metabolic status during the clinical workup of a patient.

Another topic at the cross-section of imaging and metabolic disease relates to the relatively recent discovery of functional brown adipose tissue (BAT) in adults.[9] Interscapular BAT was actually found as an artefact in ^{18}F-fluorodeoxyglucose (^{18}F-FDG) scans performed at cold temperatures (at which BAT is activated and consumes glucose). Consequently, the roles of BAT in T1D and T2D have been studied intensively, both in physiological development and as a drug target (Figure 9.6).[18] F-FDG, as an established marker of glucose metabolism, has also been used to investigate tissue-specific insulin resistance[10] and hypometabolism in the brain in individuals with diabetes.

Clearly, there are many clinically relevant areas where biomedical imaging can increase scientific understanding. The next sections focus on two specific topics, where development of novel PET techniques may provide unique insights by enabling direct and quantitative *in vivo* studies: beta cell imaging and imaging of gut incretin hormones.

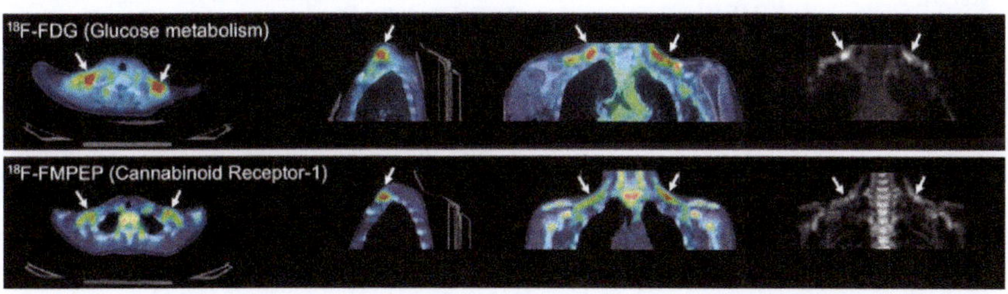

Figure 9.6 PET/CT imaging of interscapular brown adipose tissue (BAT) in individuals with diabetes. Here, BAT is visualised both by cold-stimulated glucose uptake (^{18}F-FDG, top) and by the expression of cannabinoid receptor 1 (*via* radioligand ^{18}F-FMPEP, bottom).

9.4 Beta Cell Imaging

The beta cells in the islets of Langerhans in the pancreas are the human body's sole source of insulin. It is remarkable that this specialised cell type, constituting only around 1 g of the pancreas, is responsible for control of the crucial glucose metabolism in the entire human organism (100 000 g). There are, of course, many fallback systems assisting glucose homeostasis, but the importance of the beta cells is evident in that total beta cell failure (*e.g.*, in severe T1D) is fatal if treatment is not initiated.

As outlined in the Introduction, the fate of individual beta cells in human development and the progression of diabetes are unknown due to the lack of longitudinal and non-invasive techniques for studying the pancreas. Beta cell function can be assessed using, *e.g.*, the insulin/C-peptide stimulation test, as the function is not tightly linked to actual viable beta cell mass. For example, beta cells can be exhausted and degranulated, but remain viable, in the case of chronic hyperglycaemia. This would correspond to poor function in stimulation tests (and potentially even being negative for insulin staining after biopsy).

PET has been proposed as a potential technique for quantitative imaging of beta cells.[11] This would require the identification of both a suitable surface marker on the beta cells and a suitable ligand for the marker (itself with a molecular structure that enables PET radiolabelling).

The last two decades have seen progress in this area and several PET radiopharmaceuticals have been evaluated in early-phase clinical imaging trials. Thus far, the trials have often been designed for assessing the potential of the imaging technique. Examples include controlled trials comparing the binding in individuals with long-standing, C-peptide-negative T1D compared with that in matched non-diabetes individuals, from which it has been inferred that the lower binding in T1D pancreas is related to the lower viable beta cell mass. However, recently, longitudinal studies have been performed, following PET tracer binding in individuals. The hope is that these techniques, in the future, can be used for assessment in drug and vaccination trials, for example to quantitatively measure a decreased beta cell loss in individuals on active treatment compared with that in controls.

9.4.1 Target Identification in Beta Cell Imaging

The development of beta cell PET markers (or any kind of PET agent) has several distinct stages. First, one must identify a target, most often a protein, in the tissue to be studied. In the case of beta cell imaging, the target density must be high in beta cells, while being negligible or absent from the rest of the pancreas. The reason is the limited resolution of clinical PET scanners, currently in the range of a few millimetres. In a voxel size of ~30 mm^3, one will find both islets of Langerhans, surrounding acinar exocrine tissue, and ducts and capillaries. Given the highly different phenotypes of the endocrine and exocrine cells, including the hormone secretion machinery, it is quite straightforward to identify markers specific to the islets of Langerhans.

Importantly, the diameter of most individual islets of Langerhans is in the range of ∼50–150 μm.[12] They contain of a mixture of endocrine cell types, mainly beta cells (∼60%–70%) and glucagon-secreting alpha cells (∼20%–30%), with the remaining 5–10% constituting delta, pp and other cell types.[13] To accurately image the beta cells, the target protein density must be negligible in the other endocrine subtypes. This makes target identification somewhat complicated, given that beta cells and, *e.g.*, alpha cells have a common neuroendocrine origin and share many characteristics and functions, such as hormone synthesis, storage, and exocytosis machinery. In fact, beta cells have been shown to be able to de-differentiate into glucagon-positive cells, demonstrating the similarity of these phenotypes. Thus, many putative target proteins, such as monoamine transporters or receptors, have non-negligible density in all islet cell types. It is worth mentioning that target receptor expression in nearby tissues (such as spleen, liver, stomach or gastrointestinal tract) is not usually an issue in the context of pancreatic imaging. The distance between these tissues is sufficiently large in comparison to the PET scanner resolution that spill-over signals from such receptor-positive tissues are negligible or can be corrected for.

Lastly, one must consider the quantitative aspect of protein density. It is not sufficient to demonstrate selective expression in the beta cells (compared within in other endocrine cells and the remaining pancreas), if the protein expression levels are too low for detection. PET is a highly sensitive technique, potentially enabling detection of femtomolar concentrations of protein. However, at some expression levels, the radioactive counts from radioligand binding to a receptor are insufficient to distinguish them from background noise. Here, one must also consider that a beta cell-specific signal will be diluted by a factor of ∼50–100, considering the islet/exocrine cell ratio in a given pancreatic volume.

Thus, in conclusion, an ideal target protein has high levels, but beta cell-specific localisation, in the pancreas – while expression in neighbouring tissues is permitted.

9.4.2 Ligand Identification and Optimisation in Beta Cell Imaging

The next step in the development of an imaging ligand is the identification of suitable molecules for radiolabelling. The crucial property is obviously the interaction between the ligand and the selected target protein. For PET agents, sensitivity increases with higher affinity, *i.e.*, very high affinity is required for imaging of receptors with low density. As a rule of thumb, PET radioligands should have an affinity in the nanomolar range, though poorer affinity may be acceptable for targets with very high amounts of binding sites (B_{max}). This relationship is often expressed using the binding potential (BP) formula (eqn (9.1)):

$$\text{BP (Binding Potential)} = \frac{B_{max}\,(\text{receptor density})}{K_d(\text{affinity})} \quad (9.1)$$

In the case of beta cell imaging, one can suppose that the expected receptor density B_{max} will be low, due to the dilution of tissue in pancreas, making it important to have high receptor affinity.

Selectivity is another property of particular importance in PET radioligand development. For pharmaceutical drugs, off-target effects and background non-specific binding may not preclude success. However, for a PET tracer, ensuring very little non-specific binding to other cell compartments may be crucial. The reason is of course that the PET scanner cannot distinguish between signals arising from specifically and non-specifically bound radioligand – meaning that a background signal can easily mask the target-specific signal. Thus, PET radioligands are usually not strongly lipophilic, as this can contribute to unacceptable non-specific binding in tissue.

Radioligands should normally have small molecular size, to allow imaging with short-lived radionuclides within hours of administration. Another aspect to consider is the cellular localisation of the target protein. In the case of an intracellular localisation, the PET ligand must pass over the cellular membrane to allow target binding. Active transport using membrane channels is sometimes possible, especially for small organic molecules. Passive transport is theoretically possible for non-polar lipophilic compounds, but as mentioned such ligands can lead to unwanted non-specific background binding. In the case of novel peptide scaffolds, enabling passage over the cell membrane might be a challenge Thus, for this particular class of radioligands, one must restrict the possible target proteins to those expressed on the outside of the cellular membrane, *i.e.*, that are accessible for binding.

Finally, the ligand must incorporate nuclides suitable for isotopic radiolabelling (*e.g.*, carbon-11, fluorine-18), or groups not involved in receptor binding but available for functionalisation – for example a chelator for radiometal incorporation.

With the above concepts of target protein and ligand specificity in mind, we now focus on two PET approaches for imaging of beta cells that have undergone significant clinical evaluation.

9.4.3 GLP-1R Targeting by Radiolabelled Exendin-4 Analogues

The first exendin-based radioligands were developed around 2004 and several analogues have been subject to clinical studies over the years. The main impetus for initial clinical studies with radiolabelled exendin has been the diagnosis and localisation of insulinoma, a rare form of neuroendocrine tumour derived from pancreatic beta cells.[14] However, strong binding can often be seen in the non-affected part of the pancreas, confirming expression of the glucagon-like peptide-1 receptor (GLP-1R) (see Figure 9.3). This observation, in line with the results of preclinical studies, has prompted the evaluation of radiolabelled exendin-4 as a marker for beta cells in humans. In a cross-sectional study, indium-111-labelled exendin-4 (^{111}In-NODAGA-exendin-4) demonstrated strong binding in human pancreas in non-diabetic individuals, and decreased, but at a group level overlapping, binding also in C-peptide-negative individuals with longstanding T1D.[15] The authors interpreted the remaining binding as a sign of significant residual, but "silent" beta cell mass in the individuals with T1D.

^{68}Ga-DO3A-VS-exendin-4 was recently used to evaluate GLP-1R binding in the pancreas of individuals with T2D, in the setting of a target occupancy study (more on this below).[16] The binding in the pancreas was high, but variable, indicating some variation in GLP-1R density in individuals with T2D. The correlation between radiolabelled exendin-4 and GLP-1R density in pancreas was established by demonstrating a progressively blockable signal.

A correlation between GLP-1R density and the BCM in the pancreas is yet to be conclusively demonstrated, but a significant part of the pancreatic signal is expected to arise from binding to beta cells.

9.4.4 ^{11}C-5-HTP for Beta Cell Imaging

The association between monoamine signal substances and the regulation of neuroendocrine cells, including beta cells, is well established. Furthermore, many neuroendocrine tumours (NETs), including insulinomas of beta cell origin, exhibit the molecular machinery for serotonin biosynthesis and metabolism. This has prompted carbon-11 radiolabelling of 5-hydroxytryptophan (5-HTP), the endogenous precursor for 5-HT (*i.e.*, serotonin) for imaging and diagnosis of NETs.[17] ^{11}C-5-HTP also, in an incidental finding, showed significant binding in the non-diseased pancreas in humans.

The mechanism for ^{11}C-5-HTP uptake, retention, and metabolism in pancreas has been explored in several animal models. Results indicate that ^{11}C-5-HTP is a general marker for the endocrine pancreas,[18] of which the islets of Langerhans – and, by extension, beta cells – are the major constituents. In a proof-of-concept cross-sectional clinical study, ^{11}C-5-HTP showed decreased binding in the pancreas of individuals with longstanding T1D compared with in non-diabetic individuals.[19] The decrease was on the order of 65%, suggesting that the residual binding corresponds to the remaining alpha cells in the beta cell-deficient pancreas. Further clinical studies demonstrated that ^{11}C-5-HTP binding in liver was higher in individuals with T1D who had received an intrahepatic islet transplantation than in non-transplanted individuals.[20] Interestingly, ^{11}C-5-HTP binding in the pancreas of individuals at different stages of T2D showed a limited decrease in uptake, also in patients expected to have significant beta cell loss based on peripheral biomarkers.[21] This was interpreted as retained islet mass paralleling beta cell deficiency, perhaps explained by the recently described de-differentiation mechanism. A recent longitudinal study followed individuals with recently debuted T1D for up to 2 years using imaging techniques. ^{11}C-5-HTP was one of the assessments used and demonstrated only minor loss of endocrine cells over the first 24 months of T1D, in line with the other assessments employed in the study (pancreas volume, beta cell function).[22] These findings speak against an acute and outright destruction of beta cells early in T1D.

9.4.5 Summary: Beta Cell Imaging

Here, we have discussed the challenges and opportunities of beta cell imaging. Additionally, we have reviewed two of the approaches that have been extensively evaluated in clinical practice to date (targeting the GLP-1R and ^{11}C-5-HTP). Apart from these techniques, there are several approaches targeting, *e.g.*, the vesicular amine transporter 2 (VMAT2). However, beta cell restriction of this transporter and any associated PET radioligand has not been conclusively demonstrated in any animal models, and its clinical value is unclear.[11] Several novel beta cell-specific targets are being evaluated in the context of beta cell imaging, such as CRTH2/GPR44,[23] and clinical data on these next-generation beta cell probes should become available in the next few years.

9.5 Imaging Gut Hormone Receptors

Several endogenous gut hormones and their receptors are crucial for glucose homeostasis, food intake, and energy metabolism. The most well-known are insulin and the ubiquitously expressed insulin receptor. Insulin itself has been radiolabelled, as have other ligands targeting the insulin receptor[24] – mainly by iodinisation – in order to enable *in vitro* detection of competing ligands in pharmaceutical development. Furthermore, the downstream effects of glucose uptake by insulin action in tissue have been extensively studied through PET over the last few decades, using the PET-labelled glucose analogue ^{18}F-FDG. Using this technique, one can directly observe insulin action and sensitivity in individual tissues, for example by performing ^{18}F-FDG-PET under glycaemic clamp. The *M* value (whole body measurement of insulin sensitivity) correlated well with the insulin sensitivity in skeletal muscle, which is expected as muscle tissue is the main sink of glucose in humans.[10] On the other hand, ^{18}F-FDG-PET examination may provide additional information on individual tissues not available from the global *M* value assessment.

While insulin action has been studied using this technique for some time, there have not, until recently, been any PET markers available for studies of other important gut hormones, such as glucagon (GCG), glucagon-like peptide-1 (GLP-1), and gastric inhibitory polypeptide (GIP).[25]

Despite many of the aforementioned gut hormones being known for several decades, there is still much uncertainty regarding their relative importance in human metabolism. This is probably due to crucial unknowns, such as the tissue distribution of their respective receptors (GCGR for GCG, GLP-1R for GLP-1 and GIPR for GIP). In other words: in what tissues are these receptors expressed (pancreas, gastrointestinal tract, CNS, *etc.*) and in what relative amounts? Current knowledge on gut hormone receptor tissue distribution is primarily derived from qualitative immunohistological staining of post-mortem tissue sections and transcriptomic analyses. The receptor populations in different tissues are probably involved in various downstream effects related to metabolic status. For example, GLP-1Rs in the pancreas is thought to be responsible for the incretin effect (boosting insulin secretion during hyperglycaemia), while GLP-1Rs in the brain or CNS is postulated to be involved in appetite control, and the delay in gastric emptying is considered to be mediated by GLP-1Rs in the vagal nerve. However, there are no direct studies confirming these hypotheses. Furthermore, it is not known what local hormone concentrations or receptor occupancies are required for *in vivo* receptor activation.

These questions are of utmost importance, not only for improved understanding of the basic molecular basis of metabolic homeostasis, but also for development of novel anti-diabetic treatments. Multiple gut hormone receptor agonists (mostly GLP-1R agonists) are approved for treatment of T2D, with several new classes of similar drugs in clinical development. Of particular interest is the new class of unimolecular multi-agonists simultaneously targeting two or more gut hormone receptors in different tissues. Combinations include dual GLP-1/GCG receptor agonists,[26] dual GLP-1/GIP receptor agonists,[27] and even triple GLP-1/GCG/GIP receptor agonists.[28]

Since agonism with several of these receptors has overlapping pharmacology (effect on blood glucose, weight loss, *etc.*), it is quite difficult to demonstrate whether the occasionally impressive effect demonstrated by unimolecular multi-agonists in clinical trials is due to one receptor or another – *e.g.*, is the weight loss induced by a GLP-1R/GCGR agonist due to the potent GLP-1R-mediated effect on appetite, or is it due to increased energy expenditure by agonism on the GCGR?

9.5.1 Imaging the GLP-1R

The GLP-1R is established as an important target for treatment of T2D due to the involvement of the incretin system in insulin release, control of gastric emptying, and appetite. As described above, it has also been suggested as a BCM imaging marker due to the perceived association between the GLP-1R expression in the pancreas and the beta cells. However, we will see that the GLP-1R is involved in many processes beyond insulin secretion. The endogenous GLP-1 peptide is sensitive to enzymatic degradation and subject to inactivation within minutes in the bloodstream. However, degradation-resistant GLP-1 analogues have been discovered, including the exendin-3 and exendin-4 peptides found in the venom of the Gila monster.[29] Exendin-4 and its analogues have therefore been radiolabelled and evaluated as imaging agents for several indications, including pancreatic NETs, such as insulinomas, and beta cell imaging, as described above.

Various analogues of exendin-3 and exendin-4 were used in the development of imaging agents for SPECT using ^{111}In[15,30–37] or ^{99m}Tc[30,31,33–35,38–44] and PET using ^{68}Ga,[33,34,45–50] ^{64}Cu,[47,51–54] ^{18}F,[55–66] ^{89}Zr,[67] or ^{125}I.[68–72] Clinical value has been demonstrated in both SPECT[41–43] and PET.[46,73–75] More attention is given to PET agents, due to its intrinsic advantages over SPECT technology.[76] The greatest interest is seen for the combination of ^{68}Ga and PET/CT. Various analogues, differing in the sequence of exendin and the chelator moiety, have been developed and investigated preclinically, with ^{68}Ga-labelled agents demonstrating advantages over ^{99m}Tc- and ^{111}In-labelled agents in terms of dosimetry and uptake in tumour xenografts and pancreas, beyond the general technological advantages of PET over SPECT.[33,34]

Formation of radiolytic by-products was observed when methionine was present in the exendin sequence, compromising the use of the radioactive analogues, since the oxidised counterpart would present changes in biodistribution.[38] The suppression of radiolysis was achieved by fine-tuned optimisation of a combination of radical scavengers and reaction temperature in order to achieve the required clinical product quality.[45,50,77–80] The elimination of methionine by replacement with norleucine, its isoelectric analogue, improved not only stability but also binding capability. However, relatively high blood and healthy organ uptake was observed, with the highest values for GLP-1R-positive lung and kidney, 1 h post injection in rats.[38,47,81] The structure–activity relationship and enzymatic degradation stability were thoroughly investigated for various GLP-1 analogues.[82] The amino acid residues sensitive to degradation by DPP-IV and NEP were substituted with *D*-amino acids, β-amino acids, and alkylated, glycosylated, or halogenated amino acids. The aims of these structural modifications were improvement of *in vivo* stability and renal clearance, as well as functionalisation for subsequent radioactive labelling. The position for conjugation of the chelator moiety (NODAGA, BnDTPA-Ahx) to the peptide *via* a Lys residue influenced the biological activity of the ^{68}Ga-labelled agent. Positions 12 and 40 proved preferable,[37,53] while functionalisation at position 27 led to deterioration of the binding specificity.[53,61] The kidney uptake could be considerably reduced by adding a Met-Val-Lys sequence to the Cys^{40} of an exendin-4 analogue with leucine at position 14 (NOTA-MVK-Cys^{40}-Leu^{14}-exendin-4).[83] The use of desferrioxamine B as a chelator moiety instead of macrocyclic chelators such as tetraazacyclododecantetraacetic acid (DOTA) and 1,4,7-triazacyclononane-*N*,*N*′,*N*″-triacetic acid (NOTA) also resulted in an agent with nanomolar receptor affinity.[67]

^{68}Ga-DO3A-VS-exendin-4 was evaluated in cynomolgus monkeys and pigs as regards binding in pancreas at different doses, and it was found that the binding could be dose-dependently outcompeted by co-injection of

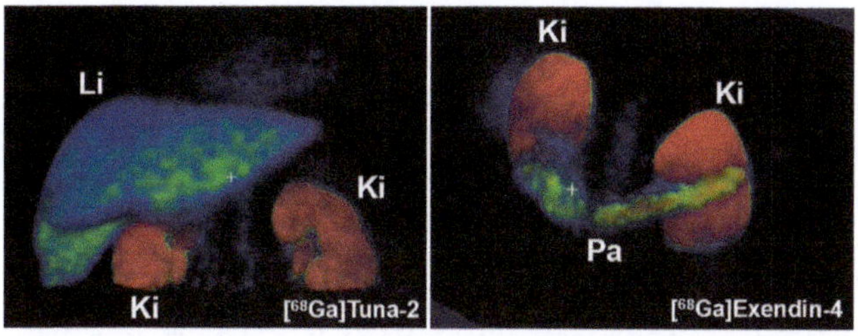

Figure 9.7 PET imaging of the GCGR and the GLP-1R in the same subject, 4 days apart. Reproduced from ref. 16, https://doi.org/10.1038/s41598-020-73815-5, under the terms of the CC BY 4.0 license https://creativecommons.org/licenses/by/4.0/.

unlabelled peptides.[45,78] This observation opened the possibility of measuring the target engagement of GLP-1R-targeting drugs, by assuming that the decrease in ^{68}Ga-DO3A-VS-exendin-4 binding, from baseline to on treatment, would be due to decreased GLP-1R availability *via* drug binding. Consequently, a clinical study using this design could demonstrate a GLP-1R occupancy of around 50% of the dual GLP-1R/GCGR dual agonist SAR425598 at clinically relevant oral dosing of the drug (Figure 9.7).[16] One could speculate that this kind of study design could be used to directly benchmark and compare different GLP-1R agonists with respect to their *in vivo* target engagement. Furthermore, it is likely feasible to measure such target engagement not only in the pancreas, but also in other tissues where the GLP-1R is known to be expressed, such as in the CNS, the intestine, and the heart.

One should also note that there are independent non-imaging-based clinical markers and signs of GLP-1R target engagement. For example, up-titration of GLP-1 dosing in patients usually leads to gastrointestinal adverse events, including nausea, which is known to be mediated *via* the GLP-1R. Thus, PET-based target engagement measurements are mainly of interest in the context of quantification of the occupancy, or as part of the pharmacology of dual or triple incretin hormone receptor agonists.

9.5.2 Imaging the GCGR

Glucagon, a 29-amino-acid residue peptide, is a hormone. A radioligand with ^{125}I is commercially available and has been used extensively for various studies of GCGR biology since the 1970s. Although this is a valuable tool for basic research studies of biological processes *in vitro*, it is not suitable for *in vivo* clinical use with PET. The feasibility of non-invasive imaging of the GCGR was demonstrated preclinically using gamma-emitting ^{67}Ga-labelled GCG.[84] Today, extensive preclinical and clinical studies have been performed using novel GCG analogues and ^{68}Ga,[16,85–88] presenting intrinsic advantages of PET over SPECT imaging, as mentioned above. These studies have developed and used two GCG analogues with improved properties as compared with endogenous hormone in terms of metabolic stability.

Both analogues consisted of 39 amino acid residues and included an uncommon amino acid, 3-thiazolylalanine, at their N-termini, and a 1,4,7,10-tetraazacyclododecane-1,4,7-triacetic acid (DO3A) chelator moiety coupled to cysteine *via* a diethyl sulfone linker at their C-terminus.[88] The analogues differed in seven amino acid residues, at positions 2, 15–17, 20, and 29. One of the analogues, [^{68}Ga]Ga-DO3A-S01-GCG, contained a lactam bridge, stabilising the alpha-helical fold of the peptide backbone. These differences resulted in higher affinity and selectivity, as well as improved metabolic stability,[88] and the analogue was used for the assessment of GCGR occupancy *in vivo* using PET in non-human primates[86] and humans.[85] The ^{68}Ga-labelling of the analogues was conducted manually for the preclinical studies and demonstrated high efficiency and reproducibility.[86,88] The radiopharmaceuticals exhibited radiolytic degradation, most likely due to the presence of a tryptophan amino acid residue. The radiolysis was suppressed by addition of ethanol as a radical scavenger, providing a radiochemical yield of over 99%. The production for clinical use was automated, to meet good manufacturing practice requirements and to reduce the radiation exposure of the operator. Under those conditions, dihydroxybenzoic acid was needed, as well as ethanol, for the suppression of radiolysis. A high success rate of the production allowed for a complex outline of the clinical study, with repeated examinations within a single day.[16,85]

[^{68}Ga]Ga-DO3A-S01-GCG was evaluated in cynomolgus monkeys using PET/CT, and was found to bind strongly and irreversibly in liver.[86] Importantly, the binding could be dose-dependently inhibited by co-injection of unlabelled precursor peptide. Pre-treatment with a high subcutaneous dose of an acylated GCGR agonist inhibited the binding of [^{68}Ga]Ga-DO3A-S01-GCG by up to 70% in liver, providing a first measurement of *in vivo* occupancy of the GCGR.

[^{68}Ga]Ga-DO3A-S01-GCG, also known as [^{68}Ga]DO3A-VS-Tuna-2, was used in a clinical trial together with [^{68}Ga]Ga-DO3A-exendin-4 (as described in the previous chapter) to evaluate the target engagement of the dual agonist SAR425598 (Figure 9.7). While SAR425598 exhibited strong interactions with the GLP-1R, the occupancy of the GCGR was modest and varied from 0% to 20% in the studied patients.[16,85] This study design can potentially be used to evaluate the target engagement of other GCGR agonists and antagonists currently under development, at relevant clinical doses.

By comparison, it is more difficult to identify a non-imaging biomarker that specifically corresponds to activation of the GCGR, especially in the setting of unimolecular multi-agonists. For example, strong occupancy at the GCGR is expected to increase blood glucose levels (by glycogenolysis). This could serve as a marker for GCGR agonism alone, but not when the drug simultaneously acts on the GLP-1R or the GIPR in pancreas and reduces the blood glucose level. Several peripheral markers have been suggested as biomarkers for activation of the hepatic GCGR, including ketones, 3-hydroxybutyrate, and fibroblast growth factor 21,[89] but the sensitivity and accuracy of such assessments remain unclear.

9.5.3 Imaging the GIPR

Gastric inhibitory polypeptide (or glucose-dependent insulinotropic polypeptide) is a 42-amino-acid residue peptide produced by enteroendocrine K cells. Its major function is stimulation of insulin secretion from pancreatic cells after food ingestion *via* binding to GIPRs.[90] Radiolabelled GIP ligands hold interest for clinical applications, *e.g.*, diagnostic imaging of NETs, and a number of analogues have been developed. The native GIP(1–42) peptide was truncated to GIP(1–30) with maintained affinity[91] and has been used to develop a series of radiolabelled analogues for *in vivo* targeting of GIPR-positive tumours.[92] A GIP(1–30) functionalised with DOTA bifunctional chelator *via* AHx-linker and lysine amino acid residue at the N-terminus and introduced with norleucine at position 14 was labelled with ^{111}In. The resulting radioligand (^{111}In-EG4; [Nle,^{14}Lys^{30}(Ahx-DOTA)]GIP(1–30)NH_2) demonstrated good affinity and selectivity for GIPR expressed on transfected INR1G9-hGIPr cells *in vitro*. The analogue was also labelled with ^{68}Ga and confirmed the feasibility of imaging in INR1G9-hGIPr xenograft-bearing mice. The stability of the full-length peptide, GIP(1–42), was improved by acetylation of the N-terminal tyrosine residue.[93] The analogue labelled with ^{111}In *via* DTPA coupled to a lysine residue at position 37 ([Lys^{37}(^{111}In-DTPA)]*N*-acetyl-GIP_{1-42}) demonstrated stability in serum for 24 hours, with an intact fraction of 73%. More considerable modifications with the aim of creating an analogue of high affinity, metabolic stability, and favourable for PET imaging pharmacokinetics were introduced to hGIP(1–30), with 10 amino acid residues exchanged at positions 2, 10, 14, 16, 20, 21, 23, 24, 29, and 30, respectively.[94] Aminoisobutyric acid substituted alanine at position 2 to improve structuration and resistance to enzymatic degradation. An additional glycine (position 29) was introduced and lysine at the C-terminus was coupled to a DOTA chelator moiety for labelling with ^{68}Ga. The fraction of intact radioligand ([^{68}Ga]Ga-S02-GIP-T4) after 90 min incubation in non-human primate plasma was over 93%, and the potency values (EC_{50}, [pM]) determined in human, rat, and non-human primate cells/tissues were either improved or at the same level as for native GIP(1–42).

[^{68}Ga]Ga-S02-GIP-T4 was evaluated *in vivo* in cynomolgus monkeys, in a dose escalation-binding paradigm similar to that described above for certain GLP-1R and GCGR PET tracers.[94] The binding in the pancreas was low, as may be expected based on the reportedly low GIPR density in the pancreas compared with, *e.g.*, GLP-1R density, and could be dose-dependently competed off by an unlabelled tracer precursor. Furthermore, pancreatic binding could be abolished by subcutaneous administration of a high dose of a stable GIPR agonist, demonstrating sufficient sensitivity of the probe for assessment of GIPR occupancy *in vivo*. [^{68}Ga]Ga-S02-GIP-T4 has not yet entered clinical studies.

GIPR activation has been associated with bone turnover and thus certain peripheral markers of bone resorption and formation have been proposed as putative markers of GIPR target engagement. Levels of C-telopeptide of cross-linked type 1 collagen (CTX) correlate with bone mineral density, and CTX has therefore been suggested as a marker also of GIPR agonism in clinical drug trials.[95] Currently, the sensitivity of CTX as a marker of GIPR activation is unknown.

9.5.4 Further Imaging Exploration of the Gut Hormones and Their Receptors

Here, we have discussed recent progress in the development of novel PET radiopharmaceuticals for the GLP-1R, the GIPR, and the GCGR – all of which are in clinical trials or close to translation. However, in addition to these important incretin gut hormones, there are other hormones that are gaining increasing interest in the context of pharmaceutical drug development, including but not limited to glucagon-like peptide-2 (GLP-2),[96] peptide YY,[97] and amylin (also known as islet amyloid polypeptide, IAPP).[98] At the moment, there are no published imaging markers mimicking these peptides or targeting their respective receptors. However, given the recent experience in the development of PET tracers for the GLP-1R, the GIPR, and the GCGR, based on rational redesign and stabilisation of the endogenous ligands, we may expect to see increasing activity in this field in the coming years. The state-of-the-art for the development and radiolabelling of novel peptide-based PET tracers will be explored in the next section.

9.6 Advances in PET Radiochemistry

PET imaging involves the administration of radiolabelled compounds (radiopharmaceuticals), to track their fate, metabolism, and binding to macromolecules inside the living body. The technique is therefore limited by the existence of suitable radiopharmaceuticals, since radiolabelling of small molecules, peptides, and proteins is non-trivial – especially under good manufacturing practice conditions and considering the short radioactive half-life of common PET radionuclides. However, the arsenal of chemical techniques for site-specific and fast radiolabelling of compounds is evolving rapidly, enabling novel PET radiopharmaceutical development. In this section, we focus on the recent technical developments of reproducible and reliable radiolabelling of peptides and proteins, which have provided access to large numbers of biologics and endogenous peptide hormones for use in PET imaging.

9.6.1 Radiolabelling of Peptides

Several radiolabelled low- and high-molecular-weight ligands have been studied for imaging of various biological processes and targets expressed on beta cells.[11] In this chapter, we focus on peptide-based ligands, since the majority of new diabetes drugs are peptides. Moreover, endogenous peptide hormones are important in diabetes development and management.

Molecular targets like the GLP-1R, the GCGR, and the GIPR have been considered, as described above. Due to how T2D is treated, the GLP-1R has been the most explored target since the discovery of the agonist exendin-4, an *in vivo* stable peptide, and its analogues.[11,30,31,40,99] Other peptide hormones involved in processes such as glucose homeostasis and appetite control include peptide YY, IAPP, and glucagon-like peptide-2. Crucially, their receptors and their biodistribution in the human body are not completely mapped; PET imaging may play a role here in the future.

Peptide-based imaging agents commonly comprise radionuclides from chemical classes such as halogens (*e.g.*, ^{18}F or $^{124/125/131}I$) and metals (*e.g.*, ^{64}Cu, ^{68}Ga, ^{89}Zr, ^{111}In, ^{99m}Tc, or ^{177}Lu) depending on the radionuclide availability, decay mode, labelling chemistry, and application of the resulting radioligands. Labelling with radiometals is commonly chelator-mediated. Various chelators conjugated with the peptides form stable complexes with relevant radiometals. Here, we have included: DOTA, NOTA, 2-(4,7-bis(carboxymethyl)-1,4,7-triazonan-1-yl) pentanedioic acid (NODAGA) for interaction with $^{68}Ga(III)$; DOTA and diethylenetriamine-pentaacetic acid (DTPA) for interaction with $^{111}In(III)$; 6-aminohexanoic acid-hydrazinonicotinamide (Ahx-HYNIC) for interaction with ^{99m}Tc; *N*′-[5-(acetyl-hydroxy-amino)pentyl]-*N*-[5-[3-(5-aminopentyl-hydroxy-carbamoyl) propanoylamino]pentyl]- *N*-hydroxy-butane diamide (desferriox-amine B, DFO) for interaction with ^{68}Ga and ^{89}Zr; bicyclic cage-like chelator (sarcophagine, Sar) for interaction with $^{64}Cu(II)$.

As a gamma emitter for SPECT, ^{99m}Tc has advantages over ^{111}In in the form of lower radiation burden and higher resolution. However, both ^{111}In and ^{99m}Tc have disadvantages such as relatively high radiation burden, low spatial resolution and sensitivity, as well as poor quantification, which can be overcome by using positron-emitting ^{68}Ga. ^{68}Ga also presents advantages over other positron emitters, such as ^{64}Cu and ^{89}Zr, in terms of lower radiation dose, possibility of repeated examinations, and higher positron abundance, resulting in higher sensitivity and shorter scanning time. The half-life of 110 min and a decay mode with 97% positron emission make ^{18}F an attractive radionuclide. A potential advantage of using ^{18}F as a radiolabel is the faster washout by the renal system. However, relatively high non-specific uptake in the liver and intestines in animal studies is a drawback.[55,60–63] Peptides can be labelled by either a covalent interaction with an ^{18}F-bearing prosthetic group (click chemistry) or a coordinate interaction of a chelator (*e.g.*, NOTA) with $[Al^{18}F]^{2+}$. However, ^{18}F requires a cyclotron and hot cell for radiation protection and the labelling chemistry is complex. ^{68}Ga is produced from a simple generator system and the straightforward labelling chemistry can be accomplished using minimal radiation shielding.

The most critical aspects that must be addressed in developing peptide-based imaging agents are binding affinity, structure–activity relationship, specific activity, radiolytic stability, *in vitro* and *in vivo* stability, kidney uptake, and physiological potency of the peptide ligands. The importance of high affinity and specific activity (MBq/μg) is dictated by the low target receptor density requiring the lowest amount of the peptide ligand possible, in order to avoid saturation of the target by the unlabelled peptide counterpart and to provide sufficiently high radiation for detection, as well as to avoid a possible pharmacological effect. The structure–activity relationship in terms of peptide sequence, chelator, prosthetic group moiety, and radionuclide can be studied in order to modulate and optimise affinity, stability, pharmacokinetics, and biodistribution.

9.6.2 Discovery and Generation of Novel Peptides for Imaging

Sometimes, when receptors of interest are identified, there may not be any known endogenous or synthetic ligands. In other cases, only antibodies or unstable endogenous peptides are available.

Antibodies can have excellent specificity; this is obvious from the growing class of biologics in pharmaceutical drug development. However, for imaging purposes, their slow biological clearance (related to their large molecular size) is a challenge. Normally, several days' wait is required for antibody receptor targeting and blood clearance to allow high contrast imaging when using a radiolabelled antibody. This, in turn, requires the radiolabel to have a radioactive half-life in the range of days, which will yield a significant deposited dose to the individual. This may be motivated in cases where antibody imaging may lead directly to diagnosis (*i.e.*, clinical decisions), or in cases that are already exposed to radiological examinations in their regular clinical routine (*e.g.*, in oncology). However, neither of these is normally applicable to PET imaging in metabolic diseases, which often affect young individuals (for instance in T1D).

Recently, many new peptide scaffolds smaller than full antibodies have emerged, allowing for the discovery and design of novel binders for many conceivable targets and receptors. Examples include the Affibody scaffold (58-mer peptides)[100] and single-domain antibodies (nanobodies).[101] Peptides in that size range usually have a shorter clearance time (hours instead of days), thereby allowing for radiolabelling with gallium-68 and fluorine-18, leading to lower radiation dose.

Endogenous hormone peptides, such as GLP-1, insulin, GCG, *etc.*, are usually unstable in the circulation due to, *e.g.*, endopeptidases and DPPIV, often being degraded or deactivated within minutes. This is, of course, a feature "by design" of many endogenous peptide hormones, allowing downstream regulation in remote tissues with a short duration. The endogenous peptide binder is often a suitable starting point for the development of an imaging agent for the system in question, but the biological half-life must be greater than a few minutes. Thus, there is a challenge similar to that in the case of antibodies, but due to a short biological half-life instead of a long one. Nowadays, precise peptide engineering in combination with knowledge on the mechanism of action for many peptidases can be used to improve the stability and plasma half-life of peptides. This can involve simple modifications such as C-terminal amidation or more complex techniques such as protecting vulnerable residues by cyclisation.

9.7 Conclusions and Future Directions

9.7.1 Improved Understanding of Diabetes by Beta Cell Imaging and Hormone Receptor Imaging

The examples on putative beta cell imaging agents as well as the hormone receptor agents show how imaging techniques can potentially be used to further our understanding of diabetes progression. Currently, several PET radioligands are in clinical use and could conceivably be used for longitudinal assessment of, *e.g.*, GLP-1R density in the years before and after diabetes diagnosis.

As discussed above, many of the molecular mechanisms underlying the onset and pathophysiological development of T1D and T2D remain unclear, especially in humans, due to the lack on non-invasive tools for studying the mode of action *in vivo*.

One can also foresee more direct clinical benefits from implementing more advanced imaging in patient management. For example, GLP-1R imaging may provide clues for predicting the efficacy of GLP-1R agonist therapy in the individual patient – *e.g.*, will an individual with very low GLP-1R density in the pancreas respond to this type of treatment? However, since PET scans carry the disadvantage of ionising radiation (albeit at low levels), there must be a clear benefit for the individual before such examinations are deployed broadly.

In the near future, it will be easier to see the strong potential for these agents in the drug development process. Quantitative PET imaging of beta cells or the main gut hormone receptors could clearly open the way for novel and previously unavailable biomarkers for evaluating drug effects and drug modes of action at a very early stage of the clinical phase.

9.7.2 Other Emerging Applications

In this chapter, we have provided an overview of emerging PET probes for the study of beta cells and incretin hormones. Obviously, there are many other disease processes crucially involved in diabetes pathophysiology and complications, not least inflammation and fibrosis in both the pancreas (mainly T1D) and other tissues, such as the liver (T2D). For example, fibrosis – the excessive deposition of connective tissue components – is an outcome of dysregulated tissue repair following injury or during chronic inflammatory disorders. Fibrosis is commonly found in tissues affected by diabetic complications like diabetic nephropathy, cardiomyopathy, or liver dysfunction.[102]

In recent years, a variety of probes have been developed that target fibrosis-specific or fibrosis-associated processes. A promising direction includes targeting of activated fibroblasts, which are the key cellular mediators of fibrosis. Targets for PET/SPECT imaging of fibroblasts in the liver include: fibroblast activation protein,[103] platelet-derived growth factor receptor beta,[104] translocator protein,[105] and αvβ3 integrin.[106] A different approach involves the deposition of extracellular matrix (ECM) proteins, including type I collagen, fibronectin, and elastin, which form the fibrous tissue in an injured area. Collagens are the most abundant proteins in the ECM, with type I collagen being the most abundant among them. The feasibility of *in vivo* imaging of fibrosis *via* targeting of type I collagen has been evaluated using CBP8[107] and collagelin[108] radiolabelled peptides. However, despite promising results, their use is limited by low affinity (1.5–2 μM) for type I collagen. In a more recent approach, Rosestedt *et al.* reported the development of a radiolabelled peptide-based targeting vector with 10-fold higher affinity (170 nM).[109] ^{68}Ga-labelled LRELHLNNN has demonstrated a favourable biodistribution profile, although improved proteolytic stability would be required.

The topics of PET probes for immune cells and fibrotic processes are broad enough to warrant separate reviews, but suffice it to say that such techniques could also provide further insight into disease development and the evaluation of drug modes of action.

9.7.3 Emerging Technological Advances

Finally, it is pertinent to mention the recent and ongoing developments in PET instrumentation. Above, we have described the advances in radiochemistry and PET ligand development that may open up new areas of clinical study. In parallel, there has been rapid improvement in engineering of, *e.g.*, PET detector technology and scanner modality integration. These optimisations are sometimes incremental, as in the case of the continuous improvement of detector sensitivity which over time enables drastically lowered PET and CT radiation doses. The recent progress in some areas has been dramatic, enabling conceptually new scanner designs. Integrated simultaneous PET/MRI scanners are now available, which potentially enable anatomical co-registration, with excellent soft-tissue contrast, in addition to fMRI and lower radiation dosing as MRI replaces CT. A second example is the full-body PET scanners[110] – with an almost 2-metre axial field of view – which is a great change from just a few years ago, when the scanned area was in the range of a decimetre. The full-body scanner enables simultaneous dynamic imaging of all tissues in the body, with improvements in both sensitivity and resolution.

These advances, coupled to novel PET radiopharmaceutical developments, will hopefully yield disruptive new tools for the study of diabetes in the years to come.

Acknowledgements

Olof Eriksson's lab is funded by Science For Life Laboratory, the Swedish Research Council (2020-02312), Diabetes Wellness Sverige (2409-PG), Diabetesfonden (DIA2019-411), Barndiabetesfonden, the European Foundation for the Study of Diabetes (EFSD), JDRF (1-SRA-2020-973-S-B), Magnus Bergvalls Foundation, the Family Ernfors Fund, and ExoDiab.

References

1. L. Krogvold, B. Edwin, T. Buanes, J. Ludvigsson, O. Korsgren, H. Hyöty, G. Frisk, K. F. Hanssen and K. Dahl-Jørgensen, *Diabetologia*, 2014, **57**, 841.
2. N. Maclean and R. F. Ogilvie, *Diabetes*, 1955, **4**, 367.
3. J. Rahier, Y. Guiot, R. M. Goebbels, C. Sempoux and J. C. Henquin, *Diabetes, Obes. Metab.*, 2008, **10**, 32.
4. C. Talchai, S. Xuan, H. V. Lin, L. Sussel and D. Accili, *Cell*, 2012, **150**, 1223.
5. H. Kumar, S. V. DeSouza and M. S. Petrov, *Comput. Methods Programs Biomed.*, 2019, **178**, 319.
6. S. V. DeSouza, R. G. Singh, H. D. Yoon, R. Murphy, L. D. Plank and M. S. Petrov, *Expert Rev. Gastroenterol. Hepatol.*, 2018, **12**, 757.
7. E. Ahlqvist, P. Storm, A. Käräjämäki, M. Martinell, M. Dorkhan, A. Carlsson, P. Vikman, R. B. Prasad, D. M. Aly, P. Almgren, Y. Wessman, N. Shaat, P. Spégel, H. Mulder, E. Lindholm, O. Melander, O. Hansson, U. Malmqvist, Å. Lernmark, K. Lahti, T. Forsén, T. Tuomi, A. H. Rosengren and L. Groop, *Lancet Diabetes Endocrinol.*, 2018, **6**, 361.
8. J. Virostko, J. Williams, M. Hilmes, C. Bowman, J. J. Wright, L. Du, H. Kang, W. E. Russell, A. C. Powers and D. J. Moore, *Diabetes Care*, 2019, **42**, 248.
9. K. A. Virtanen, M. E. Lidell, J. Orava, M. Heglind, R. Westergren, T. Niemi, M. Taittonen, J. Laine, N.-J. Savisto, S. Enerbäck and P. Nuutila, *N. Engl. J. Med.*, 2009, **360**, 1518.
10. E. Johansson, M. Lubberink, K. Heurling, J. W. Eriksson, S. Skrtic, H. Ahlström and J. Kullberg, *Radiology*, 2018, **286**, 271.
11. O. Eriksson, M. Laughlin, M. Brom, P. Nuutila, M. Roden, A. Hwa, R. Bonadonna and M. Gotthardt, *Diabetologia*, 2016, **59**, 1340.
12. B. Hellman, *Nature*, 1959, **184**, 1498.
13. A. Pisania, G. C. Weir, J. J. O'Neil, A. Omer, V. Tchipashvili, J. Lei, C. K. Colton and S. Bonner-Weir, *Lab. Investig.*, 2010, **90**, 1661.
14. E. Christ, D. Wild, S. Ederer, M. Béhé, G. Nicolas, M. E. Caplin, M. Brändle, T. Clerici, S. Fischli, C. Stettler, P. J. Ell, J. Seufert, B. Gloor, A. Perren, J. C. Reubi and F. Forrer, *Lancet Diabetes Endocrinol.*, 2013, **1**, 115.
15. M. Brom, W. Woliner-Van Der Weg, L. Joosten, C. Frielink, T. Bouckenooghe, P. Rijken, K. Andralojc, B. J. Göke, M. De Jong,

D. L. Eizirik, M. Béhé, T. Lahoutte, W. J. G. Oyen, C. J. Tack, M. Janssen, O. C. Boerman and M. Gotthardt, *Diabetologia*, 2014, **57**, 950.

16. O. Eriksson, T. Haack, Y. Hijazi, L. Teichert, V. Tavernier, I. Laitinen, J. E. Berglund, G. Antoni, I. Velikyan, L. Johansson, S. Pierrou, M. Wagner and J. Tillner, *Sci. Rep.*, 2020, **10**, 16758.
17. H. Örlefors, A. Sundin, H. Ahlström, P. Bjurling, M. Bergström, A. Lilja, B. Långström, K. Öberg and B. Eriksson, *J. Clin. Oncol.*, 1998, **16**, 2534.
18. O. Eriksson, R. K. Selvaraju, L. Johansson, J. W. Eriksson, A. Sundin, G. Antoni, J. Sörensen, B. Eriksson and O. Korsgren, *J. Nucl. Med.*, 2014, **55**, 460.
19. O. Eriksson, D. Espes, R. K. Selvaraju, E. Jansson, G. Antoni, J. Sorensen, M. Lubberink, A. R. Biglarnia, J. W. Eriksson, A. Sundin, H. Ahlström, B. Eriksson, L. Johansson, P. O. Carlsson and O. Korsgren, *Diabetes*, 2014, **63**, 3428.
20. O. Eriksson, R. Selvaraju, T. Eich, M. Willny, T. B. Brismar, L. Carlbom, H. Ahlström, G. Tufvesson, T. Lundgren and O. Korsgren, *Diabetes*, 2016, **65**, 2482.
21. L. Carlbom, D. Espes, M. Lubberink, M. Martinell, L. Johansson, H. Ahlström, P. O. Carlsson, O. Korsgren and O. Eriksson, *Diabetes*, 2017, **66**, 1286.
22. D. Espes, P. O. Carlsson, R. K. Selvaraju, M. Rosestedt, P. Cheung, H. Ahlström, O. Korsgren and O. Eriksson, *Diabetes*, 2021, **70**, 966.
23. O. Eriksson, P. Johnström, Z. Cselenyi, M. Jahan, R. K. Selvaraju, M. Jensen-Waern, A. Takano, M. S. Winzell, C. Halldin, S. Skrtic and O. Korsgren, *Diabetes*, 2018, **67**, 182.
24. J. Jiráček, L. Žáková and A. Marek, *J. Labelled Compd. Radiopharm.*, 2020, **63**, 576.
25. Y. Seino, M. Fukushima and D. Yabe, *J. Diabetes Invest.*, 2010, **1**, 8.
26. J. Tillner, M. G. Posch, F. Wagner, L. Teichert, Y. Hijazi, C. Einig, S. Keil, T. Haack, M. Wagner, M. Bossart and P. J. Larsen, *Diabetes, Obes. Metab.*, 2019, **21**, 120.
27. J. P. Frias, M. A. Nauck, J. Van, M. E. Kutner, X. Cui, C. Benson, S. Urva, R. E. Gimeno, Z. Milicevic, D. Robins and A. Haupt, *Lancet*, 2018, **392**, 2180.
28. B. Finan, B. Yang, N. Ottaway, D. L. Smiley, T. Ma, C. Clemmensen, J. Chabenne, L. Zhang, K. M. Habegger, K. Fischer, J. E. Campbell, D. Sandoval, R. J. Seeley, K. Bleicher, S. Uhles, W. Riboulet, J. Funk, C. Hertel, S. Belli, E. Sebokova, K. Conde-Knape, A. Konkar, D. J. Drucker, V. Gelfanov, P. T. Pfluger, T. D. Müller, D. Perez-Tilve, R. D. DiMarchi and M. H. Tschöp, *Nat. Med.*, 2015, **21**, 27.
29. J. Eng, P. C. Andrews, W. A. Kleinman, L. Singh and J. P. Raufman, *J. Biol. Chem.*, 1990, **265**, 20259.
30. M. Gotthardt, G. Lalyko, J. van Eerd-Vismale, B. Keil, T. Schurrat, M. Hower, P. Laverman, T. M. Behr, O. C. Boerman, B. Göke and M. Béhé, *Regul. Pept.*, 2006, **137**, 162.

31. D. Wild, M. Béhé, A. Wicki, D. Storch, B. Waser, M. Gotthardt, B. Keil, G. Christofori, J. C. Reubi and H. R. Mäcke, *J. Nucl. Med.*, 2006, **47**, 2025.
32. A. Wicki, D. Wild, D. Storch, C. Seemayer, M. Gotthardt, M. Behe, S. Kneifel, M. J. Mihatsch, J. C. Reubi, H. R. Mäcke and G. Christofori, *Clin. Cancer Res.*, 2007, **13**, 3696.
33. M. Brom, W. J. G. Oyen, L. Joosten, M. Gotthardt and O. C. Boerman, *Eur. J. Nucl. Med. Mol. Imaging*, 2010, **37**, 1345.
34. D. Wild, A. Wicki, R. Mansi, M. Béhé, B. Keil, P. Bernhardt, G. Christofori, P. J. Ell and H. R. Mäcke, *J. Nucl. Med.*, 2010, **51**, 1059.
35. M. Brom, L. Joosten, W. J. G. Oyen, M. Gotthardt and O. C. Boerman, *Contrast Media Mol. Imaging*, 2012, **7**, 160.
36. H. Kimura, H. Matsuda, Y. Ogawa, H. Fujimoto, K. Toyoda, N. Fujita, K. Arimitsu, K. Hamamatsu, Y. Yagi, M. Ono, N. Inagaki and H. Saji, *Bioorg. Med. Chem.*, 2017, **25**, 1406.
37. H. Kimura, N. Fujita, K. Kanbe, H. Matsuda, H. Watanabe, K. Arimitsu, H. Fujimoto, K. Hamamatsu, Y. Yagi, M. Ono, N. Inagaki and H. Saji, *Bioorg. Med. Chem.*, 2017, **25**, 5772.
38. B. Janota, U. Karczmarczyk, E. Laszuk, P. Garnuszek and R. Mikołajczak, *Nucl. Med. Rev.*, 2016, **19**, 104.
39. D. Seo, B. L. Faintuch, E. Aparecida de Oliveira and J. Faintuch, *Nucl. Med. Biol.*, 2017, **49**, 57.
40. M. Gotthardt, M. Fischer, I. Naeher, J. B. Holz, H. Jungclas, H. W. Fritsch, M. Béhé, B. Göke, K. Joseph and T. M. Behr, *Eur. J. Nucl. Med.*, 2002, **29**, 597.
41. D. Wild, H. Mäcke, E. Christ, B. Gloor and J. C. Reubi, *N. Engl. J. Med.*, 2008, **359**, 766.
42. E. Christ, D. Wild, F. Forrer, M. Brändle, R. Sahli, T. Clerici, B. Gloor, F. Martius, H. Maecke and J. C. Reubi, *J. Clin. Endocrinol. Metab.*, 2009, **94**, 4398.
43. A. Sowa-Staszczak, D. Pach, R. Mikołajczak, H. Mäcke, A. Jabrocka-Hybel, A. Stefańska, M. Tomaszuk, B. Janota, A. Gilis-Januszewska, M. Małecki, G. Kamiński, A. Kowalska, J. Kulig, A. Matyja, C. Osuch and A. Hubalewska-Dydejczyk, *Eur. J. Nucl. Med. Mol. Imaging*, 2013, **40**, 524.
44. V. Medina-García, B. E. Ocampo-García, G. Ferro-Flores, C. L. Santos-Cuevas, L. Aranda-Lara, R. García-Becerra, D. Ordaz-Rosado and L. Melendez-Alafort, *Nucl. Med. Biol.*, 2015, **42**, 911.
45. R. K. Selvaraju, I. Velikyan, L. Johansson, Z. Wu, I. Todorov, J. Shively, F. Kandeel, O. Korsgren and O. Eriksson, *J. Nucl. Med.*, 2013, **54**, 1458.
46. O. Eriksson, I. Velikyan, R. K. Selvaraju, F. Kandeel, L. Johansson, G. Antoni, B. Eriksson, J. Sörensen and O. Korsgren, *J. Clin. Endocrinol. Metab.*, 2014, **99**, 1519.
47. M. Kirsi, C. Bin Yim, F. Veronica, I. Tamiko, E. Viki-Veikko, R. Johan, J. Jori, S. Tiina, T. Tuula, T. Marko, G. Eleni, B. Martin, G. Martin, R. J. Claude, M. Helmut, R. Anne, S. Olof and N. Pirjo, *Mol. Imaging Biol.*, 2014, **16**, 255.

48. A. Rydén, G. Nyman, L. Nalin, S. Andreasson, I. Velikyan, O. Korsgren, O. Eriksson and M. Jensen-Waern, *Nucl. Med. Biol.*, 2016, **43**, 742.
49. A. Rydén, G. Nyman, L. Nalin, S. Andreasson, O. Korsgren, O. Eriksson and M. Jensen-Waern, *Nucl. Med. Biol.*, 2016, **43**, 397.
50. O. Eriksson, U. Rosenström, R. K. Selvaraju, B. Eriksson and I. Velikyan, *Acta Diabetol.*, 2017, **54**, 1039.
51. Z. Wu, I. Todorov, L. Li, J. R. Bading, Z. Li, I. Nair, K. Ishiyama, D. Colcher, P. E. Conti, S. E. Fraser, J. E. Shively and F. Kandeel, *Bioconjugate Chem.*, 2011, **22**, 1587.
52. C. Bin Yim, K. Mikkola, V. Fagerholm, V. V. Elomaa, T. Ishizu, J. Rajander, J. Schlesinger, A. Roivainen, P. Nuutila and O. Solin, *Nucl. Med. Biol.*, 2013, **40**, 1006.
53. A. Jodal, B. Lankat-Buttgereit, M. Brom, R. Schibli and M. Béhé, *EJNMMI Res.*, 2014, **4**, 1.
54. Z. Wu, S. Liu, I. Nair, K. Omori, S. Scott, I. Todorov, J. E. Shively, P. S. Conti, Z. Li and F. Kandeel, *Theranostics*, 2014, **4**, 770.
55. H. Gao, G. Niu, M. Yang, Q. Quan, Y. Ma, E. N. Murage, J. M. Ahn, D. O. Kiesewetter and X. Chen, *Mol. Pharming*, 2011, **8**, 1775.
56. E. J. Keliher, T. Reiner, G. M. Thurber, R. Upadhyay and R. Weissleder, *ChemistryOpen*, 2012, **1**, 177.
57. K. Mikkola, C. Bin Yim, P. Lehtiniemi, S. Kauhanen, M. Tarkia, T. Tolvanen, P. Nuutila and O. Solin, *EJNMMI Res.*, 2016, **6**, 91.
58. H. Kimura, Y. Ogawa, H. Fujimoto, E. Mukai, H. Kawashima, K. Arimitsu, K. Toyoda, N. Fujita, Y. Yagi, K. Hamamatsu, T. Murakami, A. Murakami, M. Ono, Y. Nakamoto, K. Togashi, N. Inagaki and H. Saji, *Bioorg. Med. Chem.*, 2018, **26**, 463.
59. D. O. Kiesewetter, N. Guo, J. Guo, H. Gao, L. Zhu, Y. Ma, G. Niu and X. Chen, *Theranostics*, 2012, **2**, 999.
60. D. O. Kiesewetter, H. Gao, Y. Ma, G. Niu, Q. Quan, N. Guo and X. Chen, *Eur. J. Nucl. Med. Mol. Imaging*, 2012, **39**, 463.
61. Y. Wang, K. Lim, M. Normandin, X. Zhao, G. W. Cline and Y. S. Ding, *Nucl. Med. Biol.*, 2012, **39**, 167.
62. Z. Wu, S. Liu, M. Hassink, I. Nair, R. Park, L. Li, I. Todorov, J. M. Fox, Z. Li, J. E. Shively, P. S. Conti and F. Kandeel, *J. Nucl. Med.*, 2013, **54**, 244.
63. H. Wu, S. Liang, S. Liu, Y. Pan, D. Cheng and Y. Zhang, *Nucl. Med. Commun.*, 2013, **34**, 701.
64. Y. Xu, D. Pan, Q. Xu, C. Zhu, L. Wang, F. Chen, R. Yang, S. Luo and M. Yang, *J. Cancer Res. Clin. Oncol.*, 2014, **140**, 1479.
65. X. Yue, X. Yan, C. Wu, G. Niu, Y. Ma, O. Jacobson, B. Shen, D. O. Kiesewetter and X. Chen, *Mol. Pharming*, 2014, **11**, 3875.
66. Q. Xu, C. Zhu, Y. Xu, D. Pan, P. Liu, R. Yang, L. Wang, F. Chen, X. Sun, S. Luo and M. Yang, *J. Drug Targeting*, 2015, **23**, 813.
67. A. Bauman, I. E. Valverde, C. A. Fischer, S. Vomstein and T. L. Mindt, *J. Nucl. Med.*, 2015, **56**, 1569.

68. E. Mukai, K. Toyoda, H. Kimura, H. Kawashima, H. Fujimoto, M. Ueda, T. Temma, K. Hirao, K. Nagakawa, H. Saji and N. Inagaki, *Biochem. Biophys. Res. Commun.*, 2009, **389**, 523.
69. B. Waser and J. C. Reubi, *Eur. J. Nucl. Med. Mol. Imaging*, 2014, **41**, 1166.
70. B. Waser and J. C. Reubi, *Eur. J. Nucl. Med. Mol. Imaging*, 2011, **38**, 1054.
71. J. Lv, Y. Pan, X. Li, D. Cheng, S. Liu, H. Shi and Y. Zhang, *PLoS One*, 2014, **9**, e96833.
72. T. Läppchen, R. Tönnesmann, J. Eersels, P. T. Meyer, H. R. Maecke and S. N. Rylova, *PLoS One*, 2017, **12**, 170435.
73. K. Antwi, M. Fani, G. Nicolas, C. Rottenburger, T. Heye, J. Claude Reubi, B. Gloor, E. Christ and D. Wild, *J. Nucl. Med.*, 2015, **56**, 1075.
74. Y. Luo, M. Yu, Q. Q. Pan, W. Wu, T. Zhang, D. O. Kiesewetter, Z. Zhu, F. Li, X. Chen and Y. Zhao, *Eur. J. Nucl. Med. Mol. Imaging*, 2015, **42**, 531.
75. I. Velikyan, T. N. Bulenga, R. Selvaraju, M. Lubberink, D. Espes, U. Rosenström and O. Eriksson, *Am. J. Nucl. Med. Mol. Imaging*, 2015, **5**, 109.
76. I. Velikyan and O. Eriksson, *Theranostics*, 2020, **10**, 437.
77. I. Velikyan, U. Rosenstrom and O. Eriksson, *Am. J. Nucl. Med. Mol. Imaging*, 2017, **7**, 111.
78. L. Nalin, R. K. Selvaraju, I. Velikyan, M. Berglund, S. Andréasson, A. Wikstrand, A. Rydén, M. Lubberink, F. Kandeel, G. Nyman, O. Korsgren, O. Eriksson and M. Jensen-Waern, *Eur. J. Nucl. Med. Mol. Imaging*, 2014, **41**, 1800.
79. R. K. Selvaraju, I. Velikyan, V. Asplund, L. Johansson, Z. Wu, I. Todorov, J. Shively, F. Kandeel, B. Eriksson, O. Korsgren and O. Eriksson, *Nucl. Med. Biol.*, 2014, **41**, 471.
80. R. K. Selvaraju, T. N. Bulenga, D. Espes, M. Lubberink, J. Sörensen, B. Eriksson, S. Estrada, I. Velikyan and O. Eriksson, *Am. J. Nucl. Med. Mol. Imaging*, 2015, **5**, 259.
81. T. Sako, K. Hasegawa, M. Nishimura, Y. Kanayama, Y. Wada, E. Hayashinaka, Y. Cui, Y. Kataoka, M. Senda and Y. Watanabe, *Biochem. Biophys. Res. Commun.*, 2013, **442**, 79.
82. B. Manandhar and J. M. Ahn, *J. Med. Chem.*, 2015, **58**, 1020.
83. M. Zhang, O. Jacobson, D. O. Kiesewetter, Y. Ma, Z. Wang, L. Lang, L. Tang, F. Kang, H. Deng, W. Yang, G. Niu, J. Wang and X. Chen, *Bioconjugate Chem.*, 2019, **30**, 1745.
84. A. R. Jalilian, M. Jouiaei, A. R. Doroudi, F. Bolourinovin and J. Garousi, *Nukleonika*, 2010, **55**, 219.
85. O. Eriksson, I. Velikyan, T. Haack, M. Bossart, I. Laitinen, P. J. Larsen, J.-E. Berglund, G. Antoni, L. Johansson, S. Pierrou, J. Tillner and M. Wagner, *J. Nucl. Med.*, 2021, **62**, 833.
86. O. Eriksson, I. Velikyan, T. Haack, M. Bossart, A. Evers, I. Laitinen, P. J. Larsen, O. Plettenburg, A. Takano, C. Halldin, G. Antoni, L. Johansson, S. Pierrou and M. Wagner, *Sci. Rep.*, 2019, **9**, 14960.

87. M. Wagner, J. G. Doverfjord, J. Tillner, G. Antoni, T. Haack, M. Bossart, I. Laitinen, L. Johansson, S. Pierrou, O. Eriksson and I. Velikyan, *Pharmaceuticals*, 2020, **13**, 1.
88. I. Velikyan, T. Haack, M. Bossart, A. Evers, I. Laitinen, P. Larsen, O. Plettenburg, L. Johansson, S. Pierrou, M. Wagner and O. Eriksson, *EJNMMI Res.*, 2019, **9**, 17.
89. T. D. Müller, B. Finan, C. Clemmensen, R. D. Di Marchi and M. H. Tschöp, *Physiol. Rev.*, 2017, **97**, 721.
90. K. E. Mayo, L. J. Miller, D. Bataille, S. Dalle, B. Göke, B. Thorens and D. J. Drucker, *Pharmacol. Rev.*, 2003, **55**, 167.
91. T. Yaqub, I. G. Tikhonova, J. Lättig, R. Magnan, M. Laval, C. Escrieut, C. Boulègue, C. Hewage and D. Fourmy, *Mol. Pharmacol.*, 2010, **77**, 547.
92. E. Gourni, B. Waser, P. Clerc, D. Fourmy, J. C. Reubi and H. R. Maecke, *J. Nucl. Med.*, 2014, **55**, 976.
93. S. M. A. Willekens, L. Joosten, O. C. Boerman, M. Brom and M. Gotthardt, *Sci. Rep.*, 2018, **8**, 2948.
94. O. Eriksson, I. Velikyan, T. Haack, M. Bossart, A. Evers, K. Lorenz, I. Laitinen, P. J. Larsen, O. Plettenburg, L. Johansson, S. Pierrou and M. Wagner, *Diabetes*, 2021, **70**, 842.
95. M. Christensen, L. Vedtofte, J. J. Holst, T. Vilsbøll and F. K. Knop, *Diabetes*, 2011, **60**, 3103.
96. R. Suzuki, G. A. Brown, J. A. Christopher, C. C. G. Scully and M. Congreve, *J. Med. Chem.*, 2020, **63**, 905.
97. T. Tang, C. Hartig, Q. Chen, W. Zhao, A. Kaiser, X. Zhang, H. Zhang, H. Qu, C. Yi, L. Ma, S. Han, Q. Zhao, A. G. Beck-Sickinger and B. Wu, *Nat. Commun.*, 2021, **12**, 737.
98. M. C. Riddle, R. Nahra, J. Han, J. Castle, K. Hanavan, M. Hompesch, D. Huffman, P. Strange and P. Ohman, *Diabetes Care*, 2018, **41**, 2346.
99. D. Tornehave, P. Kristensen, J. Rømer, L. B. Knudsen and R. S. Heller, *J. Histochem. Cytochem.*, 2008, **56**, 841.
100. S. Ståhl, T. Gräslund, A. Eriksson Karlström, F. Y. Frejd, P. Å. Nygren and J. Löfblom, *Trends Biotechnol.*, 2017, **35**, 691.
101. S. Muyldermans, *FEBS J.*, 2021, **288**, 2084.
102. C. R. Ban and S. M. Twigg, *Vasc. Health Risk Manage.*, 2008, **4**, 575.
103. A. J. Lay, H. E. Zhang, G. W. McCaughan and M. D. Gorrell, *Front. Biosci.-Landmark*, 2019, **24**, 1.
104. J. Lambrecht, S. Verhulst, I. Mannaerts, J. P. Sowa, J. Best, A. Canbay, H. Reynaert and L. A. van Grunsven, *EBioMedicine*, 2019, **43**, 501.
105. A. Hatori, J. Yui, L. Xie, K. Kumata, T. Yamasaki, M. Fujinaga, H. Wakizaka, M. Ogawa, N. Nengaki, K. Kawamura, F. Wang and M. R. Zhang, *Sci. Rep.*, 2015, **5**, 17327.
106. T. Shao, Z. Chen, V. Belov, X. Wang, S. H. Rwema, V. Kumar, H. Fu, X. Deng, J. Rong, Q. Yu, L. Lang, W. Lin, L. Josephson, A. E. Samir, X. Chen, R. T. Chung and S. H. Liang, *J. Hepatol.*, 2020, **73**, 161.

107. P. Désogère, L. F. Tapias, L. P. Hariri, N. J. Rotile, T. A. Rietz, C. K. Probst, F. Blasi, H. Day, M. Mino-Kenudson, P. Weinreb, S. M. Violette, B. C. Fuchs, A. M. Tager, M. Lanuti and P. Caravan, *Sci. Transl. Med.*, 2017, **9**, eaaf4696.
108. J. Muzard, L. Sarda-Mantel, S. Loyau, A. Meulemans, L. Louedec, C. Bantsimba-Malanda, F. Hervatin, J. Marchal-Somme, J. B. Michel, D. Le Guludec, P. Billiald and M. Jandrot-Perrus, *PLoS One*, 2009, **4**, e5585.
109. M. Rosestedt, I. Velikyan, U. Rosenström, S. Estrada, O. Åberg, J. Weis, C. Westerlund, S. Ingvast, O. Korsgren, P. Nordeman and O. Eriksson, *Nucl. Med. Biol.*, 2021, **93**, 54.
110. R. D. Badawi, H. Shi, P. Hu, S. Chen, T. Xu, P. M. Price, Y. Ding, B. A. Spencer, L. Nardo, W. Liu, J. Bao, T. Jones, H. Li and S. R. Cherry, *J. Nucl. Med.*, 2019, **60**, 299.

CHAPTER 10

Invasive and Implantable Glucose Sensors: Perspective for the Artificial Pancreas

OMAR DIOURI[a,b] AND ERIC RENARD*[a,b]

[a] Department of Endocrinology, Diabetes, Nutrition, Montpellier University Hospital, Montpellier, France; [b] Institute of Functional Genomics, CNRS, INSERM, University of Montpellier, Montpellier, France
*Email: e-renard@chu-montpellier.fr

10.1 Introduction

The advent and increased use of continuous glucose monitoring (CGM) devices over the last decade have led to a significant evolution in the therapeutic management of diabetes thanks to the tremendous amount of data continuously collected throughout the day. The timely detection of glucose variations has enabled the development of hybrid closed-loop insulin delivery systems, in which an insulin pump is driven by a control algorithm receiving CGM data, which entered the commercialization phase in the last few years. However, the root causes for keeping these closed-loop systems as "hybrid" are the persistent limitations in the performance of glucose sensing and those of insulin kinetics.[1,2] These limitations kept the need for user announcement of meals and physical activity in order to overcome delays in glucose sensing and insulin action. The road toward a fully automated insulin delivery system without user intervention is clearly dependent on improving glucose sensor performances. Indeed, the detection of rapid

Nanoscience & Nanotechnology Series No. 59
Nanotechnology for Diabetes Management
Edited by Amar Abderrahmani, Sabine Szunerits, Rabah Boukerroub and Abdelfattah El Ouaamari

Published by the Royal Society of Chemistry, www.rsc.org

glucose variations with a delay of between 10 and 20 min[3] makes controllers inefficient to automatically react to a rapid glucose rise due to a meal consumption, or a sudden drop caused by a physical activity.[4] This sensing delay combined with that of insulin action from so-called fast-acting insulin analogues[5] leads to mandatory user interactions with the system. Improved sensor accuracy is also of utmost importance for the advent of fully automated closed loops, because measurement errors, specifically at low and high glucose levels, have a direct impact on closed-loop performances.[6,7] Besides performances, usability, ease of use, costs, and pain at insertion, are among the multiple aspects that limit the wider use of continuous glucose monitoring devices (CGM) amongst people with type 1 diabetes, and the future adoption of closed-loop systems.

In this chapter, we present the latest research progress made on invasive and implantable glucose sensors, highlighting the techniques and strategies that could help in the next few years to overcome these barriers in terms of performance, sensor delay, and acceptability of the devices, which are essential in the technological road towards an artificial pancreas.

10.2 Needle Coating Strategies

Minimally invasive needle-type sensors represent most of the commercially available CGM devices. These devices use the glucose oxidase (GOx) reaction to measure interstitial glucose,[8] by generating a current on an electrode which is highly correlated to the interstitial glucose level thanks to a linear relation ranging from 2 $mmol\,L^{-1}$ (40 $mg\,dl^{-1}$) up to 40 $mmol\,L^{-1}$ (400 $mg\,dl^{-1}$).[9] Different algorithms have been developed to treat the sensor signal in order to reduce calibration needs and increase glucose-sensing performances.[10,11] With the sensor lag time being a major constraint of interstitial glucose monitoring, some algorithms also have been explored to reduce it.[12]

In terms of new sensor prototypes, efforts were mainly oriented toward new coating strategies, that can mitigate foreign body response, which is the main limitation to long-term use of actual devices.

A first prototype used a double coating to increase GOx needle performances.[12] A first porous polyvinylidene fluoride membrane was used to filter excessive glucose reaching the enzyme layer and to extend the sensor linear range, and a second nanosphere Nafion membrane allowed interferences to be reduced. This strategy allowed a glucose linear range from 0 to 20 $mmol\,L^{-1}$, which remains insufficient to catch higher values for type 1 diabetes. *In vitro* experiments showed excellent linearity, sensitivity, and selectivity. Sensitivity was also evaluated over time using a bovine serum, and showed good stability, with a reduction to 75% of sensitivity after 21 days, while it dropped below 50% for a non-coated sensor. An *in vivo* test was also performed, showing a good response to blood glucose variations after glucose and insulin injections.

Burugapalli *et al.* proposed different prototypes[13] with polyurethane and gelatin coaxial membranes, to prevent fibrous encapsulation and increase sensor lifetime. *In vivo* testing was performed in 12 rats over 9 weeks, with a continuous measurement compared to a blood glucometer. The results showed that the proposed coating prevented fibrous capsule formation and allowed good sensor sensitivity for up to 3 weeks. The authors also investigated the sensor delay response, which increased from 10 minutes at the first week's test, to around 15 minutes at the third week. This increase could be due to tissue formation around the sensor. The best prototype with gelatin also showed the best stability over time, with a controlled reduction of sensitivity compared to others.

Another prototype[14] consisting of polyaniline nanofibers for enzyme immobilization coated by polyurethane/epoxy-enhanced polyurethane double membranes was presented. The sensitivity was evaluated by keeping the sensors in bovine serum for 66 days, and showed an increase of 30% in the first 13 days, remaining stable for the following days. Three prototypes were then implanted in three rats, and showed similar sensitivity. Sensor delay was evaluated to 8 minutes, with 1 minute ascribable to the sensor, and succeeded to work continuously for at least 26 days, indicating its potential for long-term continuous glucose monitoring.

Cu nanoflowers were also used to produce a flexible needle prototype,[15] with a Nafion layer and polyurethane membrane like in the sensors described above that were deposited over a Pt microelectrode. The designed sensor was evaluated *in vivo* in bovine serum and achieved good selectivity and an increase in sensitivity in the first 6 days, then remained stable up to 20 days. *In vivo* evaluation was made in rats for only a few hours: after a 2-hour stabilization period, a glucose injection was performed. The sensor exhibited a good response to glucose variation, with a delay of 5 minutes compared to the measurements performed by a glucometer.

Shoenfisch *et al.*[16] investigated in a series of works the role of nitric oxide to reduce the foreign body response to needle-type sensors insertion. In their last work, the sensors were evaluated in a swine model for 28 days, where the sensors released nitric oxide over 14 or 30 days. The active release of nitric oxide succeeded in maintaining a good glucose monitoring accuracy for 3 weeks, with a mean absolute relative difference (MARD) below 15%. This metrics is frequently used to assess the performances of CGM devices and represents the average absolute error between all sensor values and reference values, and is between 9% and 15% for commercialized CGM devices.

These developments give us a glimpse of what the upgrades are that major CGM companies are working on to improve their enzymatic sensors. New strategies of coating may help to increase sensor lifetime from 7 to 14 days for available devices, to longer periods of up to 1 month. This lifetime extension helps to reduce the number of sensors used per year and the painful insertions, reducing the burden for patients and the economic and ecologic impacts of such devices. While algorithms were developed to reduce the number of calibrations needed and to compensate for sensor delay, these developments show a possibility of increasing the sensor stability over time, indicating that calibration can be avoided, like in the currently available

Freestyle Libre (Abbott, USA) or Dexcom G6[17] (Dexcom, USA). Even though the different techniques presented above can help to reduce the processing time and the delay between fluid capture and voltage measurement, the physiological delay between interstitial glucose and blood glucose seems difficult to reduce with subcutaneous needles. The lag-time reduction observed in some experiments needs to be explored with human subjects and compared to actual devices before reaching the final conclusions.

10.3 Microneedles

While flexible needles are commonly used in CGMs, microneedles may be revealed to be game-changers. In contrast to the typical CGMs that are based on needles inserted in the subcutaneous tissue, microneedles address the dermis as illustrated in Figure 10.1, where interstitial fluid (ISF) is more abundant and matches blood glucose dynamics better than subcutaneous tissue.[18] Furthermore, the reduced invasiveness of the devices could increase their acceptability, and also minimize issues due to foreign body response which reduce sensor lifetime. The abundance of interstitial fluid in dermis and the reduced size of the sensor may also help to reduce sensor delay.

A three-electrode enzymatic sensor has been designed[19,20] with a microneedle length of under 1 mm, allowing dermis interstitial glucose measurement, and resulting in an almost 40-fold size reduction compared to actual sensors. *In vitro* evaluation revealed good sensitivity, an excellent linearity up to 200 $mg\,dL^{-1}$ and specificity. *In vivo* evaluation on one subject was performed with a prototype without coated membranes. After insertion, a glucose tolerance test was performed. This proof-of-concept sensor showed good linearity and a measurement delay around 10 min compared to capillary blood glucose measurement.

This microneedle prototype was also designed with a novel coating technique that allowed mixing a GOx layer, a tetrathiafulvalene layer, and a glutaraldehyde layer.[21] This innovative technology has been reported as cost-effective and compatible with mass production, and more sensitive compared to other coating methods. An *in vivo* testing with a patch composed of 32 microneedles of 1.5 mm in length dispatched on a surface of 2×2 cm area was performed on healthy volunteers and showed good linearity with capillary blood glucose measurements, however there was an observed 20-min lag time.

Teymourian *et al.*[22] addressed another challenge, by combining parallel continuous monitoring of glucose and ketone bodies, or lactate and ketone bodies. Measuring ketone bodies alongside glucose could be of the greatest interest, since it allows to detect a lack of insulin. A systematic detection of ketone bodies could provide an additional input to closed-loop algorithms and inform patients in the case of catheter occlusion while discriminating hyperglycemic situations resulting from meals. A prototype consisting of four hollow microneedles of 1 mm in length was evaluated *in vitro* and showed an excellent accuracy for ketone bodies and glucose detection,

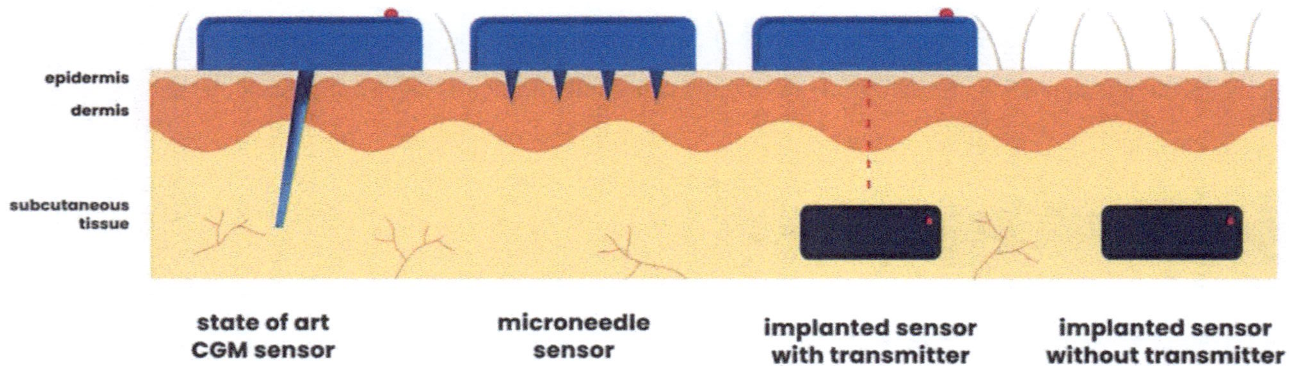

Figure 10.1 Representation of different types of glucose sensors.

alongside stability against pH variations and acetyl acetate that occur during diabetic ketoacidosis events.

A second multi-analyte microneedle sensor for lactate and glucose monitoring was evaluated *in vitro* using artificial interstitial fluid.[23] This sensor featured good accuracy and selectivity and the capacity to simultaneously measure both lactates and glucose, with however a small glucose linear range up to 5 mmol L^{-1}.

More recently, a silk/polyols/GOD sensor was prototyped,[24] with an 800 μm microneedle length. *In vitro* evaluation on serum showed good signal stability for 24 hours, a storage stability up to 5 weeks, and a linear range from 1.7 to 10 mmol L^{-1}.

Another enzyme-free device using a porous Pt-black microneedle electrode with nafion coating was evaluated.[25] The sensor array had a size of 600×100×150 μm, and was based on a direct electrochemical oxidation of glucose on the nanoporous surface of the sensor. *In vitro* assessment on artificial interstitial fluid showed an extended linear range up to 30 mmol L^{-1}, suitable for type 1 diabetes monitoring, with good stability and selectivity.

As already mentioned, only some of these prototypes were subject to preliminary *in vivo* testing on animals or healthy subjects. These evaluations failed to show a reduction in sensor lag time as was expected, but these first evaluations are not decisive, and must be compared during the same experiments and in the same subjects to a classical CGM sensor. However, microneedle technology has been explored by some companies such as PkVitality[26] (France) or Biolinq[27] (USA) that may release their first commercial devices in a few years. As shown in the papers described above, microneedle sensors could increase CGM use thanks to their reduced invasiveness, cost reduction, and increased lifetime. These characteristics are key for users' and payers' acceptability.

10.4 Fully Implantable Sensors

A first fully implantable glucose sensor was recently launched in the market by Senseonics[28] (USA), using a fluorescence technique. This first device was composed of one implantable part inside subcutaneous tissue, and a receiver outside. The device was approved for a lifetime of 6 months. It required minimally invasive surgery for implant and explant procedures, and succeeded to obtain sensor performances similar to those of usual CGM devices.[29] Implantable strategies have been widely explored in the last few years, with devices aiming at a longer lifetime and a smaller size to reduce negative impacts of foreign body response. This evolution may also help to increase sensor performances as observed in following studies.

Dou *et al.*[30] proposed a new approach, by immobilizing a nano-boronic acid-hydrogel membrane onto a quartz crystal, then using the nature of quartz crystal microbalance that allows measuring glucose, at a detection range from 1 to 30 mmol L^{-1}. A series of evaluations were made, including *in vivo* tests in rats, that showed good selectivity against other types of molecules, good linearity, and no cytotoxicity.

One of the most interesting papers sponsored by the company Integrated Medical Sensors[31] (USA) presents a subcutaneous electrochemical sensor with a size of 600 μm×3,000 μm×100 μm.[32] The size of this sensor is smaller than that of currently implanted sensors in the market. The reduced size provides a big advantage as it allows easy insertion through a needle, reduces the foreign body response, and minimizes microvascular damage, resulting in a reduced blood–ISF lag. The sensor seeds communicate with an external watch receiver through Bluetooth. Extended evaluation of the device was performed in swine and rats, with a comparison to a Dexcom G5 and an Abbott Freestyle Libre. Experiments showed a reduced lag time of below 5 minutes, while it was between 12 and 15 minutes for commercial devices. The prototype also displayed an excellent accuracy even under 40 $mg\,dL^{-1}$, and a MARD of 12% *vs.* 13% for G5 and Freestyle Libre, and a lifetime of 1 month. These performances, specifically accuracy at low glycemia level and reduced lag time, are key to increasing the closed-loop performance and the ability of controllers to react rapidly to acute glycemic variations.

A non-enzymatic glucose sensor was also proposed, using a fluorescent dye hydrogel, combined with a LED that excites the glucose-responsive fluorescent dye, and a photodiode that detects the response.[33] The prototype succeeded in maintaining its performance for 45 days, a duration that can be extended by improving the electric circuit. The implantable sensor, using a relatively similar technology to the Eversense (Senseonics, USA) device, achieved a better MARD in rats and swine models than commercial CGM devices.

A polymer-dot transducer was proposed to detect oxygen consumption resulting from Gox reaction, allowing the measurement of optically interstitial glucose through a subcutaneous injection.[34–36] *In vivo* performance and safety tests were performed in mice and showed excellent linearity with a large dynamic range up to 20 $mmol\,L^{-1}$, and a lifetime of 30 days with only a slight intensity reduction. After 30 days, mice were sacrificed and no organ damage or inflammation was observed, as well as the lack of nanoparticle sensor diffusion outside the injection site.

Profusa (USA) presented in a paper published in 2015 a first preclinical evaluation of its fully implantable sensor.[37] This device encapsulates a Gox sensor and long-lifetime phosphors inside poly(HEMA-*co*-acrylamide) hydrogel. The implanted sensors generate luminescence depending on oxygen coming from Gox reduction, which is detected by an external optical reader. The first pre-clinical evaluation performed in mini-pigs showed the ability of the prototype to monitor the blood glucose level from 0 to 400 $mg\,dL^{-1}$, with however response differences between the sensors implanted, indicating that the location of the sensor may have an impact on its performance.

A few years later, Profusa (USA) continued its search for methods to produce an implantable device[38] and presented its latest results during the

ATTD 2021 meeting.[39] The permanently implanted sensor with a size of 2.5×0.4×0.3 mm was evaluated in 12 patients with diabetes over 3 months, each of them receiving two sensors. Preliminary results after 10 weeks of sensor use revealed a MARD around 10.1% and Parkes Error Grid Analysis with 87.3% of values in zone A and 12.7% in zone B, where zones A and B represent the grid glucose measurement errors that have the lowest clinical impact.

These recent works show that implantable sensors may succeed in reducing the time lag and increasing the sensitivity of devices for low blood sugar levels, which are some of the main requirements for closed-loop systems to increase their performances. The miniaturization of devices and the reduced foreign body response help to maintain good microvascularization around the sensor, which plays a role in reducing the physiological lag time, combined to measurement techniques that allow faster sensing.[40] Many companies are working on implantable sensors, including Senseonics (USA), Profusa (USA) and Integrated Medical Sensors (USA), as earlier presented, or Glysens (USA),[41] Glusense (Israel),[42] Indigo Diabetes (Belgium),[43] and Life-Care (Norway),[44] however, these devices are currently under development and almost no information is available on their performances.

10.5 Intraperitoneal Glucose Measurement

As presented above, recent sensor prototypes with dermal or subcutaneous glucose measurement that were evaluated on lag time against blood measurement hardly succeeded in reducing this gap. However, glucose kinetics in peritoneal space is substantially faster,[45] which makes it the best option for sensor location in order to improve sensor performances towards better accuracy in acute situations with a rapid increase or decrease of glucose, as was shown in some experiments where commercial CGMs were placed in intraperitoneal space.[46]

However, despite their potential advantages in terms of lag time reduction, the development of devices specifically for intraperitoneal glucose measurement is so far not under the scope of academic and industrial research teams in contrast with other types of sensors. Only two papers were found in the last 5 years on this topic.

Theranova (USA) presented a proof-of-concept study,[47] using a fluorescent device that was implanted in sheep models in subcutaneous tissue or intraperitoneal space. The device was adapted to allow flushing with saline solution to clean the sensor surface and increase its lifetime, which can only be done in intraperitoneal space. The sensors were implanted for 3 months, and multiple glucose response challenges were performed. Implantation in the intraperitoneal space showed a consistent responsiveness, with a time to maximum between 11 to 17 min, while the subcutaneous implant lost its responsiveness with a time to maximum of between 9 and 72 min. If the protocol does not allow to demonstrate directly the impact of intraperitoneal measurement on the reduction of lag time, it shows the possibility to propose a long-term sensor that maintains its performances. However, the road to an intraperitoneal device for glucose sensing seems longer than that to other technologies.

More recently, Jernelv *et al.* used infrared measurement to evaluate glucose in extracted peritoneal fluid compared to blood glucose, in a preliminary *in vitro* experiment.[48] The results showed that this technique can measure glucose in peritoneal fluid with the same precision. The authors indicated that future work will include *in vivo* measurements performed in living animals.

10.6 Conclusion

The achievement of more performant closed-loop systems will rely greatly on CGM technological development. The integration of nanotechnology in these sensors has opened different paths to improve and explore new approaches to measure blood glucose. Table 10.1 summarizes each main category of CGM device identified in this review.

We have seen in the last few years a continuous development of commercial CGM systems based on flexible needles and GOx enzymatic sensors with longer duration of activity, reaching 14 days for some models. The performances of these systems were also improved in terms of MARD between 8% and 10% in human subjects, and stability thanks to coating membranes and algorithms that suppressed the need for calibration. Most of the work presented here showed an extension of sensor lifetime, thanks to the use of novel coating techniques that mitigate foreign body response.

In parallel to flexible sensors, microneedles have gained a lot of interest, with some startups that may enter in a commercialization phase in the next 2–3 years. The authors have presented many advantages of microneedle use, including reduced invasiveness and skin trauma, and a prolonged sensor lifetime reaching 1 month, and potentially a reduction of costs. Along with these advantages for users, these devices may potentially achieve the same performances as current devices, and eventually reduce slightly the lag time thanks to their smaller size, allowing a reduction of fluid diffusion and analysis time. These devices could also be able to detect simultaneously multiple analytes, like lactates or ketones. Such performance could enhance closed-loop algorithm performances during physical activity, or during hyperglycemic periods.

Senseonics (USA) proposed the first implantable sensor in the market, which is also the only sensor using a technique other than enzymatic reactions. This chapter shows that implantable sensors offer the possibility to exploit different sensing techniques, that may offer faster glucose measurement than the classical enzymatic ones, allowing a reduction in the lag time. Nanotechnologies help also to reduce sensor size, which will allow to propose permanent sensors with implantation without surgery. The size reduction of implanted sensors is key to preserving the microvascularization

Table 10.1 Main characteristics of invasive and implantable glucose-monitoring sensors.

	Flexible coated needles	Microneedles	Implanted sensors	Intraperitoneal sensors
Performance and accuracy	Equivalent to actual devices	Equivalent to actual devices	Increased sensitivity for low values. Better MARD	Increased sensitivity for low values
Lag time	No improvement expected	No reduction measured during first trials. Slight reduction may be expected	First results show a reduction of lag time	Reduced lag time
Sensor lifetime	Increased lifetime up to 3 or 4 weeks	Prolonged lifetime that could be above 1 month	Very extended lifetime, up to a permanent sensor	—
Level of invasiveness	Percutaneous needle up to 6 mm	Less than 1 millimeter under the skin. Very fast skin recovery	Implantable by percutaneous needle, with outside receiver	Highly invasive
Level of development	New coatings to reduce foreign body reactions	Performance evaluations on human volunteers	First pre-clinical trial with patients	Proof-of-concept studies
Industrial development	Most CGMs based on flexible needles	First commercial versions expected in less than 5 years	One implanted sensor already present. New devices expected in less than 5 years	No visibility on future commercial devices

of the subcutaneous tissue near the sensor, which results in better fluidic exchange between interstitial and blood glucose.

Finally, the peritoneal space remains underinvestigated because of the difficulty of access compared to subcutaneous tissue.

It is therefore clearly expected that, in the next few years, new devices based upon microneedles or fully implantable sensors may change the market for CGM devices, as many companies are working on these devices currently.

References

1. R. A. Lal, L. Ekhlaspour, K. Hood and B. Buckingham, *Endocr. Rev.*, 2019, **40**, 1521–1546.
2. V. Gingras, N. Taleb, A. Roy-Fleming, L. Legault and R. Rabasa-Lhoret, *Diabetes Obes. Metab.*, 2018, **20**, 245–256.
3. A. Basu, S. Dube, S. Veettil, M. Slama, Y. C. Kudva, T. Peyser, R. E. Carter, C. Cobelli and R. Basu, *J. Diabetes Sci. Technol.*, 2015, **9**, 63–68.
4. D. P. Zaharieva, K. Turksoy, S. M. McGaugh, R. Pooni, T. Vienneau, T. Ly and M. C. Riddell, *Diabetes Technol. Ther.*, 2019, **21**, 313–321.
5. T. Heise, E. Zijlstra, L. Nosek, T. Rikte and H. Haahr, *Diabetes Obes. Metab.*, 2017, **19**, 208–215.
6. D. Rodbard, *J. Diabetes Sci. Technol.*, 2014, **8**, 980–985.
7. E. Zijlstra, T. Heise, L. Nosek, L. Heinemann and S. Heckermann, *Diabetes Obes. Metab.*, 2013, **15**, 130–135.
8. J. E. Lane, J. P. Shivers and H. Zisser, *Curr. Opin. Endocrinol. Diabetes Obes.*, 2013, **20**, 106–111.
9. A. Heller and B. Feldman, *Chem. Rev.*, 2008, **108**, 2482–2505.
10. Z. Mahmoudi, M. H. Jensen, M. Dencker Johansen, T. F. Christensen, L. Tarnow, J. S. Christiansen and O. Hejlesen, *Diabetes Technol. Ther.*, 2014, **16**, 667–678.
11. G. Acciaroli, M. Vettoretti, A. Facchinetti, G. Sparacino and C. Cobelli, *IEEE Trans. Biomed. Eng.*, 2018, **65**, 587–595.
12. A. Facchinetti, *Sensors* 2016, **16**, 2093.
13. D. Chen, C. Wang, W. Chen, Y. Chen and J. X. J. Zhang, *Biosens. Bioelectron.*, 2015, **74**, 1047–1052.
14. K. Burugapalli, S. Wijesuriya, N. Wang and W. Song, *J. Biomed. Mater. Res., Part A*, 2018, **106**, 1072–1081.
15. L. Fang, B. Liang, G. Yang, Y. Hu, Q. Zhu and X. Ye, *Biosens. Bioelectron.*, 2017, **97**, 196–202.
16. Y. Fang, S. Wang, Y. Liu, Z. Xu, K. Zhang and Y. Guo, *Biosens. Bioelectron.*, 2018, **110**, 44–51.
17. M. J. Malone-Povolny, E. P. Merricks, L. E. Wimsey, T. C. Nichols and M. H. Schoenfisch, *ACS Sens.*, 2019, **4**, 3257–3264.
18. R. P. Wadwa, L. M. Laffel, V. N. Shah and S. K. Garg, *Diabetes Technol. Ther.*, 2018, **20**, 395–402.

19. W. Groenendaal, K. A. Schmidt, G. Von Basum, N. A. W. Van Riel and P. A. J. Hilbers, *Diabetes Technol. Ther.*, 2008, **10**, 283–293.
20. F. Ribet, G. Stemme and N. Roxhed, *Biosens. Bioelectron.*, 2017, **90**, 577–583.
21. F. Ribet, G. Stemme and N. Roxhed, *Biomed. Microdevices*, 2018, **20**, 1–10.
22. S. Samavat, J. Lloyd, L. O'dea, W. Zhang, E. Preedy, S. Luzio and K. S. Teng, *Biosens. Bioelectron.*, 2018, **118**, 224–230.
23. H. Teymourian, C. Moonla, F. Tehrani, E. Vargas, R. Aghavali, A. Barfidokht, T. Tangkuaram, P. P. Mercier, E. Dassau and J. Wang, *Anal. Chem.*, 2020, **92**, 2291–2300.
24. P. Bollella, S. Sharma, A. E. G. Cass and R. Antiochia, *Electroanalysis*, 2019, **31**, 374–382.
25. L. Zhao, Z. Wen, F. Jiang, Z. Zheng and S. Lu, *RSC Adv.*, 2020, **10**, 6163–6171.
26. S. R. Chinnadayyala and S. Cho, *Nanomaterials*, 2021, **11**, 1–15.
27. PKvitality - Bio-wearables Health & Sport Home page - PKVitality, https://www.pkvitality.com/ (accessed 11 July 2021).
28. Biolinq. Intelligent Continuous Glucose Monitoring System, https://www.biolinq.me/ (accessed 11 July 2021).
29. Eversense Continuous Glucose Monitoring | Long-term Continuous Glucose Monitor, https://www.eversensediabetes.com/ (accessed 11 July 2021).
30. G. Cappon, M. Vettoretti, G. Sparacino and A. Facchinetti, *Diabetes Metab. J.*, 2019, **43**, 383–397.
31. Q. Dou, D. Hu, H. Gao, Y. Zhang, A. K. Yetisen, H. Butt, J. Wang, G. Nie and Q. Dai, *RSC Adv.*, 2017, **7**, 41384–41390.
32. Integrated Medical Sensors, http://www.integratedmedicalsensors.com/ (accessed 11 July 2021).
33. M. Mujeeb-U-Rahman, M. Honarvar, M. Sencan, W. Van Antwerp, M. H. Nazari, M. Sencan and W. Van Antwerp, *Sci. Rep.*, 2019, **9**, 17421.
34. J. Sawayama and S. Takeuchi, *Adv. Healthcare Mater.*, 2020, **2001286**, 1–7.
35. K. Sun, Y. Tang, Q. Li, S. Yin, W. Qin, J. Yu, D. T. Chiu, Y. Liu, Z. Yuan, X. Zhang and C. Wu, *ACS Nano*, 2016, **10**, 6769–6781.
36. K. Sun, Y. Yang, H. Zhou, S. Yin, W. Qin, J. Yu, D. T. Chiu, Z. Yuan, X. Zhang and C. Wu, *ACS Nano*, 2018, **12**, 5176–5184.
37. K. Sun, Z. Ding, J. Zhang, H. Chen, Y. Qin, S. Xu, C. Wu, J. Yu and D. T. Chiu, *Adv. Healthcare Mater.*, 2021, **10**, 2001019.
38. R. M. Unruh, J. R. Roberts, S. P. Nichols, S. Gamsey, N. A. Wisniewski and M. J. McShane, *J. Diabetes Sci. Technol.*, 2015, **9**, 985–992.
39. L. V. Le, G. S. Chendke, S. Gamsey, N. Wisniewski and T. A. Desai, *J. Diabetes Sci. Technol.*, 2020, **14**, 204–211.
40. A. Ceriello and F. Cameron, *Diabetes Technol. Ther.*, 2021, **23**, A1–A206.
41. K. L. Helton, B. D. Ratner and N. A. Wisniewski, *J Diabetes Sci. Technol.*, 2011, **5**, 632–646.
42. GlySens Incorporated – The Next Generation Continuous Glucose Monitor, http://glysens.com/ (accessed 11 July 2021).

43. GluSense – Implantable Glucose Sensor, https://glusensemedical.com/ (accessed 11 July 2021).
44. Hassle-free glucose monitoring with our next-gen sensor – Indigo, https://indigomed.com/ (accessed 11 July 2021).
45. Lifecare | Sencell – Continous glucose monitoring, https://www.lifecare.no/ (accessed 11 July 2021).
46. M. K. Åm, K. Kölle, A. L. Fougner, I. Dirnena-Fusini, P. C. Bösch, R. Ellingsen, D. R. Hjelme, Ø. Stavdahl, S. M. Carlsen and S. C. Christiansen, *PLoS One*, 2018, **13**, e0205447.
47. L. M. Huyett, R. Mittal, H. C. Zisser, E. S. Luxon, A. Yee, E. Dassau, F. J. Doyle and D. R. Burnett, *J. Diabetes Sci. Technol.*, 2016, **10**, 1192–1194.
48. I. L. Jernelv, D. R. Hjelme and A. Aksnes, *Biomed. Opt. Express*, 2020, **11**, 3818.

CHAPTER 11

Artificial Intelligence for Diabetes-related Complications: The Eye as a Window to the Systemic Health

MICHELLE YIP,[a,b,†] ZHAORAN WANG,[a,b,c,†] LAURA GUTIERREZ,[a,b] VALENCIA FOO,[a,b] JANE LIM,[a,b] GILBERT LIM,[a] DINESH GUNASEKARAN,[a,b] TIEN YIN WONG[a,b] AND DANIEL SHU WEI TING*[a,b]

[a] Singapore Eye Research Institute, Singapore; [b] Singapore National Eye Center, 11 Third Hospital Avenue, 168751, Singapore; [c] Duke-NUS Medical School, Singapore
*Email: daniel.ting.s.w@singhealth.com.sg

11.1 Introduction

Diabetes mellitus (DM) is one of the world's fastest growing chronic diseases and a leading cause of acquired vision loss.[1] The number of people with diabetes has almost tripled from 171 million (2.8%) in 2000 to 463 million (8.3%) in 2019, and this number is projected to double over the next few decades to 700 million in 2045.[1,2] DM has widespread repercussions in

[†]First authors, contributed equally.

Nanoscience & Nanotechnology Series No. 59
Nanotechnology for Diabetes Management
Edited by Amar Abderrahmani, Sabine Szunerits, Rabah Boukerroub and Abdelfattah El Ouaamari

Published by the Royal Society of Chemistry, www.rsc.org

various body systems including micro- and macrovascular complications of poor glycaemic control.[3]

Diabetic retinopathy (DR), one of the microvascular complications of DM affecting almost 93 million people in 2012, is the leading cause of blindness in the working population around the world.[4] The number of patients with DR is postulated to grow to 191 million in 2030 and, of these, 56.3 million would have vision-threatening DR.[5] Peripheral neuropathy in diabetic patients is one of the most common microvascular complications, occurring in approximately half of patients, with progression to ischaemic and neurotrophic ulcers in the skin and the cornea.[6] Diabetic kidney disease is estimated to affect 40% of all diabetes, with a large proportion of them progressing to end-stage renal failure, eventually requiring renal replacement therapies.[7] DM has also been well documented to increase the risk of cardiovascular diseases, with the Framingham Study showing an increased risk of clinically atherosclerotic diseases by twofold to threefold.[8] Besides the heart, other macrovascular complications of DM can occur in the brain, resulting in increased risk of cerebrovascular disease and neurodegenerative diseases such as vascular dementia and Alzheimer's disease in later life.[9]

With this rapid rate of increase of the disease burden, there is undoubtedly a need for a tool to identify these diseases accurately in the early stages, implement treatment, and manage glycaemic control.[10,11] One such tool that is gaining popularity is artificial intelligence (AI). AI has been widely utilized in many fields for automation, design-making, logistics and, ultimately, to improve the efficiency and accuracy of tasks.[12,13] AI has transformed our way of living and will continue to, paving the way for the Fourth Industrial Revolution with the fusion of various technologies and the blurring of the lines of physical, biological, and digital spheres.[13–15] It has had success in various sectors, including in marketing to reach target audiences, in banking for evaluating loan applications, in industry for preventive and condition-based maintenance, and the list goes on.[16–18] Developments of AI in medicine, assisting management of DM and its systemic complications, could help prevent its progression and reduce the resulting morbidity and mortality.

11.2 Artificial Intelligence for Diabetic Retinopathy and Diabetic Macular Oedema

11.2.1 Initial Success with Fundus Imaging

The automated detection of DR using standard fundus images has been widely explored by many studies worldwide, demonstrating outstanding diagnostic performance with deep learning (DL) technologies (Table 11.1). By evaluating 24 studies with 48 different patient cohorts, a recent meta-analysis reported an area under the receiver operating characteristic curve (AUC) of 0.939 (95% CI 0.920–0.958) for detecting all, referable, or vision-threatening DR from fundus photographs.[19] The development of such DL

Table 11.1 Summary table for the different artificial intelligence systems used in ophthalmology and systemic diseases.[a]

Year	Authors	Disease	Test data sets	Modality	Test sample size (n)	AI algorithm	AUC/ICC	Sensitivity (%)	Specificity (%)
Diabetic retinopathy (DR)									
Year	Authors	Disease	Test data sets	Modality	Test sample size (n)	AI algorithm	AUC	Sensitivity (%)	Specificity (%)
2021	Scheetz *et al.*[30]	DR (referable)	Australia	Fundus photos	236 Patients	CNN (Inception-v3)	0.92	96.9	87.7
2021	Heydon *et al.*[29]	DR (referable)	DESP	Fundus photos	30,405	CNN EyeArt v2.1	—	95.7	54.0
2021	Tang *et al.*[42]	DR (referable)	Hong Kong UK India Argentina	UWF-SLO	9392 Images	CNN (ResNet50)	0.981	94.9	87.2
2021	Nagasawa *et al.*[41]	DR (detecting no DR versus DR)	Japan	Wide-field fundus photos, OCT, OCTA	491	CNN (VGG16)	0.790 (Optos) 0.883 (OCTA) 0.847 (Optos OCTA)	80.9 83.9 (OCTA) 78.6 (Optos OCTA)	55.0 (Optos) 71.6 (OCTA) 69.8 (Optos OCTA)
2020	Bhuiyan *et al.*[28]	DR (referable)	KG-set MESSIDOR	Fundus photos	88,702 1748	CNN (Xception, Inception-v3, Inception-Resnet-V2)	—	82.6	93.7
2021	Olvera-Barrios *et al.*[40]	DR (staging)	NDESP	Fundus photos UWF-SLO	1257 Patients	CNN (EyeArt V.2.1.0)	—	Wide-field: 92.3 (any retinopathy) 99.0 (vision-threatening retinopathy) Fundus photo: 92.3 (any retinopathy) 100.0 (vision-threatening retinopathy)	For no retinopathy: Wide-field: 53.0 Fundus photos: 74.0

Table 11.1 (*Continued*)

Year	Authors	Disease	Test data sets	Modality	Test sample size (n)	AI algorithm	AUC	Sensitivity (%)	Specificity (%)
2019	Gulshan *et al.*[32]	DR (referable)	India	Fundus photos	3049 Patients	CNN	0.963–0.980	88.9–92.1	92.2–95.2
2019	Bellemo *et al.*[31]	DR (referable)	SiDRP, Chingola, Zambia	Fundus photos	76,370	CNN (VGGNet)	0.973	92.3	89.0
2019	Ruamviboonsuk *et al.*[33]	DR (referable)	Thailand	Fundus photos	25,326	CNN (Inception-v4)	0.987	97.0	96.0
2018	Abramoff *et al.*[26]	DR (referable)	USA	Fundus photos, OCT	819 Patients	CNN (IDx-DR)	—	87.2	90.7
2018	Li *et al.*[24]	Vision-threatening referable DR	ZhongShan	Fundus photos	8000	CNN (Inception-v3)	0.989	92.5	98.5
			NIEHS		7181		0.955		
			SIMES		15,679				
			AusDiab		12,341				
2017	Ting *et al.*[22]	DR (referable)	SiDRP 14-15	Fundus photos	71,896	CNN (VGG-19)	0.936	90.5	91.6
			Guangdong		15,798		0.949	98.7	81.6
			SIMES		3052		0.889	97.1	82.0
			SINDI		4512		0.917	99.3	73.3
			SCES		1936		0.919	100.0	76.3
			BESAFEDS		1052		0.929	94.4	88.5
			RVEEH		1968		0.980	98.8	86.5
			Mexican		2302		0.983	98.9	92.2
			CUHK		1172		0.950	91.8	84.8
			HKU		1254		0.948	99.3	83.1
					7706		0.964	100.0	81.3
2017	Gargeya *et al.*[23]	DR (referable)	Eye-PACS	Fundus photos	75,137	CNN, gradient boosting classifiers	0.970	94.0	98.0
			MESSIDOR 2		1748		MESSIDOR 2 – 0.94	93.0	87.0
			E-Ophtha		463		E-Ophtha – 0.995	90.0	94.0
2018	Keel *et al.*[27]	DR (referable)	Melbourne, Australia	Fundus photos	96 Patients	CNN (Inception-v3)	—	92.3	93.7
2016	Abramoff *et al.*[20]	DR (diagnosis)	Messidor-2	Fundus photos	1748	CNN (AlexNet/VGG)	0.980	96.8	87.0

2016	Gulshan *et al.*[21]	DR (referable)	Messidor-2 Eye-PACS	Fundus photos	1748 9963	CNN (Inception-V3)	0.990 0.991	96.1 97.5	93.9 93.4
2021	Bora *et al.*[36]	DR (predictive risk of development)	Eye-PACS Thailand	Fundus photos	3678 2345	CNN (Inception-v3)	0.700	—	—
2020	Varadarajan *et al.*[37]	Predicting Ci-DME	Eye-PACS Thailand	Fundus photos	554 697	CNN (Inception-v3)	0.890	85.0	80.0
2019	Arcadu *et al.*[34,35]	Predicting diabetic macular thickening	RIDE/RISE	Fundus photos	17,997	CNN (Inception-v3)			
		Central subfield thickness >250 μm					0.970	87.0	96.4
		Central foveal thickness >250 μm					0.910	80.0	85.0
2021	Mirshahi *et al.*[47]	Segmentation of the fovea avascular zone	Iran	OCTA	131 Eyes	CNN (Detectron2)	—	—	—
2021	Zang *et al.*[193]	DR (referable)	Oregon, USA	OCT OCTA	303 Eyes	CNN (DcardNet)	—	96.0	90.5
2020	Sandhu *et al.*[59]	NPDR (diagnosis)	USA	Clinical bio-markers, OCT, OCTA	111 Patients	Random forest	0.960	100.0	94.0
2020	Le *et al.*[57]	DR (diagnosis)	Chicago, USA	OCTA	131 Images	CNN (VGG 16)	0.960	83.7	90.8
2019	Heisler *et al.*[58]	DR (referable)	Canada	OCTA	380 Eyes	CNN (VGG19, ResNet 50, DenseNet) and class activation maps	—	90.4 (ensemble network)	93.3 (ensemble network)
2018	De Fauw *et al.*[49]	DR (referable)	UK	OCT	14 884	CNN (U-Net)	0.992	—	—

Table 11.1 (*Continued*)

Year	Authors	Disease	Test data sets	Modality	Test sample size (*n*)	AI algorithm	AUC	Sensitivity (%)	Specificity (%)
Diabetic macular oedema (DME) and other macular diseases									
2020	Orlando *et al.*[46]	Quantifying photoreceptor alteration in DME and RVO	Vienna	OCT	40 OCT datasets	FCNN (U-Net, BRU-Net, All-Dropout, U2-Net)	0.959 (Ensemble for CSF)	85.4 (BRU-Net for CSF)	—
2020	Bhatia *et al.*[51]	DME	Iran	OCT	148 OCT volumes	CNN (VGG16)	0.980	96.0	91.0
			Greece		100 OCT volumes		0.998	96.0	96.0
2020	Rasti *et al.*[54]	DME (predicting effect of anti-VEGF treatment)	US	OCT	127 Patients	CNN (VGG)	0.866	80.1	85.0
2019	Hecht *et al.*[52]	Differentiating DME from pseudophakic cystoid macular oedema	Israel	OCT	153 Patients	SVM, decision trees, random forest, gradient-boosted trees	0.937–0.987	94.0–98.0	94.0–95.0
2018	Kermany *et al.*[48]	Diagnosing choroidal neovascularisation or DME	Guangzhou, China	OCT	1000 Images	CNN (Inception-V3)	99.9 (differentiating DME from normal)	96.8 (differentiating DME from normal)	99.6 (differentiating DME from normal)
2018	Schlegl *et al.*[44]	Detecting and quantifying macular fluid in DME, RVO, AMD	Vienna	OCT	1200 Images	CNN	0.940 (IRC) 0.920 (SRF)	—	—

Year	Authors	Disease	Test data sets	Modality	Test sample size (*n*)	AI algorithm	AUC	Specificity
Diagnosing diabetes mellitus								
2020	Aslam *et al.*[53]	Distinguishing diabetic status	UK	OCTA	182 Eyes	Random forest	0.800	49.0

Year	Authors	Application (diagnosis/ prognosis)	Outcome Measures	Study Population	Datasets	Sample size	AI algorithm	AUC	Sensitivity (%)	Specificity (%)
In Vivo Confocal Microscopy (IVCM) and Diagnosing Diabetic Peripheral Neuropathy (DPN)										
2021	Yildiz *et al.*[101]	Quantification of nerve fibre properties: 1. Differentiate DPN+ from DPN- 2. Differentiate healthy from diabetics	CNFL, mean length per segment, no. of branch points, no. of tail points, no. of nerve segments, fractal number	Healthy eyes and diabetic eyes	Turkey	3835	U-Net CNN with data augmentation	1. 0.826 2. 0.806	1. 67.7 2. 78. 4	1. 86.7 2. 72.2
2021	Salahouddin *et al.*[99]	Quantification of cornea nerves 1. Differentiate DPN+ from DPN 2. Differentiate healthy from diabetics	CNFL	Healthy eyes and diabetic eyes	Qatar/UK	708	U-Net CNN with ANFIS	1. 0.860 2. 0.950	1. 84.0 2. 92.0	1. 71.0 2. 80.0
2020	Scarpa *et al.*[100]	Distinguish healthy from DPN+ eyes Considers simultaneous different non-overlapping images without building mosaics for analysis	Classification of healthy *vs.* pathological (diabetic) eyes	Healthy eyes and Diabetic eyes	Italy	100	CNN	0.830	98.0	94.0

Table 11.1 (*Continued*)

Year	Authors	Application (diagnosis/ prognosis)	Outcome Measures	Study Population	Datasets	Sample size	AI algorithm	AUC	Sensitivity (%)	Specificity (%)
2020	Williams *et al.*[98]	Quantification of nerve fibre properties Differentiate DPN+ from DPN-	CNFL, mean length per segment, no. of branch points, no. of tail points, no. of nerve segments, fractal number	Healthy eyes and diabetic eyes	BioImLab, Italy, Beijing, ENA group, UK	3835	U-Net CNN with data augmentation	0.830	68.0	87.0
2016	Al-Fahdawi *et al.*[194]	To propose a corneal nerve segmentation system and morphometric parameter quantification system to differentiate healthy from DPN (absent, mild DPN, and moderate DPN) eyes	CNFL, CNFD, nerve thickness, tortuosity coefficient,	Healthy eyes and Diabetic eyes	Qatar, UK	498	Three-step segmentation (preprocessing, morphological operations, edge detection)	—	—	—
2017	Chen *et al.*[97]	To propose a fully automatic nerve fibre detection and quantification system	CNFL, CNFD, CNBD, CNFWH, CNFOH	Healthy eyes and diabetic eyes	ENA group, UK	1088	Feature description and classifier training	0.770	72.0	72.0
2014	Petropoulos *et al.*[79]	Distinguish healthy from DPN eyes	CNFD, CNFL, CNBD	Healthy eyes and diabetic eyes	UK	1506	Dual model feature descriptor combined with a neural network classifier	0.640–0.840	0.60–0.80	0.56–0.74
2012	Ferreira *et al.*[195]	Automated extraction and segmentation of nerve morphometric parameters for DPN diagnosis	Tortuosity coefficient, CNFL, CNFD	Healthy eyes and diabetic eyes	Portugal	47	Local equalization, phase symmetry, and nerve reconstruction using morphologic operations	—	90.0	—
2010	Ferreira *et al.*[196]	Automated extraction and segmentation of nerve morphometric parameters for DPN diagnosis	Tortuosity coefficient, CNFL	Healthy eyes and diabetic eyes	Portugal	15	Local equalization, phase symmetry, histogram processing, nerve reconstruction	—	66.6	99.6

Year	Authors	Disease	Test data sets	Modality	Test sample size (n)	AI algorithm	AUC	R^2
Systemic biomarkers								
2020	Rim et al[112]	Prediction of systemic biomarkers and body composition indices	Seoul, South Korea BES SEED UK Biobank	Fundus photos	236,257	CNN (VGG16)	Sex – 0.800–0.960	Height 0.08–0.42 Weight 0.04–0.36 Serum creatinine 0.01–0.38 Blood pressure 0.17–0.31 Haemoglobin 0.06–0.56

Year	Authors	Disease	Test data sets	Modality	Test sample size (n)	AI algorithm	AUC/CC	Sensitivity (%)	Specificity (%)
Adverse event prediction									
2014	Sudharsan *et al.*[127]	Hypoglycaemia prediction	Maryland, USA	Self-monitored blood glucose readings	56 000 Data points	Random forest, SVM, K-NN, Naïve Bayes	—	92.0	70.0
2013	Georga *et al.*[126]	Hypoglycaemic events prediction (diagnosis)	Patras, Greece	Clinical data	15 Patients	SVR	—	94.0	—
2017	Albers *et al.*[123]	Personalized glucose forecasting	New York, USA	Self-monitored blood glucose readings	5 Patients	Data assimilation (Bayesian modelling)	—	—	—
2015	Zecchin *et al.*[122]	Real-time prediction of glucose concentration	Padova, Italy	Serum glucose concentration	10 Patients	Jump neural network	—	—	—
2015	Zarkogianni *et al.*[197]	Glucose prediction model	METABO	Self-monitored blood glucose readings	10 Patients	Feedforward neural network	CC 84.2–97.9	—	—

Year	Authors	Disease	Test data sets	Modality	Test sample size (n)	AI algorithm	AUC/ICC	Sensitivity (%)	Specificity (%)
Cardiovascular disease									
2021	Cheung *et al.*[135]	Risk of cardiovascular disease (diagnosis)	SP2 Dunedin HKCES AHES RICP IRED CUHK-STDR GUSTO SiDRP CVD screening study	Fundus photos	70 000	CNN (SIVA-DLS)	Intraclass correlation coefficients 0.82–0.95	—	—

Table 11.1 (*Continued*)

Year	Authors	Disease	Test data sets	Modality	Test sample size (*n*)	AI algorithm	AUC/ICC	Sensitivity (%)	Specificity (%)
2021	Rim *et al.*[137]	Cardiovascular risk stratification using coronary artery calcium scores (diagnosis)	SEED CMERC-HI UK Biobank	Fundus photos	216, 152	RetiCAC (Efficientnet)	0.742	—	—
2020	Chang *et al.*[136]	Carotid artery atherosclerosis (diagnosis)	HPC-SNUH	Fundus photos, clinical data	15 408 Images, 32 227 Patients	DL-FAS (Xception, ImageNet)	0.713	56.9	58.3
2018	Poplin *et al.*[110]	Predicting cardiovascular risk factors (diagnosis)	UK Biobank EyePACS	Fundus photos, clinical data	284 335 Patients	Inception-v3 neural network	Major adverse cardiovascular events 0.70 Gender 0.97 Smoking status 0.71	—	—
2014	Berikol *et al.*[132]	Acute coronary syndrome (diagnosis)	Karaman, Turkey	Clinical data, ECG, echocardio-graphy	228 Patients	SVM	—	98.2	100.0
1996	Baxt *et al.*[130]	Myocardial infarction (diagnosis)	California, USA	Clinical data, ECG	1070 Patients	ANN	—	96.0	96.0
2006	Green *et al.*[131]	Acute coronary syndrome (diagnostic)	Lund, Sweden	Clinical data, ECG	634 Patients	ANN	0.800	95.0	41.0
1991	Baxt *et al.*[129]	Myocardial infarction (diagnosis)	California, USA	Clinical data, ECG	331 Patients	ANN	—	97.2	96.2
Chronic kidney disease									
2020	Sabanayagam *et al.*[146]	Chronic kidney disease (diagnosis)	SEED	Fundus photos, Risk factors	5188 Patients	CondenseNet	0.938	84.0	85.0
			SP2		3735 Patients		0.810	74.0	75.0
			BES		1538 Patients		0.858	79.0	79.0
2019	Kanda *et al.*[142]	Identified patients at high risk of chronic kidney disease (progression)	Yamagata, Japan	Clinical data, CKD stages	7465 Patients	Bayesian network, SVM	—	—	—
2016	Chen *et al.*[140]	Chronic kidney disease potential risk assessment (diagnosis)	UCI machine learning repository	Clinical data	400 Patients	KNN, SVM, SIMCA	—	KNN 99.6–100 SVM 99.8–100.0 SIMCA 81.6–90.4	KNN 99.3–96.1 SVM 99.3–97.6 SIMCA 99.8–100.0

2013 Liu *et al.*[198]	GFR estimation	Guangzhou, China	Serum creatinine, age, sex	1200 Patients	ANN	—	—	—
Cerebrovascular and neuro-degenerative diseases								
2021 Tian *et al.*[158]	Alzheimer's disease (diagnosis)	UK Biobank	Fundus photos	175 831	U-net with SVM	—	84.8	79.2
2020 Wisely *et al.*[157]	Alzheimer's disease (diagnosis)	Durham, North Carolina	Fundus photos OCT OCTA FAF UWF SLO	346 Eyes	ResNet	0.836	100.0	75.0
2019 Palomar *et al.*[199]	Multiple sclerosis	Zaragoza, Spain	Swept source OCT	258 Patients	Decisions trees, multilayer perceptron, SVM	0.995–0.998	95.5–97.2	95.2–97.9
2017 Jeena *et al.*[173]	Ischaemic stroke (diagnosis)	Trivandrum, India	Fundus photos	50 Patients	ANN	—	100.0	96.0
2021 Vasseneix *et al.*[156]	Papilloedema severity (progression)	BONSAI	Fundus photos	2103	U-Net, VGGNet, ImageNet	0.930	91.8	86.2
2020 Biousse *et al.*[155]	Optic disc classifications (normal/ papilloedema/other abnormalities) (diagnosis)	BONSAI	Fundus photos	14 341	BONSAI-DLS (U-Net, DenseNet)	0.970	91.0	93.3
2020 Milea *et al.*[154]	Papilloedema (diagnosis)	BONSAI	Fundus photos	14 341	U-Net, DenseNet	0.960	96.4	84.7

[a]ACC, accuracy; AFEDS, African American Eye Disease Study; AHES, Australian Heart Eye Study; AMD, age-related macular degeneration; ANFIS, adaptive neuro-fuzzy inference system; ANN, artificial neural network; AUC, area under the curve; AusDiab, Australian Diabetes, Obesity and Lifestyle study; BES, Beijing Eye Study; CC, correlation coefficient; Ci-DME, centre-involving diabetic macular oedema; CMERC-HI, Cardiovascular and Metabolic Disease Etiology Research Center-High Risk; CNBD, corneal nerve branch density; CNFD, corneal nerve-fibre density; CNFL, corneal nerve-fibre length; CNFOH, cornea nerve-fibre orientation histogram; CNFWH, cornea nerve-fibre width histogram; CNN, convolutional neural network; CSF, central subfield thickness; CUHK-STDR, Chinese University of Hong Kong Sight-Threatening Diabetic Retinopathy Study; CVD screening study, cardiovascular disease screening using retinal vascular imaging study; DESP, Diabetic Eye Screening Programme; DL-FAS, deep learning-funduscopic atherosclerosis score; ECG, electrocardiogram; ENA, Early Neuropathy Assessment; FAF, fundus autofluorescence; FCNN, fully convolutional neural networks; GFR, glomerular filtration rate; GUTSO, Growing Up in Singapore Towards Healthy Outcomes Study; HKCES, Hong Kong Children Eye Study; HKU, Hong Kong University; HPC-SNUH, Health Promotion Centre of Seoul National University Hospital; ICC, intraclass correlation coefficient; IRC, intraretinal cystoid fluid; IRED, Retinal Imaging in Renal Disease Study; KG, Kaggle data set; KNN, K-nearest neighbour; NDESP, English National Diabetic Eye Screening Programme; NIEHS, National Indigenous Eye Health Survey; OCT, optical coherence tomography; OCTA, optical coherence tomography angiography; RetiCAC, deep-learning retinal coronary artery calcium scores; RICP, Retinal Imaging in Chest Pain Study; RVEEH, Royal Victorian Eye and Ear Hospital; RVO, retinal vein occlusion; SCES, Singapore Chinese Eye Study; SEED, Singapore Epidemiology of Eye Diseases Study; SiDRP, Singapore Integrated Diabetic Retinopathy Program; SIMCA, soft independent modelling of class analogy; SIMES, Malays from the Singapore Malay Eye Study; SINDI, Singapore Indian Eye Study; SP2, Singapore Prospective Study Program; SRF, subretinal fluid; SVM, support vector machine; SVR, support vector for regression; UCI, UC Irvine machine learning repository; UWF SLO, ultra wide-field colour scanning laser ophthalmoscopy; VEGF, vascular endothelial growth factor.

systems first started by using retrospective datasets. Abramoff *et al.*,[20] Gulshan *et al.*,[21] Ting *et al.*,[22] Gargeya *et al.*,[23] Li *et al.*,[24] and Krause *et al.*[25] have reported sensitivities of over 90%, specificities of over 85%, and AUC of more than 0.90 in detecting referable DR (moderate non-proliferative DR or worse). When tested on independent external validation datasets collected from multi-centre and multi-ethnic groups, the diagnostic performance remained high.[22,24] Following this, Abramoff *et al.* enrolled 900 patients with diabetes but no history of DR to join a clinical trial for autonomous detection of more than mild DR, which achieved a sensitivity of 87.2% and specificity of 90.7%. Their DL system was subsequently authorized by the FDA as the first AI diagnostic system in medicine.[26] Keel *et al.* tested a DL model in two endocrinology outpatient clinics, which demonstrated a sensitivity of 92.3% and specificity of 93.7% for correct referral for DR. In addition, of the 96 participants, 96% were either satisfied or very satisfied with the automatous screening model and 78% reported their preference of the DL system over manual screening.[27] Using prospectively obtained retinal images from 974 patients with diabetes in a primary care setting, the DL screening system built by Bhuiyan *et al.* could not only detect referable DR at a sensitivity of 82.6% and specificity of 93.7%, but also send the diagnostic report within a minute to the primary care provider via a telemedicine platform.[28] Another prospective study by Heydon *et al.* evaluated the performance of an AI algorithm in automatous triaging of patients in three English Diabetic Eye Screening Programmes, with a sensitivity of 95.7% and specificity of 54.0%.[29] Recently, Scheetz conducted an AI-assisted DR screening in two real-world healthcare settings involving 236 participants, which achieved an AUC of 0.92, sensitivity of 96.9%, and specificity of 87.7% for referable DR. Moreover, the feedback from the study participants and clinical staff revealed high acceptance and positive experience with the AI-assisted DR screening program.[30]

11.3 Application to Real-world Target Populations

At the global health setting, DL technologies may be a promising alternative when the establishment of a manual screening programme is challenging, particularly in rural areas with reduced accessibility and limited resources. Several DL algorithms have been validated prospectively in overseas clinical settings. In Zambia, a country in Africa with a ratio of fewer than three ophthalmologists to one million population, Bellemo *et al.* tested the DL system developed by Ting *et al.*[22] on over 1,500 participants recruited from a mobile screening unit, achieving an AUC of 0.973 at a sensitivity of 92.3% and specificity of 89.04% in detecting referable DR.[31] In India, Gulshan *et al.* conducted a prospective observational study to evaluate the diagnostic performance of their DL system[21] in detecting referable DR, which was shown to be equal or even superior to manual grading, with AUC of 0.963 and 0.980, sensitivity of 88.9% and 92.1%, and specificity of 92.2% and 95.2% in two eye care centres, respectively.[32] In Thailand, Ruamviboonsuk *et al.* reported that the DL model[25] outperformed human graders in a national screening program with a higher sensitivity (0.97 *vs.* 0.74, $p<0.001$) for the detection of referable DR.[33] Despite the excellent performance of the DL frameworks in those countries with relatively limited healthcare resources, it is pivotal to establish a supporting ecosystem for the implementation of such DL models.

11.4 Use of Deep Learning for Prediction of Progression

Apart from the classification and detection of DR, several groups have studied the predictive role of DL systems. Arcadu *et al.* reported a DL algorithm for the prediction of DR progression by two-step or more worsening on the Early Treatment Diabetic Retinopathy Study (ETDRS) and Diabetic Retinopathy Severity Scale (DRSS), with mean AUC ranging from 0.524 to 0.616 within 24 months from seven-field fundus photographs.[34,35] Using fundus images of almost 600,000 eyes from two retrospective longitudinal datasets in two countries, Bora *et al.* created a DL framework for automated risk stratification of developing DR, which correctly predicted the development of DR within 2 years at an AUC of 0.79 and 0.70 with internal and external validation datasets, respectively. The performance was further improved to an AUC of 0.81 after introducing available risk factors to the datasets in the internal validation set and a similar increase in AUC was also observed in the external validation dataset.[36] Besides risk stratification, Arcadu *et al.* applied DL to predict optical coherence tomography (OCT) measurements of macular thickness from fundus photographs in patients with diabetic macular oedema (DME), achieving an AUC > 0.90 for all models and cut-off thickness values.[34] Similarly, Varadarajan *et al.* reported a DL model for predicting DME from fundus images with 0.89 AUC, 85% sensitivity, and 80% specificity.[37] These predictive DL frameworks may augment telemedicine-based DR screening by enhancing the efficiency of DME diagnosis from 2D retinal photographs and optimizing screening intervals.

11.5 Standard Fundal Imaging *Versus* Widefield Fundus Ophthalmoscopy

Compared to the most commonly used standard fundus photographs in DR screening, widefield fundus ophthalmoscopies may reduce the rates of ungradable images[38] and offer more detailed visualization of DR features that are not detected in the standard fundus images or are outside the 45 degree fields.[39] Olvera-Barrios *et al.* used a commercially available automated retinal image analysis software to detect DR from widefield retinal images and standard fundus photographs, reporting similar performance with sensitivity of 92.27% (from widefield retinal images) and 92.26% (standard fundus photography) for any retinopathy, respectively.[40] Using a smaller dataset of 491 Optos ultra-widefield fundus ophthalmoscopy images, Nagasawa reported a DL convolutional neural network (CNN) for detecting DR with 0.79 AUC, 80.9% sensitivity, and 55% specificity.[41] Using 9,392 ultra-widefield scanning laser ophthalmoscope images, a recent multicentre study by Tang *et al.* indicated that the DL system achieved good diagnostic performance on detecting referable DR and vision-threatening DR, with AUC >0.90, sensitivities and specificities >80% in three geographical external validation datasets.[42]

11.6 Utility of Optical Coherence Tomography

High-resolution spectral domain OCT has been increasingly used in ophthalmology, especially in the diagnosis and management of DME.[43] Broadly speaking, the development of OCT-based DL algorithms for DR and DME can be divided into three areas, namely the structural segmentation, disease classification, and predictive goals (Table 11.1).

For segmentation tasks, DL frameworks have been trained for accurate detection and quantification of intraretinal fluid and subretinal fluid.[44,45] Heatmaps have shown the focus of these frameworks in these pathological areas to diagnose DME (Figure 11.1A, B). These are similar to how the algorithms detect DR from fundal images as discussed earlier through pathological lesions recognized from training images (Figure 11.1C). In addition, Orlando *et al.* built an ensemble of four CNNs for the automatic segmentation and characterization of photoreceptor alteration in DME and retinal vein occlusion, achieving non-inferior performance to manual annotations by a human expert.[46] DL has also been applied for the segmentation of foveal avascular zone (FAZ) in OCT angiography (OCTA), using a retrospective dataset of 131 eyes of patients with diabetes and 32 normal eyes. The results indicated that the DL model had excellent agreement with the manual measurements and outperformed the automated measurements by the built-in software at delineation of the FAZ in diabetic eyes.[47] These segmentation DL models may play a role in evaluating the degree of visual impairment and guiding therapeutic management based on quantitative measurements of structural changes in the diseased retina.

Many studies focus on the classification function of OCT-based DL models and automated generation of referral opinions.[48–50] Trained with over 100,000 OCT images, Kermany *et al.* reported a robust DL algorithm for discriminating DME, drusen, and choroidal neovascularization (CNV) from normal OCT scans with 97.8% sensitivity, 97.4% specificity, and 96.6% accuracy. An AUC of 0.99 was achieved to distinguish urgent referrals (CNV or DME) by their DL model, comparable to the performance of experienced human experts.[48] Similarly, De Fauw *et al.* described a two-stage DL model trained with 14,884 scans by decoupling the segmentation and classification networks, reaching or even outperforming experts at detecting those who require urgent referrals with AUC > 0.90.[49] Bhatia *et al.* evaluated a DL-based clinical decision-supporting software for the detection of retinal diseases from macula OCT images with AUC > 0.98 in five independent validation

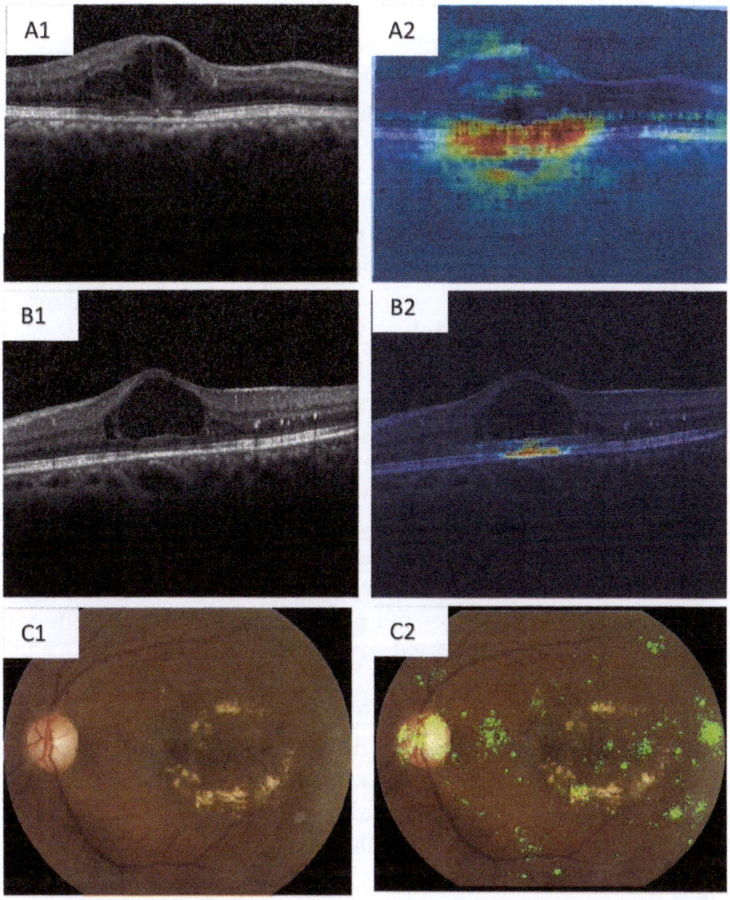

Figure 11.1 Heatmaps of different imaging modalities of diabetic retinopathy and diabetic macular oedema. (A and B) OCT images and corresponding heatmaps of diabetic macular oedema (DME). (A1) An OCT image of DME, (A2) corresponding heatmap with the red signal suggesting focus of the model in the intraretinal cysts and epithelium pigment retina layer (EPR); (B1) OCT image of DME, (B2) corresponding heatmap with red region visualizing the EPR of the retina; (C) DR fundus photo and corresponding overlay of heatmaps. (C1) Fundus photograph of DR and DME showing macular hard exudates with some scattered haemorrhages. (C2) Heatmap of the corresponding images shows in green: the AI focus is on most of the macular hard exudates and scattered haemorrhages.

datasets.[51] In addition to distinguishing diseased from normal retina, Hecht *et al.* developed a machine learning algorithm to differentiate DME from pseudophakia cystoid macular oedema in patients with diabetes after cataract surgery. Tested by cross-validation, their model had an AUC of 0.937–0.987, sensitivity of 94–98%, and specificity of 94–95% with three to six OCT parameters.[52]

In terms of predictive tasks, DL techniques were explored by Aslam *et al.* to identify diabetic status from OCTA images, with an AUC of 0.8 provided by the Random Forest classifier.[53] Another recent pilot study retrospectively trained a DL model to predict the efficacy of anti-vascular endothelial growth factor (VEGF) in DME based on pre-treatment OCT images, achieving an AUC of 0.866, sensitivity of 80.1%, and specificity of 85.0% in detecting patients who were responsive to anti-VEGF injection (defined as a more than 10% post-treatment reduction in retinal thickness).[54] Although further studies involving larger prospective datasets are warranted, such AI algorithms for predicting treatment outcome may facilitate more effective and personalized management of DME.

Recently, there has been an increasing interest in exploring the use of DL for detecting early DR changes in OCTA. Compared to colour fundus photographs and fluorescein angiography, OCTA is a non-invasive imaging modality that allows better visualization of retinal vasculatures and multiple retinal layers which may facilitate earlier detection of subtle DR changes.[55,56] The DL framework developed by Le *et al.* with transfer learning was reported to have a sensitivity of 83.76% and specificity of 90.82% for distinguishing healthy, no DR, and DR eyes.[57] Heisler *et al.* built an ensemble model using enface OCT and OCTA images of the superficial and deep capillary plexus, achieving sensitivity of 90.41% and 87.38%, and specificity of 93.33% and 92.09% in the detection of referable DR, with majority soft voting and stacking techniques, respectively.[58] Another recent study adopted a multimodal approach for the automated diagnosis of DR, combining clinical biomarkers, demographic information, OCT, and OCTA images. Compared to OCT images alone, the multimodal algorithm improved the accuracy from 76% to 96%, sensitivity from 85% to 100%, specificity from 87% to 94%, and AUC from 0.78 to 0.96.[59]

11.7 Artificial Intelligence for Corneal Neuropathy

Diabetic peripheral neuropathy (DPN) is the most common chronic complication of types 1 and 2 diabetes, affecting up to 50% of diabetics after 10 years of disease duration.[60,61] It is a length-dependent axonopathy that is characterized by sensory loss and pain, and affects small unmyelinated or thinly myelinated C and Aδ nerve fibres first, followed by large, myelinated nerve fibres.[62–64] DPN is significantly underdiagnosed,[65] leading to missed opportunities in preventing progression to severe DPN, foot ulceration, and lower limb amputations, and has a high 5-year mortality rate.[66]

Diabetic corneal neuropathy (DCN), as an ocular manifestation of DPN, affects 47–64% of patients during the clinical course of DM.[67] Due to chronic hyperglycaemia, the accumulation of advanced glycation end-products, increased polyol pathway flux, reactive oxygen species production, and protein kinase C pathway activation, all result in the triggering of apoptosis of corneal neural cells and neuronal degeneration.[68] There is progressive damage to corneal nerves that decreases corneal sensitivity and increases the risk of anterior segment pathology, ranging from dry eyes to neurotrophic ulcers and sight-threatening corneal infections, which can lead to a significant social and economic burden.[69]

Early identification as a window of opportunity and risk factor management are key to limiting the progression of DPN. Clinical standards utilize the 128-Hz tuning fork for vibration perception testing and the 10-gram monofilament,[69] but are poor at identifying those with early neuropathy.[70,71] Estimation of intraepidermal nerve fibre (IENF) density in a skin biopsy is considered to be the 'gold standard' in diagnosing small fibre neuropathy.[72] This method offers a sensitive but invasive means to identify early DPN with risks of bleeding and infection.[73] Other methods of small fibre function evaluation for DPN, such as thermal and vibration perception thresholds,[74,75] evoked potentials,[76] or laser Doppler flare,[77] have various limitations which preclude their mass implementation. Thus, there is an unmet need for robust screening methods capable of diagnosing early subclinical DPN. Over the past 2 decades however, *in vivo* confocal microscopy (IVCM) has been shown and validated to reliably quantify early axonal damage in DPN as a good surrogate marker of small fibre neuropathy[78,79] with high sensitivity and specificity,[80,81] closely correlated to the loss of IENF on skin biopsy in some studies.[80,82]

11.8 *In Vivo* Confocal Microscopy as a Tool to Detect Peripheral Neuropathy and Corneal Neuropathy

The cornea is the most densely innervated part of the human body, containing unmyelinated C and Aδ fibres.[83] IVCM is a rapid, non-invasive, and reproducible ophthalmic imaging technique that visualizes small nerve fibres in the corneal sub-basal plexus located between the Bowman layer and the basal epithelium for the clinical diagnosis of early DPN.[83,84] The use of IVCM to detect incident DPN and DCN has been well-established and has various advantages over current conventional methods of screening for DPN. These include its high sensitivity and specificity in detecting subclinical DPN, monitoring for DPN progression and stratifying patients at high risk of development and progression of DCN.[85–91] On the therapeutic fronts, IVCM can also identify corneal nerve regeneration prior to improvement in DPN symptoms.[92] IVCM has also shown value in monitoring progression post-therapeutic interventions for diabetes, such as after pancreas and kidney transplant, with an increase in cornea nerve fibre density (CNFD) and cornea nerve fibre length (CNFL) at 6 months, as well as in CNFD, CNFL, and CNBD (cornea nerve branch density) at 12 months seen. Furthermore, after 24 months, significant increases in CNFD, CNBD, and CNFL were noted in type 1 diabetics receiving continuous subcutaneous insulin infusions compared to those receiving multiple insulin injections. The role for AI could be further applied in monitoring post-therapeutics in dealing with high volumes of serial image analysis.[92,93] Given the rapid expansion of IVCM as a tool for DPN research, with recent clinical trials utilizing it as a surrogate endpoint, there is a clear expectation for the continued increased use in IVCM for DPN and DCN screening as well as monitoring of its progression.

Although manual IVCM analysis is highly accurate, it is laborious, not easily scalable for clinical implementation and requires expertise to quantify nerve pathology and distinguish it from background noise. The idea that AI may assist in automation of this analysis was adopted by the Early Neuropathy Assessment (ENA) group, who developed and validated the use of manual and automated image analysis software for corneal nerve tracing using IVCM called CCMetrics and ACCMetrics, respectively.[79,94]

11.9 Use of AI in Automating IVCM Analysis

AI has been applied mainly to the automated evaluation of the cornea nerve sub-basal plexus, with rapid and good performance of the identification, segmentation, and analysis of DCN in IVCM images. Various AI approaches, such as convolutional neural network (CNN), generative adversarial networks (GAN) (though not yet applied in DPN), and multi-scale adaptive dual model detection algorithms, have been described. These classification algorithms showed good sensitivity and specificity in distinguishing between patients with and without DPN, which is important for successful deployment of IVCM in large-scale clinical applications, multicentre studies, and clinical trials for disease-modifying therapies in DPN (Table 11.1).

11.10 Dual-model Feature Algorithms

The state-of-the-art ACCMetrics software application widely used in the study of DPN uses a dual-model feature descriptor combined with a neural network. The algorithm could distinguish nerve fibres from background noise and quantify cornea nerves with a constructed connectivity map, matching the accuracy of manual annotation, but with a processing time reduced to merely 10–22 seconds from 2–7 minutes with the manual analysis per image.[79,95–97] Chen *et al.* similarly proposed a fully automatic nerve fibre detection and quantification system in IVCM images using feature description by a dual-model filter and classifier training by a neural network or random forest.[97] In addition to the standard CNFD, cornea nerve branch density (CNBD), cornea nerve fibre length (CNFL), inferior whorl length, and tortuosity, it could produce additional features such as area, width, and orientation of the nerve fibres. The model achieved improved nerve fibre detection over previous results (sensitivity and specificity of 91.7%).

11.11 Convolutional Neural Networks

Several studies, such as those by Williams *et al.*[98] and Salahouddin *et al.*,[99] have employed a U-Net-based CNN algorithm for automated quantification of the corneal sub-basal nerve plexus in DPN patients. The intraclass correlation coefficients (ICC) for the algorithm in the former were superior to those for ACCMetrics for total CNFL (0.933 *vs.* 0.825), mean length per segment (0.656 *vs.* 0.325), number of branch points (0.891 *vs.* 0.570), number of tail points (0.623 *vs.* 0.257), number of nerve segments (0.878 *vs.* 0.504), and fractals (0.927 *vs.* 0.758). Salahouddin *et al.*[99] reported an algorithm which could estimate CNFL that was comparable to that by CCMetrics and even significantly outperformed ACCMetrics, with a higher pixel detection sensitivity (85% *vs.* 70%) and lower false-negative rate (15% *vs.* 30%). The AUC was 0.86 for distinguishing DPN from control subjects and 0.95 for distinguishing subjects with DPN from those without DPN, suggesting high accuracy in discriminating control from DPN-negative and DPN-positive subjects. Scarpa *et al.*[100] sought to overcome utilizing a single image that represents only a limited section of the cornea for IVCM analysis by proposing a CNN-based DL algorithm that simultaneously considers different multiple non-overlapping IVCM images without building a mosaic. The algorithm achieved a mean accuracy of 96% by providing a more complex analysis of three combined images to provide an overall classification of the cornea, demonstrating the potential of the CNN in identifying DPN and outperforming traditional methods using CNFD and tortuosity estimations.

11.12 Generative Adversarial Networks

Generative adversarial networks (GANs) consist of two separate networks which are generators and discriminators, and can produce images which are similar to the original ones. Yildiz *et al.*[101] created two separate DL algorithms, with the GAN-based method showing significantly higher accuracy than the U-NET method (AUC 0.9439 *vs.* 0.8934). Although GAN-based DL algorithms have not yet been reported in IVCM images from DPN subjects, there is some evidence of the superiority of the GAN-based algorithm for accurate segmentation of corneal sub-basal nerves compared to the U-Net based algorithm. This showcases the room for its potential to be explored in the diagnostic classification of various corneal pathologies.

11.13 Future Directions for AI in IVCM

The advent of automated algorithms for the various applications of IVCM in DPN and DNC have rapidly advanced and will likely continue to evolve as a powerful clinical and research tool. Expanding on previous studies, future work could focus on exploring IVCM use in staging severity to increase its utility as a prognostication and monitoring tool as to date no AI algorithms have been published on automated staging of IVCM for DPN or DCN severity. Another potential role of IVCM could be in other systemic diabetic complications such as autonomic neuropathies.[102–105] Petropoulous *et al.* also showed that degeneration of corneal nerve fibres is detectable before other microvascular complications appear, suggesting that early detection of small fibres insult on IVCM could work as a surrogate marker for systemic vascular risk stratification such as in microalbuminuria or retinopathy.[106]

However, crucial steps forward would involve tackling challenges of adoption of these algorithms in clinical practice. To mitigate the small field of view in IVCM, montaging technology with the aid of AI could allow for the numerous corneal maps obtained rapidly at regular intervals to quantify nerve parameters and study nerve migration patterns over time. This would be useful in monitoring disease progression and may lead to greater standardization of IVCM images in the future.[107] More work on the association of IVCM parameters with clinically meaningful outcomes such as clinical assessments of DPN and patient-oriented outcomes such as pain, disability, and quality of life may allow for greater acceptance and adoption by clinicians.

11.14 Artificial Intelligence for Systemic Diseases

DM is a disease due to dysfunction of the glucose homeostatic regulatory system in the body, usually manifesting with hyperglycaemia as the hallmark feature.[10] Although initial reviews debated on various methods of measuring glucose levels for diagnosis, the potential utility of HbA1c was first described in a 1985 WHO report and the use of HbA1c has since been largely the practice adopted even until now for the diagnosis of diabetes.[108,109] The patient's HbA1c level is usually measured through intravenous collection of blood to send for testing. However, with the development of AI and DL, it opens doors to possible non-invasive diagnostic techniques for the surveillance of the mass population and early detection of disease.[110] One such technique is through the use of retinal imaging as the eye has been widely suggested to be the window to systemic health.[111] Poplin *et al.* and the AI group at Google showed that a DL model was able to detect HbA1c using unsupervised learning of retinal fundus photographs and this has been built upon by others with improved accuracy.[110,112]

After diagnosis, treatment mainly aims at keeping glucose levels controlled as chronic hyperglycaemia may have long-term consequences with physiological glycosylation of cells, tissue damage, and may result in organ failure and acute derangements may precipitate life-threatening events.[3,113–115] The management of DM is challenging in monitoring and controlling blood glucose with pharmacological and lifestyle measures, recognizing and avoiding emergencies with extremes of glucose levels, and the detection of microvascular and macrovascular complications.[116–118] In recent years, the application of AI techniques to assist with DM control strategies has been garnering interest and momentum to tackle these challenges.[119,120]

The importance of monitoring blood glucose levels has led to the development of continuous glucose monitoring (CGM) devices with real-time dynamic surveillance of the patient's diurnal variation of blood glucose levels.[121] This also enables the collation of a substantial amount of data, which could be used to train AI algorithms in predicting glucose levels. Zecchin *et al.* utilized a Jump Neural Network architecture to predict glucose levels of type 1 DM patients with an accuracy of root mean square error of 16.6 mg dl^{-1} difference between the predicted and original glucose level, with an anticipatory mean time gain of 18.5 minutes to allow for therapeutic decisions to be made to circumvent glycaemic derangements.[122] Although these algorithms may provide relatively accurate predictions, in the real-world, the greater proportion of diabetics in the world have type 2 DM, the majority of whom do not use CGM devices and rely on self-monitoring blood glucose (SMBG) values which may result in sparse data.[123] Albers *et al.* identified this clinical question and utilized data assimilation to create a forecast, providing personalized medicine dynamically.[123]

In addition to monitoring of blood glucose levels, there is a role for AI to help predict adverse events and acute life-threatening DM emergencies.[124] Hypoglycaemia in insulin-dependent individuals could result in brain dysfunction and possibly be fatal.[125] Georga *et al.* utilized a support vector model (SVM) to predict upcoming hypoglycaemic episodes in type 1 DM patients based on insulin intake, recent glucose values from CGM, carbohydrate intake, and exercise, with a high sensitivity of 94% with time lags of approximately 5 minutes.[126] Type 2 DM patients are also at risk of this complication but with the issue of needing a few SMBG readings per day. Sudharsan *et al.* addressed this by computing a model that showed a high sensitivity of 92% at predicting a hypoglycaemic episode within the next 24 hours in type 2 DM individuals based on a minimum of 10 SMBG readings per week.[127] With better prediction of hypoglycaemia events, there can be avoidance of potentially fatal events, allowing for the optimization of insulin therapy.[124]

11.15 Artificial Intelligence for Cardiovascular Disease

Cardiovascular disease (CVD) is the leading cause of mortality around the world, and DM is a strong predisposing factor.[8,128] One effort to reduce mortality is to improve the accuracy of detection of acute coronary syndrome (ACS) that if missed could lead to fatal outcomes. To aid physicians in diagnosis, AI architectures have proposed DL algorithms that predict the presence of ACS utilizing clinical information.[129,130] Baxt *et al.* showed that an artificial neural network had a sensitivity and specificity of 97.2% and 96.2%, respectively, in detecting ACS, which is superior to physicians who had a sensitivity and specificity of 77.7.3% and 84.7%, respectively.[129] Green *et al.* demonstrated the importance of electrocardiogram (ECG) data in the decision-making, training a model with ECG data that could achieve an AUC of approximately 80% with no increase in performance when adding clinical data available at presentation to the emergency department.[131] Berikol *et al.* suggested the importance of various clinical data in training the classification model to assist the physician in deciding the patient disposition for less straightforward cases based on ACS risk stratification.[132]

Retinal vasculature has been characterized as the 'window' to the body's circulatory system as it is easily and non-invasively visible, and is correlated with several diseases that affect systemic micro- and macrovasculatures.

DR was found to have a significant association with coronary artery calcification (CAC), and is a non-invasive measure of subclinical coronary artery disease with an odds ratio of 1.43 in a study by Wong *et al.*[133] However, the studies by Wong *et al.* and Kromer *et al.* both showed little correlation established between retinal vessel calibre and coronary artery calcification or incident myocardial infarction, respectively.[133,134]

With the use of DL, the unsupervised learning of underlying associations can assist in accurate and non-invasive detection of CVD. Poplin *et al.* used DL to predict cardiovascular risk factors not previously thought to be detectable in input retinal photographs such as smoking status, HbA1c, and systolic blood pressure.[110] In addition, the model was able to detect major adverse cardiac events within 5 years from fundus images with an AUC of 0.70.[110] As retinal vasculopathy was postulated to be a marker for CVD risk, Cheung *et al.* devised a DL system for automated measurement of retinal vessel calibre and demonstrated that it correlated with classic CVD risk factors and incident CVD events using population-based data.[135]

A quantifiable risk of atherosclerosis was explored in a study by Chang *et al.* where a DL-fundoscopic atherosclerosis score (DL-FAS) was generated by the model for each retinal fundal examination, and the study showed that a score greater than 0.66 indicated an increased risk of CVD deaths, providing additional predictive value over the Framingham risk score for decision-making.[136] Although one of the currently validated non-invasive markers of CVD risk is CAC, access to computer tomography (CT) machines may be limited. Another group used DL based on fundus images to derive a CAC score, which was comparable to CT scan measured CAC when predicting cardiovascular events and further improved the risk stratification for patients.[137]

11.16 Artificial Intelligence for Chronic Kidney Disease

Diabetes is the main cause of chronic kidney disease (CKD) in most countries, accounting for at least 40% of CKD.[138] The Kidney Disease: Improving Global Outcomes (KDIGO) guidelines places emphasis on screening for early disease detection and monitoring for progression.[139] To assist in community-based monitoring, the use of AI to detect CKD could enable community deployment and increased accessibility to patients. Chen *et al.* used an input of 21 different clinical attributes such as serum creatinine, presence of diabetes, and electrolyte levels to train an AI model to detect CKD, and were able to achieve 99% accuracy.[140] A quantifiable estimation of renal function is also important to evaluate the severity of CKD. Many equations built on regression have been formulated to calculate the estimated glomerular filtration rate (eGFR) as a surrogate marker for renal function but are thought to be imprecise.[141] Liu *et al.* utilized AI to enhance the accuracy of eGFR calculations, comparing the model's eGFR to that from current regression equations, and found that an artificial neural network using three variables was not superior but further investigation, if adding additional variables, would improve the performance.[141] With a more accurate estimation of the GFR, it would lead to improved staging and prognostication of CKD. Early CKD stages may have a faster decline in renal function, however treatment strategies to address early CKD have not been well established. A worksite-based cohort study by Kanda *et al.* used SVM to identify CKD patients at high risk of progression, surprisingly showing that stage G1 was more progressive than stage G2.[142] It also used a Bayesian network to determine risk factors for CKD progression, identifying hypertension as a large contributing factor.[142]

The diagnosis and monitoring of CKD often requires blood tests for serum creatinine or urine tests for protein or albumin.[139] As serum or urine samples must be obtained, this may affect adherence to surveillance and may not be practical in all settings. Fundus imaging could provide a non-invasive screening tool to detect CKD based on physiological similarities in microvasculature, with earlier studies reporting associations between retinopathy and retinal arteriolar narrowing with CKD independent of other factors such as diabetes and hypertension.[143,144]

Based on these associations, Yip *et al.* utilized AI and machine learning to accurately and efficiently quantify retinal vascular parameters (such as arteriolar and venular calibre, tortuosity, fractal dimension, and branching angle) and detect retinopathy. This study demonstrated that smaller retinal arterioles, larger retinal venules, and the presence of retinopathy were associated with incident CKD, providing an automated non-invasive tool to detect the development of CKD.[145] DL has also been applied recently to the detection of CKD from retinal imaging, clinical risk attributes, and a combination of both images and risk factors.[146] Sabanayagam *et al.* showed that the accuracy of retinal imaging to detect CKD is high with an AUC of 0.911 and comparable to a model that also possessed the clinical information of the patient.[146] This suggests that it could be a useful automated tool for the detection of CKD in remote settings where information is sparse and resources are limited. Screening could be more widely accessible and at-risk individuals could be identified early and referred to a nephrologist promptly.

11.17 Artificial Intelligence for Cerebrovascular and Neuro-degenerative Diseases

As the global population ages, the impact of neurological disorders will increase. In 2015, there were 46.8 million people with dementia worldwide, and it is estimated that by 2050 the number will increase to 74.7 million.[147] In addition, stroke was ranked as the second leading cause of death worldwide,[148] with an annual mortality rate of 5.5 million.[149] Traditionally, patients with cerebrovascular disease (CVD) are diagnosed through physical examination, questionnaires, blood tests, brain imaging, lumbar puncture, or even brain biopsy (gold standard), although the latter can be considered rarely done due to the high risks implied.[150,151] As part of the central nervous system (CNS), the retina can also reveal brain health conditions in a less invasive way.[152,153] A more efficient trend has emerged that allows analysing of the retinal structures using artificial intelligence (AI) to reflect CNS disease.

The synergy between the eye–brain and AI was seen in a study by Milea *et al.* and the BONSAI group[154] that devised an algorithm to detect papilloedema in fundus photos. They used U-NET followed by DenseNet to detect the optic disc's location and showed good discriminatory ability with an accuracy of 94.8% for detecting papilloedema (AUC > 0.99). Building on prior work, Biousse and the BONSAI group[155] used the same pre-trained model to compare the diagnostic performance of AI with an expert neuro-ophthalmologist and proved there was no difference between the two groups. Vasseneix *et al.*[156] achieved an excellent DL system performance in classifying the severity of papilloedema based on Frisen scale. It successfully discriminated mild from severe papilloedema in fundus photos with accuracy, sensitivity, and specificity of 87.9%, 91.8%, and 86.2%, respectively. The AUC was 0.93, a similar performance to that of three expert neuro-ophthalmologists. Hence, the BONSAI group[154–156] created an algorithm that detected and classified papilloedema equivalent to that of expert neuro-ophthalmologists, providing a possible complementary screening tool in the emergency department that could facilitate triage and earlier management of potentially life-threatening diseases.

Neurodegenerative disorders, like Alzheimer's disease (AD), have also been studied using AI and retinal imaging.[157,158] Recent studies have found that thinning of the retinal nerve fibre layer (RNFL) thickness,[159] ganglion cell-inner plexiform layer (GC-IPL),[160,161] and subfoveal choroidal thickness[162] can be possible indicators of AD in retinal imaging. Wisely *et al.*[157] developed a CNN with multiple retinal image modalities and patient data that could detect symptomatic AD with OCTA. GC-IPL maps had an AUC of 0.809 and, combined with patient data, had the highest predictive value with an AUC of 0.841. Another neurological assessment for early AD detection was reported by Tian *et al.*,[158] utilizing retinal vessel and modular machine learning to achieve an accuracy of 0.82 and precision of 0.83.

Second to AD, Parkinson's disease (PD) is the most common neurodegenerative condition in the developed world.[163] Early diagnosis of PD is challenging but may be enhanced by the use of retinal imaging. Recent studies have demonstrated the usefulness of texture analysis in identifying AD or PD using OCT. Nunes *et al.*[164] used SVMs to develop a differentiation system of AD and PD using OCT data. Although the algorithm was trained with a limited number of patients (AD = 20 and PD = 28), it managed an accuracy of 0.82 in 10-fold cross-validation using texture data as opposed to thickness measurements. More recently, Chrysou *et al.*[165] classified PD patients and controls by using a random forest plot and achieved an AUC of

0.82 and accuracy of 0.64. AI may thus represent an innovative, inexpensive, and non-invasive tool in the identification of these neurodegenerative diseases.

While stroke mortality is declining in high-income countries, 87% of stroke-related deaths occur in low-income countries.[166] Automated identification of high-risk patients through cost-effective screening programmes could relieve the burden in economically underprivileged nations. The association between cerebrovascular disease (CeVD) and retinal vascular changes has been proven in various retrospective studies,[167–172] showing an increased risk of CeVD in the presence of retinal arterial and venous occlusion. A semi-automated computer software analysis (SIVA) described in a study by Ong *et al.*[171] demonstrated that in retinal fundus photos of ischaemic stroke patients, vessel networking was scarcer and more tortuous through vascular fractal dimension. Jeena *et al.*[173] used a naive Bayes (NB) classifier to diagnose patients with stroke from fundus photos, in comparison to normal controls, with an accuracy of 95.1%. Raveendran *et al.*[174] utilized anatomical and vascular analysis before the use of the NB classifier in detection of CeVD, that resulted in a high AUC of 0.976. Finally, with a much larger dataset ($n = 5500$ for patients who had a history of stroke), Lim *et al.*[175] used vascular calibre and five retinal features to identify affected patients with an ischaemic stroke with an AUC of 0.96.

11.18 Potential Challenges and Future Directions for Artificial Intelligence Implementation Among the Diabetes Population

This chapter has provided a brief review of the AI applications that can be applied to scale-up existing healthcare resources for eye care services. They can be adopted in existing clinical workflows through applications designed for the screening of patients within the community, identifying of patients who require therapeutic interventions such as laser treatment or surgery, and right-siting of them with referrals to tertiary ophthalmology care (Figure 11.2).[176,177] However, progress in the translation of AI in health care has faced several challenges.[178,179] These include addressing constraints of operationalizing these solutions and managing costs among many others. These challenges need to be overcome before AI and other digital health solutions can be implemented across varied healthcare settings and populations.[178]

Developing a suitable operational model for implementation of these solutions for each target clinical application is critical to avoid disrupting existing processes for their successful adoption. Described models in the literature range from the sorting conveyor model for out-of-hospital triage to the lighthouse model for patient monitoring. Ultimately, the ideal operational model for a given AI solution varies based on the clinical context, processes impacted by the solution, and relevant end users.[180]

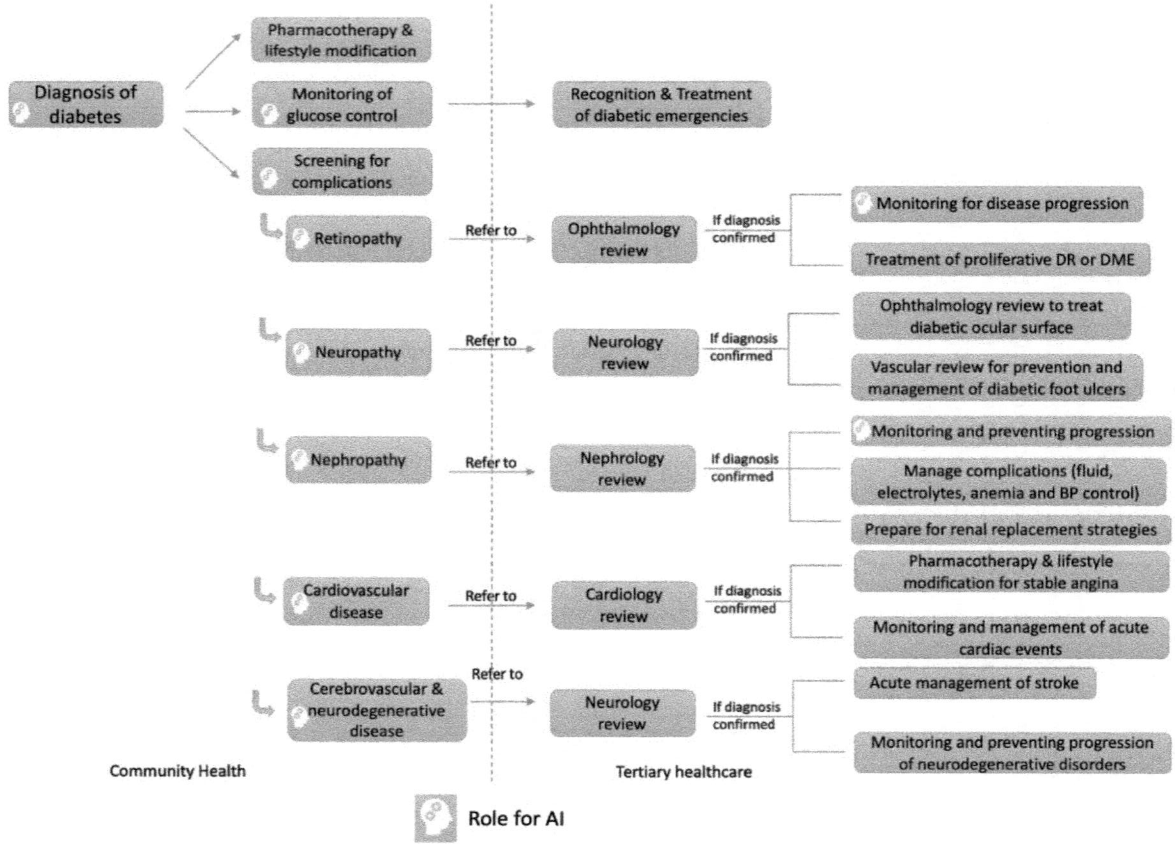

Figure 11.2 Potential AI applications within the clinical workflow. In the clinical workflow that follows after a patient is diagnosed with diabetes, there are opportunities for AI to assist in the management and screening of complications.

Other practical considerations include resource availability and cost, in order to ensure the solutions are implemented within clinical services in a financially sustainable manner. Health economic assessment (HEA) will play a critical role in the evaluation of future AI applications for health care. These include cost-minimization analysis in the context of equivalent algorithms that are being compared, as well as alternatives such as cost-effectiveness analysis, cost-utility analysis, and cost-benefit analysis for other clinical contexts, as discussed in a recent review.[181]

Finally, despite the autonomous nature of most AI healthcare applications, their success remains dependent on the acceptance and trust of its human end users, including both patients and healthcare practitioners. For AI applications intended for eye screening, end users may range from trained nurses counselling patients and performing fundus photography within the community, to ophthalmologists receiving patient referrals for follow-up tertiary care. Recent studies have highlighted how operational disruptiveness and lack of interpretability of AI solutions can contribute to barriers to adoption at scale.[182] Therefore, important future directions for AI research include engaging stakeholders to evaluate and improve the acceptance of AI technology by end users, and addressing practical considerations such as operational models and managing cost.[183,184]

11.19 Conclusion

The increasing global need to combat diabetes has spawned and encouraged the development of AI techniques to assist in tackling this disease and its various microvascular complications.[2] Automation with AI provides a reliable and accurate method to optimize screening for DM and its complications, facilitating more efficient allocation of healthcare resources and increasing the accessibility of disease screening in rural areas.[13,120] These AI models have been largely trained, tested, and built using inputs derived from clinical attributes or information collected from existing databases. A plethora of studies, as discussed earlier, have demonstrated the various associations between the eye and various systemic diseases.[111] DL has also recently garnered interest with unsupervised learning, allowing processing of large amounts of data and the extraction of meaningful patterns,[185] paving the way for more cost-effective and efficient means of health screening via different ophthalmic imaging modalities.[110,186]

With change, it might be met with resistance to the new and unknown. Technological advancements may obliterate old jobs, resulting in the fear of being obsolete; however, with the progress of information and technology, a wider and different scope of jobs is needed.[12] Whilst accuracy is key in any diagnosis, medical professionals question the impact of these algorithms on patient outcomes.[185,187] Patients are concerned that the AI might not be able to take into account the patient's unique characteristics and circumstances.[188] However, if AI was thought of as a supportive tool in providing personalized medicine, rather than one which finalized clinical decisions and medicine, this might improve consumer acceptance.[188]

Advances in AI have provided possible solutions and, at the same time, raised more issues regarding a lack of interpretability, cost, and acceptance by patients and the medical profession.[185,187,188] There is a greater emphasis in examining the impact of AI on patient outcomes and addressing the interpretability of algorithms and multimodal systems that can provide personalized medicine in patient care.[189–192] There is anticipation that in the growing global pandemic of diabetes, its treatment and complications can be better tackled with AI and its widespread adoption.

References

1. S. E. Moss, R. Klein and B. E. Klein, The 14-year incidence of visual loss in a diabetic population, *Ophthalmology*, 1998, **105**(6), 998–1003.
2. IDF Diabetes Atlas, 9th edn, 2020. https://diabetesatlas.org/en/resources/ (accessed 9 August 2021).
3. M. J. Fowler, Microvascular and macrovascular complications of diabetes, *Clin. Diabetes*, 2008, **26**(2), 77–82.
4. J. W. Yau, S. L. Rogers and R. Kawasaki, *et al.*, Global prevalence and major risk factors of diabetic retinopathy, *Diabetes Care*, 2012, **35**(3), 556–564.
5. D. S. W. Ting, G. C. M. Cheung and T. Y. Wong, Diabetic retinopathy: global prevalence, major risk factors, screening practices and public health challenges: a review, *Clin. Exp. Ophthalmol.*, 2016, **44**(4), 260–277.
6. G. Bikbova, T. Oshitari, T. Baba, M. Bikbov and S. Yamamoto, Diabetic corneal neuropathy: clinical perspectives, *Clin. Ophthalmol.*, 2018, **12**, 981.
7. American Diabetes Association, Standards of medical care in diabetes–2014, *Diabetes Care*, 2014, **37**, S14–S80.
8. W. B. Kannel and D. L. McGee, Diabetes and cardiovascular disease: the Framingham study, *JAMA*, 1979, **241**(19), 2035–2038.
9. S. M. De la Monte and J. R. Wands, Alzheimer's disease is type 3 diabetes—evidence reviewed, *J. Diabetes Sci. Technol.*, 2008, **2**(6), 1101–1113.
10. American Diabetes Association, Screening for diabetes, *Diabetes Care*, 2002, **25**(suppl 1), s21–s24.
11. J. M. G. Wilson, G. Jungner, World Health Organization, *Principles and Practice of Screening for Disease*, 1968.
12. M. Ben Ali, F. Wang, R. Boukherroub, W. Lei and M. Xia, Phytic acid-doped polyaniline nanofibers-clay mineral for efficient adsorption of copper(II) ions, *J. Colloid Interface Sci.*, 2019, **553**, 688–698.
13. N. J. Nilsson, *Principles of Artificial Intelligence*, Morgan Kaufmann, 2014.
14. K. Schwab, The Fourth Industrial Revolution: What It Means and How to Respond, 2015. https://www.foreignaffairs.com/articles/2015-12-12/fourth-industrial-revolution (accessed 22 August 2021).

15. K. Schwab, *The Fourth Industrial Revolution*: Currency, 2017.
16. J.-A. Choi and K. Lim, Identifying machine learning techniques for classification of target advertising, *ICT Express*, 2020, **6**(3), 175–180.
17. D. Reis and N. Pati, Applications of artificial intelligence to condition-based maintenance, *Rev. Admin. Empresas*, 2000, **40**, 102–107.
18. J. Turiel and T. Aste, Peer-to-peer loan acceptance and default prediction with artificial intelligence, *R. Soc. Open Sci.*, 2020, **7**(6), 191649.
19. R. Aggarwal, V. Sounderajah and G. Martin, *et al.*, Diagnostic accuracy of deep learning in medical imaging: a systematic review and meta-analysis, *NPJ Digital Med.*, 2021, **4**(1), 65.
20. M. D. Abramoff, Y. Lou and A. Erginay, *et al.*, Improved automated detection of diabetic retinopathy on a publicly available dataset through integration of deep learning, *Invest. Ophthalmol. Vis. Sci.*, 2016, **57**(13), 5200–5206.
21. V. Gulshan, L. Peng and M. Coram, *et al.*, Development and validation of a deep learning algorithm for detection of diabetic retinopathy in retinal fundus photographs, *JAMA*, 2016, **316**(22), 2402–2410.
22. D. S. W. Ting, C. Y.-L. Cheung and G. Lim, *et al.*, Development and validation of a deep learning system for diabetic retinopathy and related eye diseases using retinal images from multiethnic populations with diabetes, *JAMA*, 2017, **318**(22), 2211–2223.
23. R. Gargeya and T. Leng, Automated identification of diabetic retinopathy using deep learning, *Ophthalmology*, 2017, **124**(7), 962–969.
24. Z. Li, S. Keel and C. Liu, *et al.*, An automated grading system for detection of vision-threatening referable diabetic retinopathy on the basis of color fundus photographs, *Diabetes Care*, 2018, **41**(12), 2509–2516.
25. J. Krause, V. Gulshan and E. Rahimy, *et al.*, Grader variability and the importance of reference standards for evaluating machine learning models for diabetic retinopathy, *Ophthalmology*, 2018, **125**(8), 1264–12672.
26. M. D. Abramoff, P. T. Lavin, M. Birch, N. Shah and J. C. Folk, Pivotal trial of an autonomous AI-based diagnostic system for detection of diabetic retinopathy in primary care offices, *NPJ Digital Med.*, 2018, **1**, 39.
27. S. Keel, P. Y. Lee and J. Scheetz, *et al.*, Feasibility and patient acceptability of a novel artificial intelligence-based screening model for diabetic retinopathy at endocrinology outpatient services: a pilot study, *Sci. Rep.*, 2018, **8**(1), 4330.
28. A. Bhuiyan, A. Govindaiah and A. Deobhakta, *et al.*, Development and validation of an automated diabetic retinopathy screening tool for primary care setting, *Diabetes Care*, 2020, **43**(10), e147–e148.
29. P. Heydon, C. Egan and L. Bolter, *et al.*, Prospective evaluation of an artificial intelligence-enabled algorithm for automated diabetic retinopathy screening of 30 000 patients, *Br. J. Ophthalmol.*, 2021, **105**(5), 723–728.
30. J. Scheetz, D. Koca and M. McGuinness, *et al.*, Real-world artificial intelligence-based opportunistic screening for diabetic retinopathy in

endocrinology and indigenous healthcare settings in Australia, *Sci. Rep.*, 2021, **11**(1), 15808.
31. V. Bellemo, Z. W. Lim and G. Lim, *et al.*, Artificial intelligence using deep learning to screen for referable and vision-threatening diabetic retinopathy in Africa: a clinical validation study, *Lancet Digital Health*, 2019, **1**(1), e35–e44.
32. V. Gulshan, R. P. Rajan and K. Widner, *et al.*, Performance of a deep-learning algorithm vs manual grading for detecting diabetic retinopathy in India, *JAMA Ophthalmol.*, 2019, **137**(9), 987–993.
33. P. Ruamviboonsuk, J. Krause and P. Chotcomwongse, *et al.*, Deep learning versus human graders for classifying diabetic retinopathy severity in a nationwide screening program, *NPJ Digit. Med.*, 2019, **2**(1), 25.
34. F. Arcadu, F. Benmansour and A. Maunz, *et al.*, Deep learning predicts OCT measures of diabetic macular thickening from color fundus photographs, *Invest. Ophthalmol. Vis. Sci.*, 2019, **60**(4), 852–857.
35. F. Arcadu, F. Benmansour, A. Maunz, J. Willis, Z. Haskova and M. Prunotto, Author correction: deep learning algorithm predicts diabetic retinopathy progression in individual patients, *NPJ Digit. Med.*, 2020, **3**(1), 160.
36. A. Bora, S. Balasubramanian and B. Babenko, *et al.*, Predicting the risk of developing diabetic retinopathy using deep learning, *Lancet Digital Health*, 2021, **3**(1), e10–e19.
37. A. V. Varadarajan, P. Bavishi and P. Ruamviboonsuk, *et al.*, Predicting optical coherence tomography-derived diabetic macular edema grades from fundus photographs using deep learning, *Nat. Commun.*, 2020, **11**(1), 130.
38. P. S. Silva, M. B. Horton and D. Clary, *et al.*, Identification of diabetic retinopathy and ungradable image rate with ultrawide field imaging in a national teleophthalmology program, *Ophthalmology*, 2016, **123**(6), 1360–1367.
39. A. Olvera-Barrios, T. F. Heeren and K. Balaskas, *et al.*, Comparison of true-colour wide-field confocal scanner imaging with standard fundus photography for diabetic retinopathy screening, *Br. J. Ophthalmol.*, 2020, **104**(11), 1579–1584.
40. A. Olvera-Barrios, T. F. Heeren and K. Balaskas, *et al.*, Diagnostic accuracy of diabetic retinopathy grading by an artificial intelligence-enabled algorithm compared with a human standard for wide-field true-colour confocal scanning and standard digital retinal images, *Br. J. Ophthalmol.*, 2021, **105**(2), 265–270.
41. T. Nagasawa, H. Tabuchi and H. Masumoto, *et al.*, Accuracy of diabetic retinopathy staging with a deep convolutional neural network using ultra-wide-field fundus ophthalmoscopy and optical coherence tomography angiography, *J. Ophthalmol.*, 2021, **2021**, 6651175.
42. F. Tang, P. Luenam and A. R. Ran, *et al.*, Detection of diabetic retinopathy from ultra-widefield scanning laser ophthalmoscope images: a

multicenter deep learning analysis, *Ophthalmol. Retina*, 2021, **5**(11), 1097–1106.

43. J. Fujimoto and E. Swanson, The development, commercialization, and impact of optical coherence tomography, *Invest. Ophthalmol. Vis. Sci.*, 2016, **57**(9), Oct1–oct13.
44. T. Schlegl, S. M. Waldstein and H. Bogunovic, *et al.*, Fully automated detection and quantification of macular fluid in OCT using deep learning, *Ophthalmology*, 2018, **125**(4), 549–558.
45. Y. Guo, T. T. Hormel, H. Xiong, J. Wang, T. S. Hwang and Y. Jia, Automated segmentation of retinal fluid volumes from structural and angiographic optical coherence tomography using deep learning, *Transl. Vis. Sci. Technol.*, 2020, **9**(2), 54.
46. J. I. Orlando, B. S. Gerendas and S. Riedl, *et al.*, Automated quantification of photoreceptor alteration in macular disease using optical coherence tomography and deep learning, *Sci. Rep.*, 2020, **10**(1), 5619.
47. R. Mirshahi, P. Anvari, H. Riazi-Esfahani, M. Sardarinia, M. Naseripour and K. G. Falavarjani, Foveal avascular zone segmentation in optical coherence tomography angiography images using a deep learning approach, *Sci. Rep.*, 2021, **11**(1), 1031.
48. D. S. Kermany, M. Goldbaum and W. Cai, *et al.*, Identifying medical diagnoses and treatable diseases by image-based deep learning, *Cell*, 2018, **172**(5), 1122–1131.e9.
49. J. De Fauw, J. R. Ledsam and B. Romera-Paredes, *et al.*, Clinically applicable deep learning for diagnosis and referral in retinal disease, *Nat. Med.*, 2018, **24**(9), 1342–1350.
50. L. Wang, G. Wang and M. Zhang, *et al.*, An intelligent optical coherence tomography-based system for pathological retinal cases identification and urgent referrals, *Transl. Vis. Sci. Technol.*, 2020, **9**(2), 46.
51. K. K. Bhatia, M. S. Graham and L. Terry, *et al.*, Disease classification of macular optical coherence tomography scans using deep learning software: validation on independent, multicenter data, *Retina*, 2020, **40**(8), 1549–1557.
52. I. Hecht, A. Bar and L. Rokach, *et al.*, Optical coherence tomography biomarkers to distinguish diabetic macular edema from pseudophakic cystoid macular edema using machine learning algorithms, *Retina*, 2019, **39**(12), 2283–2291.
53. T. M. Aslam, D. C. Hoyle, V. Puri and G. Bento, Differentiation of diabetic status using statistical and machine learning techniques on optical coherence tomography angiography images, *Transl. Vis. Sci. Technol.*, 2020, **9**(4), 2.
54. R. Rasti, M. J. Allingham and P. S. Mettu, *et al.*, Deep learning-based single-shot prediction of differential effects of anti-VEGF treatment in patients with diabetic macular edema, *Biomed. Opt. Express*, 2020, **11**(2), 1139–1152.

55. J. K. Goh, C. Y. Cheung, S. S. Sim, P. C. Tan, G. S. Tan and T. Y. Wong, Retinal imaging techniques for diabetic retinopathy screening, *J. Diabetes Sci. Technol.*, 2016, **10**(2), 282–294.
56. I. A. Thompson, A. K. Durrani and S. Patel, Optical coherence tomography angiography characteristics in diabetic patients without clinical diabetic retinopathy, *Eye*, 2019, **33**(4), 648–652.
57. D. Le, M. Alam and C. K. Yao, *et al.*, Transfer learning for automated OCTA detection of diabetic retinopathy, *Transl Vis Sci Technol*, 2020, **9**(2), 35.
58. M. Heisler, S. Karst and J. Lo, *et al.*, Ensemble deep learning for diabetic retinopathy detection using optical coherence tomography angiography, *Transl. Vis. Sci. Technol.*, 2020, **9**(2), 20.
59. H. S. Sandhu, M. Elmogy and A. Taher Sharafeldeen, *et al.*, Automated diagnosis of diabetic retinopathy using clinical biomarkers, optical coherence tomography, and optical coherence tomography angiography, *Am. J. Ophthalmol.*, 2020, **216**, 201–206.
60. L. Ang, M. Jaiswal, C. Martin and R. Pop-Busui, Glucose control and diabetic neuropathy: lessons from recent large clinical trials, *Curr. Diabetes Rep.*, 2014, **14**(9), 528.
61. R. Pop-Busui, J. Lu and M. M. Brooks, *et al.*, Impact of glycemic control strategies on the progression of diabetic peripheral neuropathy in the Bypass Angioplasty Revascularization Investigation 2 Diabetes (BARI 2D) Cohort, *Diabetes Care*, 2013, **36**(10), 3208–3215.
62. Z. Iqbal, S. Azmi and R. Yadav, *et al.*, Diabetic peripheral neuropathy: epidemiology, diagnosis, and pharmacotherapy, *Clin. Ther.*, 2018, **40**(6), 828–849.
63. K. A. Sveen, B. Karimé and E. Jørum, *et al.*, Small- and large-fiber neuropathy after 40 years of type 1 diabetes: associations with glycemic control and advanced protein glycation: the Oslo Study, *Diabetes Care*, 2013, **36**(11), 3712–3717.
64. E. L. Feldman, B. C. Callaghan and R. Pop-Busui, *et al.*, Diabetic neuropathy, *Nat. Rev. Dis. Primers*, 2019, **5**(1), 41.
65. N. Tentolouris, S. Al-Sabbagh, M. G. Walker, A. J. Boulton and E. B. Jude, Mortality in diabetic and nondiabetic patients after amputations performed from 1990 to 1995: a 5-year follow-up study, *Diabetes Care*, 2004, **27**(7), 1598–1604.
66. A. Barsegian, J. Lee, M. O. Salifu and S. I. McFarlane, Corneal neuropathy: an underrated manifestation of diabetes mellitus, *J. Clin. Endocrinol. Diabetes*, 2018, **2**(1), 111.
67. M. Markoulli, J. Flanagan, S. S. Tummanapalli, J. Wu and M. Willcox, The impact of diabetes on corneal nerve morphology and ocular surface integrity, *Ocul. Surf.*, 2018, **16**(1), 45–57.
68. H. Zhao, Y. He, Y. R. Ren and B. H. Chen, Corneal alteration and pathogenesis in diabetes mellitus, *Int. J. Ophthalmol.*, 2019, **12**(12), 1939–1950.
69. NICE NifHaCE, Diabetic Foot Problems: Prevention and Management, NICE Guideline (NG19), 2015. https://www.nice.org.uk/guidance/ng19 (accessed 9 March 2020).

70. M. J. Young, J. L. Breddy, A. Veves and A. J. Boulton, The prediction of diabetic neuropathic foot ulceration using vibration perception thresholds. A prospective study, *Diabetes Care*, 1994, **17**(6), 557–560.
71. S. J. Rith-Najarian, T. Stolusky and D. M. Gohdes, Identifying diabetic patients at high risk for lower-extremity amputation in a primary health care setting. A prospective evaluation of simple screening criteria, *Diabetes Care*, 1992, **15**(10), 1386–1389.
72. I. N. Petropoulos, G. Ponirakis, A. Khan, H. Almuhannadi, H. Gad and R. A. Malik, Diagnosing diabetic neuropathy: something old, something new., *Diabetes Metab. J.*, 2018, **42**(4), 255–269.
73. M. E. Shy, E. M. Frohman and Y. T. So, *et al.*, Quantitative sensory testing: report of the Therapeutics and Technology Assessment Subcommittee of the American Academy of Neurology, *Neurology*, 2003, **60**(6), 898–904.
74. G. Gruener and P. J. Dyck, Quantitative sensory testing: methodology, applications, and future directions, *J. Clin. Neurophysiol.*, 1994, **11**(6), 568–583.
75. G. Cruccu and A. Truini, Tools for assessing neuropathic pain, *PLoS Med.*, 2009, **6**(4), e1000045.
76. A. Q. Green, S. Krishnan, F. M. Finucane and G. Rayman, Altered C-fiber function as an indicator of early peripheral neuropathy in individuals with impaired glucose tolerance, *Diabetes Care*, 2010, **33**(1), 174–176.
77. I. N. Petropoulos, U. Alam and H. Fadavi, *et al.*, Corneal nerve loss detected with corneal confocal microscopy is symmetrical and related to the severity of diabetic polyneuropathy, *Diabetes Care*, 2013, **36**(11), 3646–3651.
78. I. N. Petropoulos, T. Manzoor and P. Morgan, *et al.*, Repeatability of in vivo corneal confocal microscopy to quantify corneal nerve morphology, *Cornea*, 2013, **32**(5), e83–e89.
79. I. N. Petropoulos, U. Alam and H. Fadavi, *et al.*, Rapid automated diagnosis of diabetic peripheral neuropathy with in vivo corneal confocal microscopy, *Invest. Ophthalmol. Vis. Sci.*, 2014, **55**(4), 2071–2078.
80. U. Alam, M. Jeziorska and I. N. Petropoulos, *et al.*, Diagnostic utility of corneal confocal microscopy and intra-epidermal nerve fibre density in diabetic neuropathy, *PLoS One*, 2017, **12**(7), e0180175.
81. X. Chen, J. Graham and M. A. Dabbah, *et al.*, Small nerve fiber quantification in the diagnosis of diabetic sensorimotor polyneuropathy: comparing corneal confocal microscopy with intraepidermal nerve fiber density, *Diabetes Care*, 2015, **38**(6), 1138–1144.
82. C. Quattrini, M. Tavakoli and M. Jeziorska, *et al.*, Surrogate markers of small fiber damage in human diabetic neuropathy, *Diabetes*, 2007, **56**(8), 2148–2154.
83. A. Cruzat, Y. Qazi and P. Hamrah, In vivo confocal microscopy of corneal nerves in health and disease, *Ocul Surf*, 2017, **15**(1), 15–47.

84. P. Hossain, A. Sachdev and R. A. Malik, Early detection of diabetic peripheral neuropathy with corneal confocal microscopy, *Lancet*, 2005, **366**(9494), 1340–1343.
85. J. Burgess, B. Frank and A. Marshall, *et al.*, Early detection of diabetic peripheral neuropathy: a focus on small nerve fibres, *Diagnostics*, 2021, **11**(2), 165.
86. H. Gad, I. N. Petropoulos and A. Khan, *et al.* Corneal confocal microscopy for the diagnosis of diabetic peripheral neuropathy: a systematic review and meta-analysis, *J Diabetes Investigation*, 2021, **13**(1), 134–147.
87. B. A. Perkins, L. E. Lovblom and E. J. H. Lewis, *et al.*, Corneal confocal microscopy predicts the development of diabetic neuropathy: a longitudinal diagnostic multinational consortium study, *Diabetes Care*, 2021, **44**(9), 2107–2114.
88. C. Dehghani, N. Pritchard, K. Edwards, A. W. Russell, R. A. Malik and N. Efron, Risk factors associated with corneal nerve alteration in type 1 diabetes in the absence of neuropathy: a longitudinal in vivo corneal confocal microscopy study, *Cornea*, 2016, **35**(6), 847–852.
89. M. E. Rosenberg, T. M. Tervo, I. J. Immonen, L. J. Müller, C. Grönhagen-Riska and M. H. Vesaluoma, Corneal structure and sensitivity in type 1 diabetes mellitus, *Invest. Ophthalmol. Vis. Sci.*, 2000, **41**(10), 2915–2921.
90. A. Khan, I. N. Petropoulos and G. Ponirakis, *et al.*, Corneal confocal microscopy detects severe small fiber neuropathy in diabetic patients with Charcot neuroarthropathy, *J Diabetes Invest.*, 2018, **9**(5), 1167–1172.
91. C. Dehghani, A. W. Russell and B. A. Perkins, *et al.*, A rapid decline in corneal small fibers and occurrence of foot ulceration and Charcot foot, *J. Diabetes Complications*, 2016, **30**(8), 1437–1439.
92. M. Tavakoli, M. Mitu-Pretorian and I. N. Petropoulos, *et al.*, Corneal confocal microscopy detects early nerve regeneration in diabetic neuropathy after simultaneous pancreas and kidney transplantation, *Diabetes*, 2013, **62**(1), 254–260.
93. M. Tavakoli, P. Kallinikos and A. Iqbal, *et al.*, Corneal confocal microscopy detects improvement in corneal nerve morphology with an improvement in risk factors for diabetic neuropathy, *Diabetic Med.*, 2011, **28**(10), 1261–1267.
94. W. C. Mdicine-Qatar, *Software Licenses*, https://qatar-weill.cornell.edu/research/research-faculty/early-neuropathy-assessment-group/software-licenses (accessed 5 Sep 2021).
95. M. A. Dabbah, J. Graham, I. N. Petropoulos, M. Tavakoli and R. A. Malik, Automatic analysis of diabetic peripheral neuropathy using multi-scale quantitative morphology of nerve fibres in corneal confocal microscopy imaging, *Med. Image Anal.*, 2011, **15**(5), 738–747.
96. M. A. Dabbah, J. Graham, I. Petropoulos, M. Tavakoli and R. A. Malik, Dual-model automatic detection of nerve-fibres in corneal confocal

microscopy images, *Med. Image Comput. Comput. Assist. Intervention*, 2010, **13**(Pt 1), 300–307.

97. X. Chen, J. Graham, M. A. Dabbah, I. N. Petropoulos, M. Tavakoli and R. A. Malik, An Automatic tool for quantification of nerve fibers in corneal confocal microscopy images, *IEEE Trans. Biomed. Eng.*, 2017, **64**(4), 786–794.
98. B. M. Williams, D. Borroni and R. Liu, *et al.*, An artificial intelligence-based deep learning algorithm for the diagnosis of diabetic neuropathy using corneal confocal microscopy: a development and validation study, *Diabetologia*, 2020, **63**(2), 419–430.
99. T. Salahouddin, I. N. Petropoulos and M. Ferdousi, *et al.*, Artificial intelligence-based classification of diabetic peripheral neuropathy from corneal confocal microscopy images, *Diabetes Care*, 2021, **44**(7), e151–e153.
100. F. Scarpa, A. Colonna and A. Ruggeri, Multiple-image deep learning analysis for neuropathy detection in corneal nerve images, *Cornea*, 2020, **39**(3), 342–347.
101. E. Yildiz, A. T. Arslan and A. Yildiz Tas, *et al.*, Generative adversarial network based automatic segmentation of corneal subbasal nerves on in vivo confocal microscopy images, *Transl. Vis. Sci. Technol.*, 2021, **10**(6), 33.
102. G. A. Sivaskandarajah, E. M. Halpern and L. E. Lovblom, *et al.*, Structure–function relationship between corneal nerves and conventional small-fiber tests in type 1 diabetes, *Diabetes Care*, 2013, **36**(9), 2748–2755.
103. E. Maddaloni, F. Sabatino and R. Del Toro, *et al.*, In vivo corneal confocal microscopy as a novel non-invasive tool to investigate cardiac autonomic neuropathy in type 1 diabetes, *Diabetic Med.*, 2015, **32**(2), 262–266.
104. M. Tavakoli, P. Begum, J. McLaughlin and R. A. Malik, Corneal confocal microscopy for the diagnosis of diabetic autonomic neuropathy, *Muscle Nerve*, 2015, **52**(3), 363–370.
105. M. Tavakoli, A. J. Boulton, N. Efron and R. A. Malik, Increased Langerhan cell density and corneal nerve damage in diabetic patients: role of immune mechanisms in human diabetic neuropathy, *Contact Lens Anterior Eye*, 2011, **34**(1), 7–11.
106. I. N. Petropoulos, P. Green and A. W. Chan, *et al.*, Corneal confocal microscopy detects neuropathy in patients with type 1 diabetes without retinopathy or microalbuminuria, *PLoS One*, 2015, **10**(4), e0123517.
107. J. T. Turuwhenua, D. V. Patel and C. N. McGhee, Fully automated montaging of laser scanning in vivo confocal microscopy images of the human corneal subbasal nerve plexus, *Invest. Ophthalmol. Vis. Sci.*, 2012, **53**(4), 2235–2242.
108. World Health Organization, Diabetes mellitus: report of a WHO Expert Committee [meeting held in Geneva from 24 to 30 November 1964], 1965.

109. World Health Organization, Diabetes mellitus: report of a WHO Study Group [meeting held in Geneva from 11 to 16 February 1985], 1985.
110. R. Poplin, A. V. Varadarajan and K. Blumer, *et al.*, Prediction of cardiovascular risk factors from retinal fundus photographs via deep learning, *Nat. Biomed. Eng.*, 2018, **2**(3), 158–164.
111. P. Foster and K.-T. Khaw, The eye: window to the soul or a mirror of systemic health? *Heart*, 2009, **95**(5), 348–349.
112. T. H. Rim, G. Lee and Y. Kim, *et al.*, Prediction of systemic biomarkers from retinal photographs: development and validation of deep-learning algorithms, *Lancet Digital Health*, 2020, **2**(10), e526–e536.
113. S. M. Son, Reactive oxygen and nitrogen species in pathogenesis of vascular complications of diabetes, *Diabetes Metab. J.*, 2012, **36**(3), 190–198.
114. R. Larkins and M. Dunlop, The link between hyperglycaemia and diabetic nephropathy, *Diabetologia*, 1992, **35**(6), 499–504.
115. M. Kitada, Z. Zhang, A. Mima and G. L. King, Molecular mechanisms of diabetic vascular complications, *J. Diabetes Invest.*, 2010, **1**(3), 77–89.
116. American Diabetes Association, Nutrition recommendations and interventions for diabetes: a position statement of the American Diabetes Association, *Diabetes Care*, 2007, **30**(suppl 1), S48–S65.
117. S. R. Colberg, R. J. Sigal and J. E. Yardley, *et al.*, Physical activity/exercise and diabetes: a position statement of the American Diabetes Association, *Diabetes Care*, 2016, **39**(11), 2065–2079.
118. S. E. Inzucchi, R. M. Bergenstal and J. B. Buse, *et al.*, Management of hyperglycemia in type 2 diabetes: a patient-centered approach: position statement of the American Diabetes Association (ADA) and the European Association for the Study of Diabetes (EASD), *Diabetes Spectr.*, 2012, **25**(3), 154–171.
119. I. Contreras and J. Vehi, Artificial intelligence for diabetes management and decision support: literature review, *J. Med. Internet Res.*, 2018, **20**(5), e10775.
120. S. Ellahham, Artificial intelligence: the future for diabetes care, *Am. J. Med.*, 2020, **133**(8), 895–900.
121. M. M. Ahmadi and G. A. Jullien, A wireless-implantable microsystem for continuous blood glucose monitoring, *IEEE Trans. Biomed. Circuits Syst.*, 2009, **3**(3), 169–180.
122. C. Zecchin, A. Facchinetti, G. Sparacino and C. Cobelli, Jump neural network for real-time prediction of glucose concentration, *Artif. Neural Netw.*, 2015, 245–259.
123. D. J. Albers, M. Levine, B. Gluckman, H. Ginsberg, G. Hripcsak and L. Mamykina, Personalized glucose forecasting for type 2 diabetes using data assimilation, *PLoS Comput. Biol.*, 2017, **13**(4), e1005232.
124. UK UHSGshsa, Risk of hypoglycaemia in types 1 and 2 diabetes: effects of treatment modalities and their duration, *Diabetologia*, 2007, **50**, 1140–1147.

125. R. E. Warren and B. M. Frier, Hypoglycaemia and cognitive function, *Diabetes Obes. Metab.*, 2005, **7**(5), 493–503.
126. E. I. Georga, V. C. Protopappas, D. Ardigo, D. Polyzos and D. I. Fotiadis, A glucose model based on support vector regression for the prediction of hypoglycemic events under free-living conditions, *Diabetes Technol. Ther.*, 2013, **15**(8), 634–643.
127. B. Sudharsan, M. Peeples and M. Shomali, Hypoglycemia prediction using machine learning models for patients with type 2 diabetes, *J. Diabetes Sci. Technol.*, 2014, **9**(1), 86–90.
128. World Health Organisation, *The Top 10 Causes of Death*, 2020.
129. W. G. Baxt, Use of an artificial neural network for the diagnosis of myocardial infarction, *Ann. Intern. Med.*, 1991, **115**(11), 843–848.
130. W. G. Baxt and J. Skora, Prospective validation of artificial neural network trained to identify acute myocardial infarction, *Lancet*, 1996, **347**(8993), 12–15.
131. M. Green, J. Björk, J. Forberg, U. Ekelund, L. Edenbrandt and M. Ohlsson, Comparison between neural networks and multiple logistic regression to predict acute coronary syndrome in the emergency room, *Artif. Intell. Med.*, 2006, **38**(3), 305–318.
132. G. B. Berikol, O. Yildiz and İ. T. Özcan, Diagnosis of acute coronary syndrome with a support vector machine, *J. Med. Syst.*, 2016, **40**(4), 84.
133. T. Y Wong, N. Cheung and F. A. Islam, *et al.*, Relation of retinopathy to coronary artery calcification: the multi-ethnic study of atherosclerosis, *Am. J. Epidemiol.*, 2008, **167**(1), 51–58.
134. R. Kromer, E. Tigges, N. Rashed, I. Pein, M. Klemm and S. Blankenberg, Association between optical coherence tomography based retinal microvasculature characteristics and myocardial infarction in young men, *Sci. Rep.*, 2018, **8**(1), 1–8.
135. C. Y. Cheung, D. Xu and C.-Y. Cheng, *et al.*, A deep-learning system for the assessment of cardiovascular disease risk via the measurement of retinal-vessel calibre, *Nat. Biomed. Eng.*, 2021, **5**(6), 498–508.
136. J. Chang, A. Ko and S. M. Park, *et al.*, Association of cardiovascular mortality and deep learning-funduscopic atherosclerosis score derived from retinal fundus images, *Am. J. Ophthalmol.*, 2020, **217**, 121–130.
137. T. H. Rim, C. J. Lee and Y.-C. Tham, *et al.*, Deep-learning-based cardiovascular risk stratification using coronary artery calcium scores predicted from retinal photographs, *The Lancet Digital Health*, 2021, **3**(5), e306–e316.
138. A. S. Levey and J. Coresh, Chronic kidney disease, *Lancet*, 2012, **379**(9811), 165–180.
139. KdIGO, *KDIGO 2012 Clinical Practice Guideline for the Evaluation and Management of Chronic Kidney Disease*, 2012. https://kdigo.org/wp-content/uploads/2017/02/KDIGO_2012_CKD_GL.pdf (accessed 12 August 2021.
140. Z. Chen, X. Zhang and Z. Zhang, Clinical risk assessment of patients with chronic kidney disease by using clinical data and multivariate models, *Int. Urol. Nephrol.*, 2016, **48**(12), 2069–2075.

141. X. Liu, N.-S. Li and L.-S. Lv, *et al.*, A comparison of the performances of an artificial neural network and a regression model for GFR estimation, *Am. J. Kidney Dis.*, 2013, **62**(6), 1109–1115.
142. E. Kanda, Y. Kanno and F. Katsukawa, Identifying progressive CKD from healthy population using Bayesian network and artificial intelligence: a worksite-based cohort study, *Sci. Rep.*, 2019, **9**(1), 1–18.
143. J. E. Grunwald, J. Alexander and G.-S. Ying, *et al.*, Retinopathy and chronic kidney disease in the Chronic Renal Insufficiency Cohort (CRIC) study, *Arch. Ophthalmol.*, 2012, **130**(9), 1136–1144.
144. C. Sabanayagam, A. Shankar and D. Koh, *et al.*, Retinal microvascular caliber and chronic kidney disease in an Asian population, *Am. J. Epidemiol.*, 2009, **169**(5), 625–632.
145. W. Yip, P. G. Ong and B. W. Teo, *et al.*, Retinal vascular imaging markers and incident chronic kidney disease: a prospective cohort study, *Sci. Rep.*, 2017, **7**(1), 1–9.
146. C. Sabanayagam, D. Xu and D. S. Ting, *et al.*, A deep learning algorithm to detect chronic kidney disease from retinal photographs in community-based populations, *The Lancet Digital Health*, 2020, **2**(6), e295–e302.
147. H. Hanzouli-Ben Salah, J. Lapuyade-Lahorgue and J. Bert, *et al.*, A framework based on hidden Markov trees for multimodal PET/CT image co-segmentation, *Med. Phys.*, 2017, **44**(11), 5835–5848.
148. E. S. Donkor, Stroke in the 21st Century: A Snapshot of the Burden, Epidemiology, and Quality of Life, *Stroke Res. Treat.*, 2018, **2018**, 3238165.
149. S. S. Virani, A. Alonso and E. J. Benjamin, *et al.*, Heart disease and stroke statistics—2020 Update: a report from the American Heart Association, *Circulation*, 2020, **141**(9), e139–e596.
150. K. S. Yew and E. M. Cheng, Diagnosis of acute stroke, *Am. Fam. Physician*, 2015, **91**(8), 528–536.
151. S. B. Guze, Diagnostic and statistical manual of mental disorders, 4th ed. (DSM-IV), *Am. J. Psychiatry*, 1995, **152**(8), 1228.
152. A. London, I. Benhar and M. Schwartz, The retina as a window to the brain—from eye research to CNS disorders, *Nat. Rev. Neurol.*, 2013, **9**(1), 44–53.
153. T. E. Yap, S. I. Balendra, M. T. Almonte and M. F. Cordeiro, Retinal correlates of neurological disorders, *Ther. Adv. Chronic Dis.*, 2019, **10**, 204062231988220.
154. D. Milea, R. P. Najjar and Z. Jiang, *et al.*, Artificial intelligence to detect papilledema from ocular fundus photographs, *N. Engl. J. Med.*, 2020, **382**(18), 1687–1695.
155. V. Biousse, N. J. Newman and R. P. Najjar, *et al.*, Optic disc classification by deep learning versus expert neuro-ophthalmologists, *Ann. Neurol.*, 2020, **88**(4), 785–795.
156. C. Vasseneix, R. P. Najjar and X. Xu, *et al.*, Accuracy of a deep learning system for classification of papilledema severity on ocular fundus photographs, *Neurology*, 2021, **97**(4), e369–e377.

157. C. E. Wisely, D. Wang and R. Henao, *et al.*, Convolutional neural network to identify symptomatic Alzheimer's disease using multimodal retinal imaging, *Br. J. Ophthalmol.*, 2022, **106**, 388–395.
158. J. Tian, G. Smith and H. Guo, *et al.*, Modular machine learning for Alzheimer's disease classification from retinal vasculature, *Sci. Rep.*, 2021, **11**(1), 238.
159. Y. Lu, Z. Li and X. Zhang, *et al.*, Retinal nerve fiber layer structure abnormalities in early Alzheimer's disease: evidence in optical coherence tomography, *Neurosci. Lett.*, 2010, **480**(1), 69–72.
160. E. M. Lad, D. Mukherjee and S. S. Stinnett, *et al.*, Evaluation of inner retinal layers as biomarkers in mild cognitive impairment to moderate Alzheimer's disease, *PLoS One*, 2018, **13**(2), e0192646.
161. S. H. Choi, S. J. Park and N. R. Kim, Macular ganglion cell-inner plexiform layer thickness is associated with clinical progression in mild cognitive impairment and alzheimers disease, *PLoS One*, 2016, **11**(9), e0162202.
162. M. Bulut, F. Kurtuluş and O. Gözkaya, *et al.*, Evaluation of optical coherence tomography angiographic findings in Alzheimer's type dementia, *Br. J. Ophthalmol.*, 2018, **102**(2), 233–237.
163. L. M. de Lau and M. M. Breteler, Epidemiology of Parkinson's disease, *Lancet Neurol.*, 2006, **5**(6), 525–535.
164. A. Nunes, G. Silva and C. Duque, *et al.*, Retinal texture biomarkers may help to discriminate between Alzheimer's, Parkinson's, and healthy controls, *PLoS One*, 2019, **14**(6), e0218826.
165. A. Chrysou, Identifying de novo Parkinson's disease with optical coherence tomography of the retina: a machine learning classification approach, *Mov. Disord. Abstracts*, 2020, **35**(suppl 1), 565.
166. F. Lanas and P. Seron, Facing the stroke burden worldwide, *Lancet Glob. Health*, 2021, **9**(3), e235–e236.
167. N. Cheung, T. Mosley and A. Islam, *et al.*, Retinal microvascular abnormalities and subclinical magnetic resonance imaging brain infarct: a prospective study, *Brain*, 2010, **133**(Pt 7), 1987–1993.
168. Y. Zhou, W. Zhu and C. Wang, Relationship between retinal vascular occlusions and incident cerebrovascular diseases, *Medicine*, 2016, **95**(26), e4075.
169. Z. Khan, D. R. P. Almeida, K. Rahim, M. J. Belliveau, M. Bona and J. Gale, 10-Year Framingham risk in patients with retinal vein occlusion: a systematic review and meta-analysis, *Can. J. Ophthalmol.*, 2013, **48**(1), 40–45.e1.
170. D. A. De Silva, J. J. F. Manzano and E. Y. Liu, *et al.*, Retinal microvascular changes and subsequent vascular events after ischemic stroke, *Neurology*, 2011, **77**(9), 896–903.
171. Y.-T. Ong, D. A. De Silva and C. Y. Cheung, *et al.*, Microvascular structure and network in the retina of patients with ischemic stroke, *Stroke*, 2013, **44**(8), 2121–2127.

172. O. M. Dumitrascu and M. Koronyo-Hamaoui, Retinal vessel changes in cerebrovascular disease, *Curr. Opin. Neurol.*, 2020, **33**(1), 87–92.
173. R. S. Jeena, A. Sukesh Kumar and K. Mahadevan, Stroke diagnosis from retinal fundus images using multi texture analysis, *J. Intell. Fuzzy Syst.*, 2019, **36**(3), 2025–2032.
174. J. Raveendran Susha, A. Sukeshkumar and K. Mahadevan, Computation of retinal fundus parameters for stroke prediction, *Comput. Methods Biomech. Biomed. Eng. Imaging Vis.*, 2020, **8**(4), 374–381.
175. G. Lim, Z. W. Lim and D. Xu, *et al.*, Feature isolation for hypothesis testing in retinal imaging: an ischemic stroke prediction case study, *Proc. AAAI Conf. Artif. Intell.*, 2019, **33**, 9510–9515.
176. D. V. Gunasekeran, D. S. W. Ting, G. S. W. Tan and T. Y. Wong, Artificial intelligence for diabetic retinopathy screening, prediction and management, *Curr. Opin. Ophthalmol.*, 2020, **31**(5), 357–365.
177. M. J. Allingham, D. Mukherjee and E. B. Lally, *et al.*, A quantitative approach to predict differential effects of Anti-VEGF treatment on diffuse and focal leakage in patients with diabetic macular edema: a pilot study, *Transl. Vis. Sci. Technol.*, 2017, **6**(2), 7.
178. D. V. Gunasekeran, R. Tseng, Y. C. Tham and T. Y. Wong, Applications of digital health for public health responses to COVID-19: a systematic scoping review of artificial intelligence, telehealth and related technologies, *NPJ Digit. Med.*, 2021, **4**(1), 40.
179. M. Y. T. Yip, G. Lim and Z. W. Lim, *et al.*, Technical and imaging factors influencing performance of deep learning systems for diabetic retinopathy, *NPJ Digit. Med.*, 2020, **3**, 40.
180. D. V. Gunasekeran, Y. C. Tham, D. S. W. Ting, G. S. W. Tan and T. Y. Wong, Digital health during COVID-19: lessons from operationalising new models of care in ophthalmology, *Lancet Digit Health*, 2021, **3**(2), e124–e34.
181. Y. Xie, D. V. Gunasekeran and K. Balaskas, *et al.*, Health economic and safety considerations for artificial intelligence applications in diabetic retinopathy screening, *Transl. Vis. Sci. Technol.*, 2020, **9**(2), 22.
182. E. Beede, E. Baylor, F. Hersch, *et al.* A human-centered evaluation of a deep learning system deployed in clinics for the detection of diabetic retinopathy, *Proceedings of the 2020 CHI Conference on Human Factors in Computing Systems*, Association for Computing Machinery, 2020, pp. 1–12.
183. L. Harst, H. Lantzsch and M. Scheibe, Theories predicting end-user acceptance of telemedicine use: systematic review, *J. Med. Internet Res.*, 2019, **21**(5), e13117.
184. A. M. K. Chew, R. Ong and H. H. Lei, *et al.*, Digital health solutions for mental health disorders during COVID-19, *Front. Psychiatry*, 2020, **11**, 582007.
185. D. S. W. Ting, L. R. Pasquale and L. Peng, *et al.*, Artificial intelligence and deep learning in ophthalmology, *Br. J. Ophthalmol.*, 2019, **103**(2), 167–175.
186. Y. Xie, Q. D. Nguyen and H. Hamzah, *et al.*, Artificial intelligence for teleophthalmology-based diabetic retinopathy screening in a national

programme: an economic analysis modelling study, *Lancet Digital Health*, 2020, **2**(5), e240–e249.

187. C. J. Kelly, A. Karthikesalingam, M. Suleyman, G. Corrado and D. King, Key challenges for delivering clinical impact with artificial intelligence, *BMC Med.*, 2019, **17**(1), 1–9.
188. C. Longoni, A. Bionezzi and C. K. Morewedge, Resistance to medical artificial intelligence, *J. Consum. Res.*, 2019, **46**(4), 629–650.
189. J. Varghese, Artificial intelligence in medicine: chances and challenges for wide clinical adoption, *Visc. Med.*, 2020, **36**(6), 443–449.
190. E. H. Shortliffe and M. J. Sepúlveda, Clinical decision support in the era of artificial intelligence, *JAMA*, 2018, **320**(21), 2199–2200.
191. S. Keel, P. Y. Lee and J. Scheetz, *et al.*, Feasibility and patient acceptability of a novel artificial intelligence-based screening model for diabetic retinopathy at endocrinology outpatient services: a pilot study, *Sci. Rep.*, 2018, **8**(1), 1–6.
192. J. De Fauw, J. R. Ledsam and B. Romera-Paredes, *et al.*, Clinically applicable deep learning for diagnosis and referral in retinal disease, *Nat. Med.*, 2018, **24**(9), 1342–1350.
193. P. Zang, L. Gao and T. T. Hormel, *et al.*, DcardNet: diabetic retinopathy classification at multiple levels based on structural and angiographic optical coherence tomography, *IEEE Trans. Biomed. Eng.*, 2021, **68**(6), 1859–1870.
194. S. Al-Fahdawi, R. Qahwaji and A. S. Al-Waisy, *et al.*, A fully automatic nerve segmentation and morphometric parameter quantification system for early diagnosis of diabetic neuropathy in corneal images, *Comput. Methods Programs Biomed.*, 2016, **135**, 151–166.
195. A. Ferreira, A. M. Morgado and J. S. Silva, A method for corneal nerves automatic segmentation and morphometric analysis, *Comput. Methods Programs Biomed.*, 2012, **107**(1), 53–60.
196. A. Ferreira, A. M. Morgado and J. S. Silva *Corneal Nerves Segmentation and Morphometric Parameters Quantification for Early Detection of Diabetic Neuropathy*, Springer, Berlin Heidelberg, 2010, pp. 264–267.
197. K. Zarkogianni, K. Mitsis and E. Litsa, *et al.*, Comparative assessment of glucose prediction models for patients with type 1 diabetes mellitus applying sensors for glucose and physical activity monitoring, *Med. Biol. Eng. Comput.*, 2015, **53**(12), 1333–1343.
198. Y. Hu, T. Guo and X. Ye, *et al.*, Dye adsorption by resins: effect of ionic strength on hydrophobic and electrostatic interactions, *Chem. Eng. J.*, 2013, **228**, 392–397.
199. A. P. D. Palomar, J. Cegoñino and A. Montolío, *et al.*, Swept source optical coherence tomography to early detect multiple sclerosis disease. The use of machine learning techniques, *PLoS One*, 2019, **14**(5), e0216410.

Subject Index